Gene Silencing by RNA Interference

Technology and Application

Gene Silencing by RNA Interference

Technology and Application

EDITED BY
Muhammad Sohail

CRC PRESS

Boca Raton London New York Washington, D.C.

Library of Congress Cataloging-in-Publication Data

Gene silencing by RNA interference : technology and application / edited by Muhammad Sohail.
p. cm.
Includes bibliographical references and index.
ISBN 0-8493-2141-7 (alk. paper)
1. Gene silencing—Laboratory manuals. 2. RNA—Laboratory manuals. I. Sohail, Muhammad.

QH450. G4635 2004
660.6'5—dc22

2004051934

Visit the CRC Press Web site at www.crcpress.com

International Standard Book Number 0-8493-2141-7
Library of Congress Card Number 2004051934
Printed in the United States of America 1 2 3 4 5 6 7 8 9 0
Printed on acid-free paper

Foreword

Every so often, a new research tool emerges which reshapes an entire field of study. It has become especially clear that recent progress in the biological sciences was driven by technical innovation; Sanger's sequencing method laid the foundation for genomics and all that has followed.

RNAi is one of the most recent entries. Although described several years ago, it has only recently been widely adapted. But it has set in motion intense activity directed at discovering the function of genes and, hence, understanding the complex mechanisms of biological systems. It is a remarkable process. Who could have imagined that nature would invent a system for the destruction of single-stranded RNAs that was triggered by their double-stranded counterparts? Evidence for the system was hard won. And much work went into unravelling its complexities, which are still not fully understood.

However, its power and its usefulness are now proven. Applications will grow. It will in all likelihood become a routine technology in many areas of biology. But much needs to be done to achieve that status. It is a method that is applied to organisms, each of which will have its own quirks and foibles; i.e., the nematode worm consumes RNAi agents in its normal diet of bacteria, a system of delivery that is unlikely to work in many animals. To advance the technology, we need a deeper understanding of the molecular mechanisms that produce the effect; however, these may differ in different organisms, even in different cells of the same organism. Above all, we need established methods that work in the hands of non-specialists who want to apply them to their experiments, and in the world of high-throughput biology, we would like to have methods that can be automated to apply on the large scale.

We are fortunate to live in an age which accords to technology development the respect it deserves. Funding agencies give their support to technology development and inventors get their rewards for innovation; there are even signs that elite scientific journals are catching up with the trend.

The chapters of this volume, written by scientists at the leading edge, illustrate the present state of the art in the RNAi field. They provide the basic knowledge needed by those who want to apply the method to their own system, to those who want to understand the biology of the RNAi mechanism, and to those who want to engage in furthering this imoportant technology.

SIR EDWIN M. SOUTHERN, FRS
Oxford

Preface

Gene silencing induced by RNA interference is one of the most topical and exciting subjects in molecular biology today. It offers new ways of quickly and easily creating loss-of-function phenotypes. Therefore, molecular biologists in almost every discipline are keen to employ this tool in their model systems/organisms. DNA sequencing projects are revealing a wealth of information that needs to be analyzed and understood before it can be implemented. To do so, the functions of several thousands of genes need to be elucidated and that requires high-throughput techniques. Conventional gene knockout techniques are slow and cumbersome and are unlikely to meet the high demands of such analyses. Antisense technologies hold the potential of providing a high throughput alternative to gene knockout methods. However, there have been considerable difficulties in effectively applying conventional antisense reagents for such purpose, such as the lack of target specificity and low efficacy. The advent of RNA interference tools is likely to solve at least some of these difficulties.

RNA interference appears to be highly conserved and occurs in both plants and animals. Artificial silencing of an endogenous gene can be triggered by the introduction of short or long double-stranded RNA, a trans-gene, or a virus. Methodologies for using RNA interference in plant, insect, and mammalian cells have been developed and have been published in journal articles over the past few years. However, due to space constraints, it is not always possible for the authors to provide full details of methods, and the often sketchy methodological information can be difficult to follow. Thus, a reliable, well-researched resource that combines the essential background to the subject, with tried and tested practical techniques and troubleshooting advice was needed. This book is intended to fill the gaps, and to be useful to both beginners and experienced users in the fields, in academia, and in the commercial sector.

This volume covers a wide variety of topics — from design and production of siRNAs to their application in cultured cells, embryos, animals, and plants. Experts in the field who either invented the methods or have extensive experience in using them have written the chapters. Each chapter first provides necessary background to the subject area and then gives detailed methods in an easy-to-follow step-by-step fashion, with troubleshooting hints and useful tips. The first chapter provides a general introduction to the subject area. Chapters 2 through 8 focus on the various methods of siRNA design, chemical and enzymatic production, and plasmid- and virus-mediated intracellular expression. Chapter 9 provides valuable information on improving and monitoring exogenous delivery of siRNAs into cells. The remaining chapters deal with the application of RNAi tools in a number of biological systems including cultured cells, human parasites, embryos, model animals, and plants.

MUHAMMAD SOHAIL
Oxford

Acknowledgments

I am highly indebted to all the authors for their hard work and support in bringing this challenging project to completion and consider it an honor to have such a prestigious and world-class team. I am also grateful to my colleagues at CRC Press, Barbara Norwitz, Judith Spiegel, Erika Dery, Barbara Uetrecht-Pierre, and their teams, for their considerable patience, timely assistance and advice, and highly efficient processing of the book — from the proposal submission stages to print. Last but not least, I am most thankful to my parents for their continuous support and prayers and my children and my wife for their immense patience and support throughout the course of this project.

Editor

Muhammad Sohail is a senior researcher at the Department of Biochemistry, University of Oxford, which he joined in 1995. His major research interest is the study of nucleic acid interactions using oligonucleotide arrays, with particular focus on the design strategies of gene silencing reagents. He obtained a BSc degree in Biochemical Sciences from the University of Punjab in 1986. He then joined Quaid-i-Azam University and earned an MSc in Biological Sciences in 1989 and an MPhil in Molecular Biology in 1991. Later in 1991, Dr. Sohail moved to the University of Oxford to study for a doctoral degree and joined the Department of Biochemistry, graduating in 1994.

Contributors

Geoffrey W. Abbott
Weill Medical College of Cornell University
New York, New York

Saghir Akhtar
Cardiff University
Cardiff, United Kingdom

Dong Sung An
University of California–Los Angeles
Los Angeles, California

Peter J. Appleford
University of Oxford
Oxford, United Kingdom

Vincent C. Auyeung
California Institute of Technology
Pasadena, California

David Baltimore
California Institute of Technology
Pasadena, California

M. Barciszewska-Pacak
Adam Mickiewicz University
Poznań, Poland

Mónica Bettencourt-Dias
University of Cambridge
Cambridge, United Kingdom

Queta Boese
Dharmacon, Inc.
Lafayette, Colorado

Anke Brüning-Richardson
University of Leeds
Leeds, United Kingdom

Frank Buchholz
Max Planck Institute of Molecular Cell Biology and Genetics
Dresden, Germany

Natasha J. Caplen
National Cancer Institute
Bethesda, Maryland

Virander S. Chauhan
International Center for Genetic Engineering and Biotechnology
New Delhi, India

Irvin S.Y. Chen
University of California–Los Angeles
Los Angeles, California

Palakodeti V.N. Dasaradhi
International Center for Genetic Engineering and Biotechnology
New Delhi, India

Olivier Donzé
Apotech Corporation
Université de Genève
Genève, Switzerland

Graeme Doran
University of Oxford
Oxford, United Kingdom

David Drechsel
Max Planck Institute of Molecular Cell Biology and Genetics
Dresden, Germany

Mark E. Drew
Washington University School of Medicine
St. Louis, Missouri

Peter Dudek
Université de Genève
Genève, Switzerland

Paul T. Englund
Johns Hopkins University School of Medicine
Baltimore, Maryland

M. Figlerowicz
Institute of Bioorganic Chemistry
Polish Academy of Sciences
Poznań, Poland

Stephen P. Fox
Cardiff University
Cardiff, United Kingdom

Lisa Frenz
Cyclacel Ltd.
Dundee, United Kingdom

David M. Glover
University of Cambridge
and
Cyclacel Ltd.
Cambridge, United Kingdom

Earl Gordon
Weill Medical College of Cornell University
New York, New York

B. Grabowska
Institute of Bioorganic Chemistry
Polish Academy of Sciences
Poznań, Poland

Hidetoshi Hasuwa
Osaka University
Osaka, Japan

Andrew J. Hollins
Cardiff University
Cardiff, United Kingdom

Manzar J. Hossain
International Center for Genetic Engineering and Biotechnology
New Delhi, India

Dong-Yan Jin
The University of Hong Kong
Hong Kong, The Peoples' Republic of China

Ralf Kittler
Max Planck Institute of Molecular Cell Biology and Genetics
Dresden, Germany

Ulrich Klahre
University Heidelberg
Heidelberg, Germany

Kin Hang Kok
The University of Hong Kong
Hong Kong, The Peoples' Republic of China

Judy Lieberman
Harvard Medical School
Boston, Massachusetts

Valentine M. Macaulay
The Weatherall Institute of Molecular Medicine
Oxford, United Kingdom

Pawan Malhotra
International Center for Genetic Engineering and Biotechnology
New Delhi, India

William S. Marshall
Dharmacon, Inc.
Lafayette, Colorado

Okabe Masaru
Osaka University
Osaka, Japan

Glenn A. McConkey
University of Leeds
Leeds, United Kingdom

Frederick Meins, Jr.
The Friedrich Miescher Institute for Biomedical Research
Basel, Switzerland

Makoto Miyagishi
The University of Tokyo
Tokyo, Japan

Asif Mohmmed
International Center for Genetic Engineering and Biotechnology
New Delhi, India

James C. Morris
Clemson University
Clemson, South Carolina

Shawn A. Motyka
Johns Hopkins University School of Medicine,
Baltimore, Maryland

A. Pacak
Adam Mickiewicz University
Poznań, Poland

Deborah Palliser
Harvard Medical School
Boston, Massachusetts

Didier Picard
Université de Genève
Genève, Switzerland

Ronald H.A. Plasterk
Hubrecht Laboratory, Center for Biomedical Genetics
Utrecht, The Netherlands

Xiao-Feng Qin
UT MD Anderson Cancer Center
Houston, Texas

Angela Reynolds
Dharmacon, Inc.
Lafayette, Colorado

Martine Ruer
Max Planck Institute of Molecular Cell Biology and Genetics
Dresden, Germany

Masayuki Sano
National Institute of Advanced Industrial Science and Technology
Tsukuba Science City, Japan

Dániel Silhavy
Agricultural Biotechnology Center
Gödöllö, Hungary

Femke Simmer
Hubrecht Laboratory, Center for Biomedical Genetics
Utrecht, The Netherlands

Rita Sinka
University of Cambridge
Cambridge, United Kingdom

Muhammad Sohail
University of Oxford
Oxford, United Kingdom

Erwei Song
Harvard Medical School
Boston, Massachusetts
and
Sun-Yat-Sen University
Guangzhou, The Peoples' Republic of China

Edwin M. Southern
University of Oxford
Oxford, United Kingdom

Patrick Stern
Massachusetts Institute of Technology
Cambridge, Massachusetts

Kazunari Taira
The University of Tokyo
Hongo, Tokyo
and
National Institute of Advanced Industrial Science and Technology (AIST)
Tsukuba Science City, Japan

Luk VanParijs
Massachusetts Institute of Technology
Cambridge, Massachusetts

Renu Wadhwa
National Institute of Advanced Industrial Science and Technology (AIST)
Tsukuba Science City, Japan

Zefeng Wang
Massachusetts Institute of Technology
Cambridge, Massachusetts

Robert W. Williams
University of California
Berkeley, California

J. Wojciechowicz
University of Warmia and Mazury
Olsztyn, Poland

Alison Woollard
University of Oxford
Oxford, United Kingdom

Table of Contents

1 Gene Silencing by RNA Interference and the Role of Small Interfering RNAs

Natasha J. Caplen

CONTENTS

The completion or near completion of several whole genome sequence projects has highlighted two important issues: (1) a far higher percentage of transcription results in the generation of non-coding RNAS than was originally predicted, and (2) there is now a need to rapidly translate genomic information into functional data. The recent identification of RNA interference (RNAi) an endogenous post-transcriptional gene-silencing mechanism that requires formation of a non-coding RNA containing protein complex, and the rapid development of technologies that exploit this mechanism to determine gene function illustrates the need to characterize basic cellular processes so that these can be exploited as scientific tools. This chapter summarizes RNA-mediated epigenetic gene-silencing mechanisms, the identification and role of key players in RNAi and how this work has impacted on both our understanding of the role of non-coding RNAs in regulating endogenous gene expression and on the development of a whole new research field utilizing RNAi-generated knockdowns, even up to a genome-wide scale to study gene function.

1.1 THE IDENTIFICATION OF POST-TRANSCRIPTIONAL GENE SILENCING MECHANISMS

Post-transcriptional gene silencing (PTGS) or RNA interference (RNAi) has been observed in a wide range of organisms including plants, with *Nicotina* and *Arabidopsis* the species most widely studied; fungi, in particular *Neurospora crassa*; several

0-8493-2141-7/05/$0.00 + $.50

Metazoan and Protozoan species, for example, *Planarian*, *Hydra,* and *Trypanosoma*; and more complex organisms, particularly *Caenorhabditis elegans*, *Drosophila,* and most recently mammalian cells both human and mouse.[1–17] The term PTGS has been used mostly to indicate epigenetic post-transcriptional gene-silencing in plants and fungi (also known as co-suppression and quelling, respectively), whereas RNAi has been mostly applied to a similar epigenetic gene-silencing effect seen in invertebrates and mammals, though it has been suggested that the collective term of RNA silencing is more appropriate for all of these processes.[18] Central to PTGS and RNAi in all species studied to date is the role of small RNA molecules (20 to 25 nts in length) and the formation of a ribonucleoprotein complex called the RNA-induced silencing complex or RISC.[19–22] In plants, RNA silencing can be observed following the introduction of multiple copies of a transgene and in cells infected with cytoplasmically replicating RNA viruses.[23] Plants defective for PTGS, show increased susceptibility to infection and viruses produce proteins that block PTGS, leading to the hypothesis that PTGS acts as an antiviral response in plants.[24,25] The term RNAi was first used to describe the inhibition of gene expression seen in *C. elegans* following the direct injection into embryos of double-stranded RNA (dsRNA) of a sequence cognate to the gene that was silenced.[6] As in plants and fungi the silencing induced by dsRNA in *C. elegans* is highly potent, suggesting that an intracellular process both amplifies and spreads this effect.[6,7,26] Double-stranded RNA was quickly shown to mediate a similar inhibition of gene expression in other invertebrate species, including *Drosophila*.[8,9]

Though dsRNA generated through aberrant transcription from multiple transgenes (particularly those arranged as an inverted repeat), or from a viral intermediate, or introduced exogenously appeared to be the triggers of both PTGS and RNAi, it was unclear for some time how degradation of the target RNA was mediated. The first evidence suggesting a role for small RNA molecules in PTGS and RNAi emerged from a plant study that showed that in cells where PTGS had been triggered (by a transgene or virus) there was an accumulation of a small species of RNA estimated to be approximately 25 nucleotides (nts) in size.[19] Subsequent *in vitro* analysis of RNAi in *Drosophila* cell extracts showed that a small species of RNA, now termed small or short interfering RNAs (siRNAs), are enzymatically derived from the input double-stranded RNA.[20,22,27–29] Cleavage of the target mRNA was observed at a site that corresponded to approximately the middle of the siRNA, suggesting that the siRNA acts as guide for an enzyme-containing complex that leads to degradation of the target mRNA.[22,30] Since the identification of siRNAs, their central role in PTGS and RNAi has been more fully defined and their relationship to endogenous small non-coding RNAs has begun to be elucidated.

1.2 RNA SILENCING, siRNAS, AND RISC

siRNAs processed from dsRNA within a cell are typically of 20 to 25 nts in length, though most have been sized as 21 to 23 nt.[20,22] The enzyme responsible for the processing of siRNAs from dsRNA is an RNase III-like endonuclease termed Dicer.[29] The siRNAs are duplex RNA molecules with 2 or 3 nt 3' overhangs of ssRNA and carry a

5' phosphate group.[20,22,28] It is unclear at exactly what point an individual siRNA is unwound, a process that is ATP dependent, but stable RISC complexes appear to contain just one strand of the siRNA; ssRNA can enter the RNAi pathway but do so fair less efficiently.[20,31–33] siRNAs can be generated from all parts of the input dsRNA but there may be an accumulation of those sequences that most effectively induce silencing.[34,35] This selection may relate to certain features of the sequence of siRNAs that influence the effectiveness of a particular siRNA. An important feature may be the internal energy of the siRNA molecule, particularly the internal stability of the 5' end of the antisense strand compared to the target sequences, which in most effective siRNAs seems to be lower than other portions of the molecule; it is assumed that this allows the siRNA to be efficiently unwound.[34] The internal free energy around the predicted cleavage site also appears to need to be relatively low in comparison with adjacent regions of the molecule.[34] There is evidence from studies of RNAi in *Drosophila* cells that at least one protein, called R2D2, is involved in the transfer of siRNAs as they are generated by Dicer onto proteins that from part of the RISC complex.[36] R2D2 may influence the polarity (sense/antisense orientation) with which the siRNA interacts with components of RISC.

Several protein components of RISC have been isolated. The first to be identified, Argonaute 2 (Ago2) appears to form the core of RSIC.[37] Ago2 and Dicer are both members of the PAZ (PIWI, Argonaute, and Zwille/Pinhead) domain-containing family of proteins.[38,39] Recent structural analysis of the *Drosophila* Ago2 protein has shown that the PAZ domain can form a nucleic acid-binding fold.[40,41] The affinity of the interaction of this domain and RNA is relatively weak, but the presence of a 3' single-stranded RNA overhang appears to be critical to the interactions.[40,41] RNA, single- and double-stranded, can also bind, though in the case of duplex DNA the binding is independent of a 3' overhang.[41] Other proteins associated with RISC include in *Drosophila*, the Vasa intronic gene product (a p68 RNA helicase homolog) and the fragile X mental retardation protein, and in human cells the homolog of Ago2–E2FC and Gemin3 and Gemin4.[31,42–46] The antisense single-stranded siRNA component of RISC leads to localization of the RISC complex through perfect sequence alignment of the two RNA molecules. Cleavage of the target occurs at a position ~10 nts from the first nucleotide that represented the first based pair from the 5' end of the original siRNA.[28] The ribonuclease(s) responsible for this cleavage is not defined but an enzyme with homology to Tudor staphylococcal nuclease is associated with RISC and effective RNAi.[47]

1.3 SPECIES-SPECIFIC ASPECTS OF RNA-TRIGGERED GENE SILENCING

The identification of siRNAs as intermediate molecules in the RNAi pathway was essential to establishing the presence of RNAi in mammalian cells. Many viruses produce dsRNA intermediates during their life cycle, and thus as part of an antiviral response mammalian cells have evolved potent responses to dsRNA that result in a nonspecific down-regulation in gene expression. The best-characterized responses involve the triggering RNase L-dependent pathways through activation of 2'5'-oligoadenylate synthetase and the interferon-associated dsRNA-dependent kinase (PKR) but there is also evidence of additional dsRNA-sensitive pathways. Though only limited data exist maximal activation of these pathways may be dependent on the size and concentration of the triggering dsRNA molecule. To test if RNA molecules modeled on the structure of siRNAs could mediate sequence-specific cleavage in mammalian cells, two groups transfected siRNA molecules made from chemically synthesized oligoribonucleotides into a variety of different cell lines.[16,22] Initially, using a plasmid/siRNA co-transfection model system of RNAi sequence-specific inhibition was detected at both RNA and protein level.[16,22] Endogenous gene expression could also be downregulated using siRNAs against Lamin A/C.[22]

In the last two years these findings have been utilized and extended by many groups using cell lines derived from a very broad range of cells and against transcripts involved in a broad range of cellular processes. Double-stranded RNAs of over 30 nts can be used in cells that have an attenuated response to larger dsRNAs such as those derived from embryonic cells, but in the vast majority of studies molecules are used based on a siRNA structure.[48,49] In most cases the expression of gene expression has been determined to be specific; however, non-specific effects have been reported that may reflect some low level triggering of non-RNAi dsRNA dependent pathways, concentration-dependent saturation of RISC which may effect its cellular role, and off-target effects due to imperfect sequence matches that can trigger alternative RNA silencing mechanisms.[50–56] Mammalian cells show no amplification or spread of the inhibition of gene expression mediated by synthetic siRNAs.[30,31,54,57–59] To overcome this limitation when applying RNAi as a functional genomics tool in mammalian cells, many groups have developed siRNAs that are expressed as a single transcript that forms a stem–loop structure where the stem corresponds to the siRNA.[60–64] A key advantage of these short hairpin RNAs (shRNAs) is that they can be expressed from plasmids, usually from RNA polymerase III promoters, enabling the generation of cell lines expressing a shRNA against a particular transcript.[61,63,64]

An early and important observation unique to gene silencing in plants, *Neurospora* and *C. elegans* is the amplification and spread of the gene-silencing effect. The local (cell-to-cell plasmodesmata-based movement) and distant (phloem-based transport) spread of RNA silencing seen in plants can be observed following the use of an exogenous nucleic acid trigger and transgenic plants that induce RNA silencing in grafts.[65–67] The molecule(s) that mediate this spread have not been formally identified, but candidates include siRNAs, larger dsRNAs, and an RNA/protein complex.[68] Recent data suggests that plants generate two different species of siRNA, one of ~21 to 22 nts in length and another group of ~24 to 26 nts that are probably generated by different Dicer enzymes.[69,70] The 21 to 22 nt siRNAs appear to be critical for mediating localized RNA silencing, with one model suggesting that short siRNAs can move directly to ~10 to 15 cells but for more extensive spread synthesis of additional siRNAs is required.[71] The intracellular generation of secondary siRNAs through the activity of an RNA-dependent RNA polymerase (RdRp) has been identified in plants (*Arabidopsis*, SGS2/SDE1), *N. crassa* (QDE-1) and *C. elegans* (Ego-1, RRF-1).[72–76] It is presumed that transitive RNA silencing or transitive RNAi, as this amplification process is termed, uses the homologous ssRNA transcript as a template to produce additional dsRNA that is then subjected to Dicer to generate further siRNA. In plants, the nature of the siRNA generated following amplification is unclear. One study suggests that only siRNAs of 21 nts are produced, whereas another study observed the generation of only the longer species of siRNA.[70,71] This difference may reflect species or model system-specific variations. The role of the larger siRNAs is unclear but in one study these have been linked to expression from endogenous retrotransposons that resulted in systemic and stable silencing and it has been suggested that it is this species that could act as a mobile silencing signal through the vascular (phloem) system of plants that could then trigger the intracellular production of small siRNAs at a distant site.[69,71] Additional evidence that the systemic spread of RNA silencing in plants could be mediated in different ways depending on the nature of the initiating molecule comes from data using grafting models that showed a difference in the induction of distant RNA-silencing effects depending if the trigger was transgene or amplicon derived.[77]

In *Neurospora* the QDE-1 RdRp catalyzes at least two activities that could relate to the mediation of RNA silencing; the generation of full length RNA duplexes from ssRNA templates (with no requirement for a primer) that could then act as substrate for Dicer, and shorter <21 nt ssRNA molecules corresponding to different parts of the template that may interact directly with RISC.[78] In *C. elegans* not only is there evidence for the presence of an RdRp-dependent amplification that generates secondary siRNAs, there is also evidence for the active transport of dsRNA through the transmembrane protein SID-1 which was identified by analysis of *C. elegans* mutants that were sensitive to local RNAi but that were defective for systemic RNAi.[75,76,79–81] Expression of SID-1 as a transgene in *Drosophila* cells can induce these cells to take up dsRNA from medium that can then mediate RNAi.[81] *Drosophila* cells naturally show no cell-to-cell movement of RNA silencing and most studies suggest that there is no RdRP amplification of RNAi in *Drosophila,* though one study has reported synthesis of dsRNA in *Drosophila* cell extracts.[30,82,83]

1.4 RNAi AND EPIGENETIC ENDOGENOUS RNA SILENCING

A highly coordinated control of gene expression is essential for normal development and differentiation and for the maintenance of appropriate cellular function. In addition, there is a need for the cell to protect itself from the potential effect of aberrant expression from the large amount of selfish genetic material that is present within the genome. Until recently most research on the control of gene expression has focused on gene regulation at a genomic, transcriptional or translational level; however, with the identification of RNA silencing the role of RNA both as a mediator and as a target for controlling expression has become increasingly apparent. The first evidence for a direct link between RNAi and the control of endogenous gene expression came with the identification of a role for Dicer in the processing what are now known to be an important group of endogenous non-coding RNAs termed micro RNAS (miRNAs).[84–90] Micro RNAs are 21–23 nt ssRNA molecules that are processed by Dicer from large stem-loop single-stranded RNAs termed precursor miRNAs in an asymmetric manner.[91–94] Precursor miRNAs are themselves processed from the initially expressed transcript by another PAZ domain containing RNase III nuclease called Drosha.[95,96] Micro RNAs also interact with many, if not all, of the same components of the RISC complex that siRNAs interact with.[31,42–46] However, miRNA-RISC complexes appear to mediate silencing by a somewhat different manner in that the interaction between miRNAs and the putative target transcripts seems to require only minimal homology between the sequences of the two RNA species that leads to a structural perturbation of the mRNA and/or a locking of the RISC complex at the site of binding so that protein translation is blocked.[97–101] The action of miRNAs may be different in plants in that the miRNAs in plants have exact sequence matches with their putative target mRNAs. This suggests that here either an siRNA-like cleavage or an miRNA-like protein translational blockade could occur.[70,86,102–105] Very few target transcripts that miRNAs regulate are known, but of those that have been characterized many are associated with developmentally regulated genes which may explain, at least in part, the embryonic lethal phenotype seen in Dicer transgenic knockout mice.[105–108]

A second role for RNAi in regulating endogenous expression links it to modifications of genomic DNA that may induce repression of transcription from a variety of sources including coding sequences, heterochromatin regions and other repetitive sequences including transposons. In plants methylation of genomic DNA is often seen within the transcribed region that encodes the transcript that is the subject of RNA silencing.[109,110] This has been best studied when an RNA virus carrying a sequence homologous to the endogenous transcript triggers RNA silencing; however, it is unclear how the RNA–DNA interaction triggers methylation or which methylase is involved.[111–113] A link between heterochromatin formation and RNAi has currently only been studied in *Schizosaccharomyces pombe* where *Dcr (Dicer)*, *Ago2*, and *RdRp* knockout mutant yeast, show transcription from previously silent regions as a result of the loss of histone H3 lysine methylation, which allows for Swi6 localization and thus the spreading of heterochromatin formation.[114,115] Overlapping non-coding RNAs expressed from both strands of the repeat sequences within heterochromatin regions have been purified and siRNAs corresponding to these repeat regions have been identified.[114,116] Loss of heterochromatin RNAi in *S. pombe* leads to disruption in centromere and telomere function, leading to disruption of both mitosis and meiosis.[114,117] Another set of repeat sequences that may be silenced endogenously by RNAi is selfish genetic material such as integrated retroviral and retrotransposons. In *C. elegans* some RNAi mutants show enhanced activation of transposition and in *S. pombe* transcribed long terminal repeats can induce RNAi-dependent chromatin silencing that can extend to closely neighboring transcribed sequence.[118–120]

1.5 APPLICATIONS OF RNAi BASED TECHNOLOGIES

Sequencing and gene expression studies on a whole genome scale have generated enormous amounts of data that imply a function for a particular gene, but a direct functional analysis on a similar scale has not been possible. Determining the function of a gene through knockout transgenics has been of enormous value but this has not been applicable on a high throughput basis and is often restricted by aspects of the biology of the organism. In contrast technologies based on knockdowns generated on PTGS and RNAi have been rapidly developed on a scale from single gene to whole genome, with application both at cell culture and whole organism levels using transient and stable induction of an RNAi phenotype. Subsequent chapters in this book will describe in detail the different technologies that have been developed that exploit the RNAi pathway. Some of these technologies are species specific, whereas the use of others will be dependent on the biological question that is being asked. Key applications for RNAi-based technologies include direct gene analysis on a gene-by-gene basis, but this has also been extended to examine whole pathways and to examine broad RNAi phenotypes on a high throughput basis in whole organisms.[121–127] One advantage of RNAi-based knockdowns is that multiple gene knockdowns can be simultaneously or consecutively analyzed and, particularly in mammalian cells because different siRNA/shRNAs sequences can produce different levels of knockdown, the effect of modulating expression by inducing different degrees of inhibition can be studied.[128] In addition, RNAi has the potential to generate new model systems both cell line- and whole organism-based.[129–131] From a clinical perspective, an important application of RNAi knockdowns in mammalian cells is for drug target validation and for determining the specificity with which a particular drug interacts with a particular target molecule. In addition, as result of RNAi functional studies many new drug targets will be identified for which small molecule drugs can be developed, or it may be possible to develop the RNAi molecule itself as an inhibitor, leading to the concept of RNAi as a therapeutic approach.[132–134]

REFERENCES

1. Romano, N. and Macino, G., Quelling: transient inactivation of gene expression in *Neurospora crassa* by transformation with homologous sequences, *Mol. Microbiol.,* 6, 3343, 1992.
2. Cogoni, C., Romano, N., and Macino, G., Suppression of gene expression by homologous transgenes, *Antonie Van Leeuwenhoek,* 65, 205, 1994.
3. Cogoni, C. and Macino, G., Isolation of quelling-defective (qde) mutants impaired in posttranscriptional transgene-induced gene silencing in *Neurospora crassa*, *Proc. Natl. Acad. Sci. U.S.A.,* 94, 10233, 1997.
4. Beclin, C. et al., Infection of tobacco or *Arabidopsis* plants by CMV counteracts systemic post-transcriptional silencing of nonviral (trans)genes, *Virology,* 252, 313, 1998.
5. Ngo, H. et al., Double-stranded RNA induces mRNA degradation in *Trypanosoma brucei*, *Proc. Natl. Acad. Sci. U.S.A.,* 95, 14687, 1998.
6. Fire, A. et al., Potent and specific genetic interference by double-stranded RNA in *Caenorhabditis elegans*, *Nature,* 391, 806, 1998.
7. Montgomery, M.K., Xu, S., and Fire, A., RNA as a target of double-stranded RNA-mediated genetic interference in *Caenorhabditis elegans*, *Proc. Natl. Acad. Sci., U.S.A.,* 95, 15502, 1998.
8. Kennerdell, J.R. and Carthew, R.W., Use of dsRNA-mediated genetic interference to demonstrate that *frizzled* and *frizzled 2* act in the wingless pathway, *Cell,* 95, 1017, 1998.
9. Misquitta, L. and Paterson, B.M., Targeted disruption of gene function in *Drosophila* by RNA interference (RNA-i): a role for *nautilus* in embryonic somatic muscle formation, *Proc. Natl. Acad. Sci. U.S.A.,* 96, 1451, 1999.

10. Sanchez A.A. and Newmark, P.A., Double-stranded RNA specifically disrupts gene expression during planarian regeneration, *Proc. Natl. Acad. Sci. U.S.A.,* 96, 5049, 1999.
11. Lohmann, J.U., Endl, I., and Bosch, T.C., Silencing of Developmental Genes in Hydra, *Dev. Biol.,* 214, 211, 1999.
12. Tuschl, T. et al., Targeted mRNA degradation by double-stranded RNA *in vitro*, *Genes Dev.,* 13, 3191, 1999.
13. Vaucheret, H., Beclin, C., and Fagard, M., Post-transcriptional gene silencing in plants, *J. Cell Sci.,* 114, 3083, 2001.
14. Matzke, M.A. et al., RNA-based silencing strategies in plants, *Curr. Opin. Genet. Dev.,* 11, 221, 2001.
15. Cogoni, C., Homology-dependent gene silencing mechanisms in fungi, *Annu. Rev. Microbiol.,* 55, 381, 2001.
16. Caplen, N.J. et al., Specific inhibition of gene expression by small double-stranded RNAs in invertebrate and vertebrate systems, *Proc. Natl. Acad. Sci. U.S.A.,* 98, 9742, 2001.
17. Elbashir, S.M. et al., Duplexes of 21-nucleotide RNAs mediate RNA interference in cultured mammalian cells, *Nature,* 411, 494, 2001.
18. Baulcombe, D., RNA silencing, *Curr. Biol.,* 12, R82, 2002.
19. Hamilton, A.J. and Baulcombe, D.C., A species of small antisense RNA in posttranscriptional gene silencing in plants, *Science,* 286, 950, 1999.
20. Zamore, P.D. et al., RNAi: double-stranded RNA directs the ATP-dependent cleavage of mRNA at 21 to 23 nucleotide intervals, *Cell,* 101, 25, 2000.
21. Hammond, S.M. et al., An RNA-directed nuclease mediates post-transcriptional gene silencing in *Drosophila* cells, *Nature,* 404, 293, 2000.
22. Elbashir, S.M., Lendeckel, W., and Tuschl, T., RNA interference is mediated by 21- and 22-nucleotide RNAs, *Genes Dev.,* 15, 188, 2001.
23. Vaucheret, H. and Fagard, M., Transcriptional gene silencing in plants: targets, inducers and regulators, *Trends Genet.,* 17, 29, 2001.
24. Li, W.X. and Ding, S.W., Viral suppressors of RNA silencing, *Curr. Opin. Biotechnol.,* 12, 150, 2001.
25. Baulcombe, D., Viral suppression of systemic silencing, *Trends Microbiol.,* 10, 306, 2002.
26. Timmons, L. and Fire, A., Specific interference by ingested dsRNA, *Nature,* 395, 854, 1998.
27. Yang, D., Lu, H., and Erickson, J.W., Evidence that processed small dsRNAs may mediate sequence-specific mRNA degradation during RNAi in *Drosophila* embryos, *Curr. Biol.,* 10, 1191, 2000.
28. Elbashir, S.M. et al., Functional anatomy of siRNAs for mediating efficient RNAi in *Drosophila melanogaster* embryo lysate, *EMBO J.,* 20, 6877, 2001.
29. Bernstein, E. et al., Role for a bidentate ribonuclease in the initiation step of RNA interference, *Nature,* 409, 363, 2001.
30. Schwarz, D.S. et al., Evidence that siRNAs function as guides, not primers, in the *Drosophil*a and human RNAi pathways, *Mol. Cell,* 10, 537, 2002.
31. Martinez, J. et al., Single-stranded antisense siRNAs guide target RNA cleavage in RNAi, *Cell,* 110, 563, 2002.
32. Nykanen, A., Haley, B., and Zamore, P.D., ATP requirements and small interfering RNA structure in the RNA interference pathway, *Cell,* 107, 309, 2001.
33. Holen, T. et al., Similar behaviour of single-strand and double-strand siRNAs suggests they act through a common RNAi pathway, *Nucleic Acids Res.,* 31, 2401, 2003.
34. Khvorova, A., Reynolds, A., and Jayasena, S.D., Functional siRNAs and miRNAs Exhibit Strand Bias, *Cell,* 115, 209, 2003.

35. Schwarz, D.S. et al., Asymmetry in the Assembly of the RNAi Enzyme Complex, *Cell,* 115, 199, 2003.
36. Liu, Q. et al., R2D2, a bridge between the initiation and effector steps of the *Drosophila* RNAi pathway, *Science,* 301, 1921, 2003.
37. Hammond, S.M. et al., Argonaute2, a link between genetic and biochemical analyses of RNAi, *Science,* 293, 1146, 2001.
38. Cerutti, L., Mian, N., and Bateman, A., Domains in gene silencing and cell differentiation proteins: the novel PAZ domain and redefinition of the Piwi domain, *Trends Biochem. Sci.,* 25, 481, 2000.
39. Carmell, M.A., Xuan, Z., Zhang, M.Q., and Hannon, G.J., The Argonaute family: tentacles that reach into RNAi, developmental control, stem cell maintenance, and tumorigenesis, *Genes Dev.,* 16, 2733, 2002.
40. Song, J.J. et al., The crystal structure of the Argonaute2 PAZ domain reveals an RNA binding motif in RNAi effector complexes, *Nat. Struct. Biol.,* 10, 1026, 2003.
41. Lingel, A. et al., Structure and nucleic-acid binding of the *Drosophila* Argonaute 2 PAZ domain, *Nature*, 2003.
42. Caudy, A.A. et al., Fragile X-related protein and VIG associate with the RNA interference machinery, *Genes Dev.,* 16, 2491, 2002.
43. Ishizuka, A., Siomi, M.C., and Siomi, H., A *Drosophila* fragile X protein interacts with components of RNAi and ribosomal proteins, *Genes Dev.,* 16, 2497, 2002.
44. Mourelatos, Z. et al., miRNPs: a novel class of ribonucleoproteins containing numerous microRNAs, *Genes Dev.,* 16, 720, 2002.
45. Hutvagner, G. and Zamore, P.D., A microRNA in a multiple-turnover RNAi enzyme complex, *Science,* 297, 2056, 2002.
46. Doi, N. et al., Short-Interfering-RNA-Mediated Gene Silencing in Mammalian Cells Requires Dicer and eIF2C Translation Initiation Factors, *Curr. Biol.,* 13, 41, 2003.
47. Caudy, A.A. et al., A micrococcal nuclease homologue in RNAi effector complexes, *Nature,* 425, 411, 2003.
48. Wianny, F. and Zernicka-Goetz, M., Specific interference with gene function by double-stranded RNA in early mouse development, *Nat. Cell Biol.,* 2, 70, 2000.
49. Billy, E. et al., Specific interference with gene expression induced by long, double-stranded RNA in mouse embryonal teratocarcinoma cell lines, *Proc. Natl. Acad. Sci. U.S.A.,* 98, 14428, 2001.
50. Bridge, A.J. et al., Induction of an interferon response by RNAi vectors in mammalian cells, *Nat. Genet.*, 2003.
51. Sledz, C.A. et al., Activation of the interferon system by short-interfering RNAs, *Nat. Cell Biol.,* 5, 834, 2003.
52. Semizarov, D. et al., Specificity of short interfering RNA determined through gene expression signatures, *Proc. Natl. Acad. Sci. U.S.A.,* 100, 6347, 2003.
53. Jackson, A.L. et al., Expression profiling reveals off-target gene regulation by RNAi, *Nat. Biotechnol.,* 21, 635, 2003.
54. Chi, J.T. et al., Genome wide view of gene silencing by small interfering RNAs, *Proc. Natl. Acad. Sci. U. S. A.,* 100, 6343, 2003.
55. Doench, J.G., Petersen, C.P., and Sharp, P.A., siRNAs can function as miRNAs, *Genes Dev.,* 17, 438, 2003.
56. Zeng, Y., Yi, R., and Cullen, B.R., MicroRNAs and small interfering RNAs can inhibit mRNA expression by similar mechanisms, *Proc. Natl. Acad. Sci. U.S.A.,* 100, 9779, 2003.
57. Celotto, A.M. and Graveley, B.R., Exon-specific RNAi: a tool for dissecting the functional relevance of alternative splicing, *RNA,* 8, 718, 2002.

58. Chiu, Y.L. and Rana, T.M., RNAi in human cells: basic structural and functional features of small interfering RNA, *Mol. Cell,* 10, 549, 2002.
59. Holen, T. et al., Positional effects of short interfering RNAs targeting the human coagulation trigger Tissue Factor, *Nucleic Acids Res.,* 30, 1757, 2002.
60. Paddison, P.J. et al., Short hairpin RNAs (shRNAs) induce sequence-specific silencing in mammalian cells, *Genes Dev.,* 16, 948, 2002.
61. Paddison, P.J., Caudy, A.A., and Hannon, G.J., Stable suppression of gene expression by RNAi in mammalian cells, *Proc. Natl. Acad. Sci. U.S.A.,* 99, 1443, 2002.
62. Lee, N.S. et al., Expression of small interfering RNAs targeted against HIV-1 rev transcripts in human cells, *Nat. Biotechnol.,* 20, 500, 2002.
63. Brummelkamp, T.R., Bernards, R., and Agami, R., Stable suppression of tumorigenicity by virus-mediated RNA interference, *Cancer Cell,* 2, 243, 2002.
64. Brummelkamp, T.R., Bernards, R., and Agami, R., A system for stable expression of short interfering RNAs in mammalian cells, *Science,* 296, 550, 2002.
65. Palauqui, J.C. et al., Systemic acquired silencing: transgene-specific post-transcriptional silencing is transmitted by grafting from silenced stocks to non- silenced scions, *EMBO J.,* 16, 4738, 1997.
66. Voinnet, O. and Baulcombe, D.C., Systemic signalling in gene silencing, *Nature,* 389, 553, 1997.
67. Voinnet, O. et al., Systemic spread of sequence-specific transgene RNA degradation in plants is initiated by localized introduction of ectopic promoterless DNA, *Cell,* 95, 177, 1998.
68. Mlotshwa, S. et al., RNA silencing and the mobile silencing signal, *Plant Cell,* 14 Suppl, S289, 2002.
69. Hamilton, A. et al., Two classes of short interfering RNA in RNA silencing, *EMBO J.,* 21, 4671, 2002.
70. Tang, G. et al., A biochemical framework for RNA silencing in plants, *Genes Dev.,* 17, 49, 2003.
71. Himber, C. et al., Transitivity-dependent and -independent cell-to-cell movement of RNA silencing, *EMBO J.,* 22, 4523, 2003.
72. Cogoni, C. and Macino, G., Gene silencing in *Neurospora crassa* requires a protein homologous to RNA-dependent RNA polymerase, *Nature,* 399, 166, 1999.
73. Dalmay, T. et al., An RNA-dependent RNA polymerase gene in *Arabidopsis* is required for posttranscriptional gene silencing mediated by a transgene but not by a virus, *Cell,* 101, 543, 2000.
74. Mourrain, P. et al., *Arabidopsis* SGS2 and SGS3 genes are required for posttranscriptional gene silencing and natural virus resistance, *Cell,* 101, 533, 2000.
75. Smardon, A. et al., EGO-1 is related to RNA-directed RNA polymerase and functions in germ-line development and RNA interference in *C. elegans*, *Curr. Biol.,* 10, 169, 2000.
76. Sijen, T. et al., On the role of RNA amplification in dsRNA-triggered gene silencing, *Cell,* 107, 465, 2001.
77. Mallory, A.C. et al., The capacity of transgenic tobacco to send a systemic RNA silencing signal depends on the nature of the inducing transgene locus, *Plant J.,* 35, 82, 2003.
78. Makeyev, E.V. and Bamford, D.H., Cellular RNA-Dependent RNA Polymerase Involved in Posttranscriptional Gene Silencing Has Two Distinct Activity Modes, *Mol. Cell,* 10, 1417, 2002.
79. Alder, M.N. et al., Gene silencing in *Caenorhabditis elegans* by transitive RNA interference, *RNA,* 9, 25, 2003.
80. Winston, W.M., Molodowitch, C., and Hunter, C.P., Systemic RNAi in *C. elegans* requires the putative transmembrane protein SID-1, *Science,* 295, 2456, 2002.

81. Feinberg, E.H. and Hunter, C.P., Transport of dsRNA into cells by the transmembrane protein SID-1, *Science,* 301, 1545, 2003.
82. Roignant, J.Y. et al., Absence of transitive and systemic pathways allows cell-specific and isoform-specific RNAi in *Drosophila*, *RNA,* 9, 299, 2003.
83. Lipardi, C., Wei, Q., and Paterson, B.M., RNAi as random degradative PCR: siRNA primers convert mRNA into dsRNAs that are degraded to generate new siRNAs, *Cell,* 107, 297, 2001.
84. Lau, N.C. et al., An abundant class of tiny RNAs with probable regulatory roles in *Caenorhabditis elegans*, *Science,* 294, 858, 2001.
85. Lagos-Quintana, M. et al., Identification of novel genes coding for small expressed RNAs, *Science,* 294, 853, 2001.
86. Reinhart, B.J. et al., MicroRNAs in plants, *Genes Dev.,* 16, 1616, 2002.
87. Lim, L.P. et al., The microRNAs of *Caenorhabditis elegans*, *Genes Dev.,* 17, 991, 2003.
88. Lim, L.P. et al., Vertebrate microRNA genes, *Science,* 299, 1540, 2003.
89. Lagos-Quintana, M. et al., New microRNAs from mouse and human, *RNA,* 9, 175, 2003.
90. Ambros, V., et al. MicroRNAs and Other Tiny Endogenous RNAs in *C. elegans*, *Curr. Biol.,* 13, 807, 2003.
91. Grishok, A. et al., Genes and mechanisms related to RNA interference regulate expression of the small temporal RNAs that control *C. elegans* developmental timing, *Cell,* 106, 23, 2001.
92. Hutvagner, G. et al., A cellular function for the RNA-interference enzyme Dicer in the maturation of the let-7 small temporal RNA, *Science,* 293, 834, 2001.
93. Ketting, R.F. et al., Dicer functions in RNA interference and in synthesis of small RNA involved in developmental timing in *C. elegans*, *Genes Dev.,* 15, 2654, 2001.
94. Knight, S.W. and Bass, B.L., A role for the RNase III enzyme DCR-1 in RNA interference and germ line development in *Caenorhabditis* elegans, *Science,* 293, 2269, 2001.
95. Lee, Y. et al., MicroRNA maturation: stepwise processing and subcellular localization, *EMBO J.,* 21, 4663, 2002.
96. Lee, Y. et al., The nuclear RNase III Drosha initiates microRNA processing, *Nature,* 425, 415, 2003.
97. Lee, R.C., Feinbaum, R.L., and Ambros, V., The *C. elegans* heterochronic gene lin-4 encodes small RNAs with antisense complementarity to lin-14, *Cell,* 75, 843, 1993.
98. Wightman, B., Ha, I., and Ruvkun, G., Posttranscriptional regulation of the heterochronic gene lin-14 by lin-4 mediates temporal pattern formation in *C. elegans*, *Cell,* 75, 855, 1993.
99. Moss, e.g., Lee, R.C., and Ambros, V., The cold shock domain protein LIN-28 controls developmental timing in *C. elegans* and is regulated by the lin-4 RNA, *Cell,* 88, 637, 1997.
100. Reinhart, B.J. et al,. The 21-nucleotide let-7 RNA regulates developmental timing in *Caenorhabditis elegans*, *Nature,* 403, 901, 2000.
101. Slack, F.J. et al., The lin-41 RBCC gene acts in the *C. elegans* heterochronic pathway between the let-7 regulatory RNA and the LIN-29 transcription factor, *Mol. Cell,* 5, 659, 2000.
102. Rhoades, M.W. et al., Prediction of plant microRNA targets, *Cell,* 110, 513, 2002.
103. Llave, C., et al. Cleavage of Scarecrow-like mRNA targets directed by a class of *Arabidopsis* miRNA, *Science,* 297, 2053, 2002.
104. Park, W. et al., CARPEL FACTORY, a Dicer homolog, and HEN1, a novel protein, act in microRNA metabolism in *Arabidopsis thaliana*, *Curr. Biol.,* 12, 1484, 2002.
105. Kasschau, K.D. et al., P1/HC-Pro, a viral suppressor of RNA silencing, interferes with Arabidopsis development and miRNA unction, *Dev. Cell,* 4, 205, 2003.

106. Pasquinelli, A.E. and Ruvkun, G., Control of developmental timing by micrornas and their targets, *Annu. Rev. Cell Dev. Biol.,* 18, 495, 2002.
107. Bartel, B. and Bartel, D.P., MicroRNAs: at the root of plant development?, *Plant Physiol.,* 132, 709, 2003.
108. Bernstein, E. et al., Dicer is essential for mouse development, *Nat. Genet.,* 35, 215, 2003.
109. Jones, A.L., Thomas, C.L., and Maule, A.J., *De novo* methylation and co-suppression induced by a cytoplasmically replicating plant RNA virus, *EMBO J.,* 17, 6385, 1998.
110. Jones, L. et al., RNA-DNA interactions and DNA methylation in post-transcriptional gene silencing, *Plant Cell,* 11, 2291, 1999.
111. Jones, L., Ratcliff, F., and Baulcombe, D.C., RNA-directed transcriptional gene silencing in plants can be inherited independently of the RNA trigger and requires Met1 for maintenance, *Curr. Biol.,* 11, 747, 2001.
112. Vaistij, F.E., Jones, L., and Baulcombe, D.C., Spreading of RNA targeting and DNA methylation in RNA silencing requires transcription of the target gene and a putative RNA-dependent RNA polymerase, *Plant Cell,* 14, 857, 2002.
113. Aufsatz, W. et al., RNA-directed DNA methylation in *Arabidopsis*, *Proc. Natl. Acad. Sci. U.S.A.,* 99 Suppl 4, 16499, 2002.
114. Volpe, T.A. et al., Regulation of heterochromatic silencing and histone H3 lysine-9 methylation by RNAi, *Science,* 297, 1833, 2002.
115. Hall, I.M. et al., Establishment and maintenance of a heterochromatin domain, *Science,* 297, 2232, 2002.
116. Reinhart, B.J. and Bartel, D.P., Small RNAs correspond to centromere heterochromatic repeats, *Science,* 297, 1831, 2002.
117. Hall, I.M., Noma, K., and Grewal, S.I., RNA interference machinery regulates chromosome dynamics during mitosis and meiosis in fission yeast, *Proc. Natl. Acad. Sci. U.S.A.,* 100, 193, 2003.
118. Tabara, H. and Mello, C.C. et al., The *rde-1* gene, RNA interference, and transposon silencing in *C. elegans*, *Cell,* 99, 123, 1999.
119. Ketting, R.F. et al., *Mut-7* of *C. elegans*, required for transposon silencing and RNA interference, is a homolog of Werner syndrome helicase and RNaseD, *Cell,* 99, 133, 1999.
120. Schramke, V. and Allshire, R., Hairpin RNAs and retrotransposon LTRs effect RNAi and chromatin-based gene silencing, *Science,* 301, 1069, 2003.
121. Harborth, J. et al., Identification of essential genes in cultured mammalian cells using small interfering RNAs, *J. Cell Sci.,* 114, 4557, 2001.
122. Kamath, R.S. et al., Systematic functional analysis of the *Caenorhabditis elegans* genome using RNAi, *Nature,* 421, 231, 2003.
123. Simmer, F. et al., Genome-Wide RNAi of *C. elegans* Using the Hypersensitive rrf-3 Strain Reveals Novel Gene Functions, *PLoS Biol.,* 1, E12, 2003.
124. Lum, L. et al., Identification of Hedgehog pathway components by RNAi in *Drosophila* cultured cells, *Science,* 299, 2039, 2003.
125. Kiger, A. et al., A functional genomic analysis of cell morphology using RNA interference, *J. Biol.,* 2, 27, 2003.
126. Aza-Blanc, P. et al., Identification of modulators of TRAIL-induced apoptosis via RNAi-based phenotypic screening, *Mol. Cell,* 12, 627, 2003.
127. Chanda, S.K. et al., Genome-scale functional profiling of the mammalian AP-1 signaling pathway, *Proc. Natl. Acad. Sci. U.S.A.,* 100, 12153, 2003.
128. Hemann, M.T. et al., An epi-allelic series of p53 hypomorphs created by stable RNAi produces distinct tumor phenotypes *in vivo*, *Nat. Genet.,* 33, 396, 2003.

129. Tiscornia, G. et al., A general method for gene knockdown in mice by using lentiviral vectors expressing small interfering RNA, *Proc. Natl. Acad. Sci. U.S.A.,* 100, 1844, 2003.
130. Carmell, M.A. et al., Germline transmission of RNAi in mice, *Nat. Struct. Biol.,* 10, 91, 2003.
131. Rubinson, D.A. et al., A lentivirus-based system to functionally silence genes in primary mammalian cells, stem cells and transgenic mice by RNA interference, *Nat. Genet.,*33, 401, 2003.
132. Mccaffrey, A.P. et al., RNA interference in adult mice, *Nature,* 418, 38, 2002.
133. Lewis, D.L. et al. Efficient delivery of siRNA for inhibition of gene expression in post-natal mice, *Nat. Genet.,* 32, 107, 2002.
134. Caplen, N.J., RNAi as a gene therapy approach, *Expert. Opin. Biol. Ther.,* 3, 575, 2003.

2 Basics of siRNA Design and Chemical Synthesis

Queta Boese, Angela Reynolds, and William S. Marshall

CONTENTS

2.1 INTRODUCTION: SUCCESSFUL RNA INTERFERENCE DEPENDS ON DESIGN AND SYNTHESIS

In principle, RNA interference (RNAi) is a straightforward method for inducing sequence-specific silencing of one or more genes of interest with the simple introduction of small interfering RNAs (siRNAs). In practice, there are fundamental challenges to be considered in order to achieve successful gene silencing. Among these are siRNA design and production, as it is well known that not all siRNAs directed against a given target silence with equivalent efficiencies. Thus, identification of sequence related elements that are critical for functionality are important in the development of this new and potent technology. Similarly, there is a variety of methods available for producing the small functional duplex intermediates, the choice of which is largely driven by time, labor, and cost considerations.

0–8493–2141–7/05/$0.00 + $.50

2.2 DESIGN

Since the initial reports of RNAi in mammalian systems, numerous studies have echoed the necessity of screening multiple siRNAs to identify functional duplexes. For example, early attempts to silence vimentin failed,[1] resulting in the design of several additional duplexes before identifying one that effectively reduced gene expression.[2] Similarly, attempts to reduce or knockdown PLK1 (polo-like kinase 1),[3] POLH (DNA polymerase H),[4] and hTF (human Tissue Factor)[5] all exhibited varying degrees of success due to the wide range of functionality of the respective siRNAs (10 to 90%). Together these and other studies clearly demonstrate that the ability to identify a functional duplex is not trivial. Furthermore, these data demonstrate that siRNA selection strategies represent a critical first step that can have significant downstream implications in terms of the time and labor invested to achieve efficient siRNA-mediated gene knockdown.

The requirements for siRNA design include the availability of quality sequence information and sound guidelines for identifying functional duplexes. Recommended sources of reliable information are sequence databases hosted by NCBI (National Center for Biotechnology Information), DDBJ (DNA Data Bank of Japan), or EMBL (European Molecular Biology Laboratory) (see Table 2.1). These databases are populated with sequence information deposited from a variety of sources. Sequence entries include genomic, cDNA, expressed sequence tag (EST), or protein sequence information. The best resources are annotated databases, such as the Reference Sequence (RefSeq) or Unigene databases, which contain curated records representing alignments of multiple entries that resolve potential sequencing artifacts, errors, or single nucleotide polymorphisms (SNPs). The result is a unique "reference" sequence record for each gene.

The mRNA sequence entries derived from ESTs, e.g., RefSeq's NM _123456, are most often used because these records are supported by expression data. Information derived from whole genome sequencing efforts should be evaluated with care to

TABLE 2.1
Resources for Publicly Available Nucleotide Sequence Information and Analytical Tools

Database	URL	Contents
GenBank	ncbi.nlm.nih.gov/Genbank	Sequences assigned unique identifiers including "gi" (Gene Info) or alphanumeric codes (e.g., U12345 or AY123456)
RefSeq	ncbi.nlm.nih.gov/RefSeq	Curated sequences assigned unique alphanumeric codes (e.g., NM_123456 (mRNA specific))
Unigene	ncbi.nlm.nih.gov/UniGene	Collection of curated, non-redundant sequences
Entrez	ncbi.nlm.nih.gov/Entrez	Portal to databases and analytical tools
NCBI	ncbi.nlm.nih.gov	Collection of databases and analytical tools
EMBL	ebi.ac.uk/embl	Collection of databases and analytical tools; sequences assigned unique alphanumeric codes
DDBJ	ddbj.nig.ac.jp	Collection of databases and analytical tools; sequences assigned unique alphanumeric codes
BLAST	ncbi.nlm.nih.gov/BLAST	Analytical tool for aligning sequences

avoid selecting targets from introns, intergenic sequences, and non-coding regions. Protein sequences should also be avoided because the degeneracy of the genetic code will not allow for unambiguous nucleic acid sequence selection. Finally, it is important to remember that these sequence databases are dynamic and updated versions with new information appear periodically. Therefore, when troubleshooting a non-functional siRNA that is expected to be functional by all known criteria, it is prudent to review the original source of sequence information as well as the expression of the intended target in the host cell or model system.

Once a dependable mRNA sequence has been identified, an siRNA directed against a unique target must be designed. This can be achieved by one of several methods, including:

1. *Random design.* A method that selects 19 to 21 base targets within the mRNA without regard for sequence or positional information.
2. *Conventional design.* A method that selects 19 to 21 base targets using simple design parameters promoted by Dr. Thomas Tuschl and colleagues[6] followed by BLAST™ analysis to ensure that the target is unique.
3. *Rational design.* These methods employ algorithms that incorporate multiple parameters followed by stringent BLAST™ and extensive bioinformatic analyses (e.g., isoform inclusion or exclusion; Entrez SNP analysis) to identify functional siRNAs.

Of these three methods, the first two require screening at least three or more duplexes per target to identify the best siRNAs for consistent, reproducible silencing. The latter category eliminates the need to screen multiple duplexes because the design strategies rely on biochemical and biophysical parameters associated with siRNA functionality.

2.2.1 Random Methods of siRNA Design

Random siRNA design does not rely on sequence composition or relative position within the mRNA coding sequence or the 5' or 3' untranslated regions (UTRs) and has been used primarily to identify potential trends for selecting functional targets.[5,7,8] Figure 2.1 documents the range of silencing efficiencies for 90 siRNAs, targeting sites shifted by every other base within a select region in two genes: human cyclophilin (M60857; 193–390) and firefly luciferase (fLuc, U47298; 1434–1631).[8] In an effort to set standards for comparing functionality of different siRNAs, an "F" value designation is used throughout, where F95, F90, F80, and F70 refer to silencing or reduction in mRNA expression by 95, 90, 80, and 70% respectively. The histograms in Figure 2.1 show that there are no clear patterns or sequence preferences and the silencing efficiencies vary widely. Thus, when using random selection the probability of selecting a duplex that reduces target expression by 80% or more (≥ F80) is approximately 46% (41/90) for human cyclophilin and 53% (48/90) for firefly luciferase.

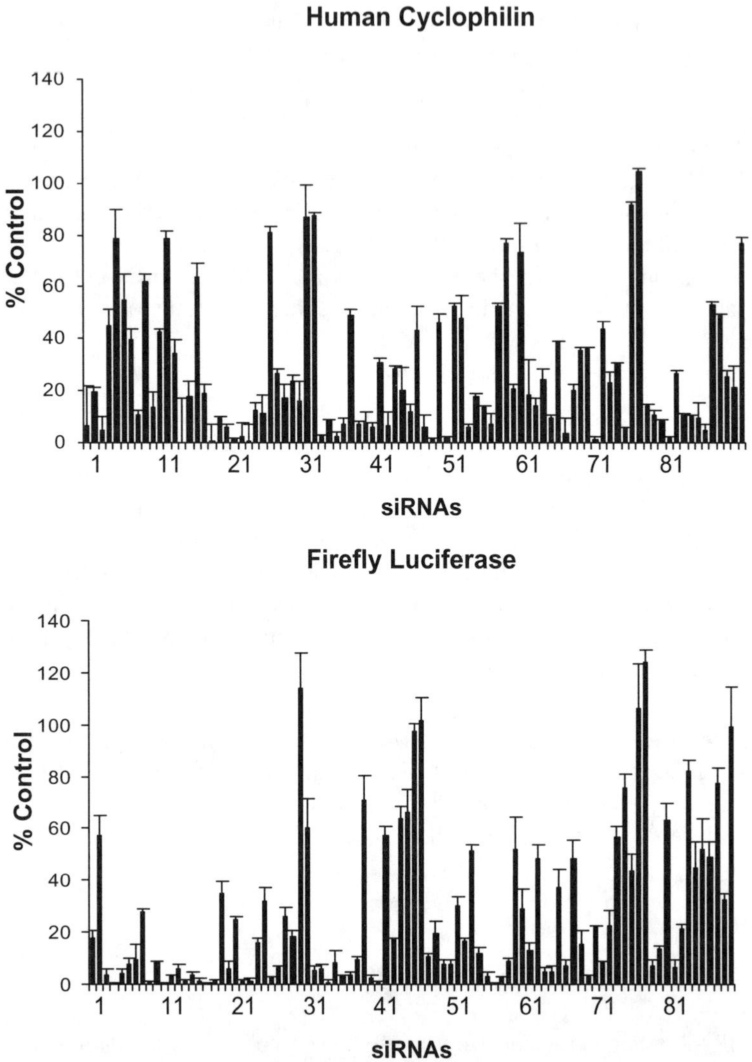

FIGURE 2.1 Silencing functionality of siRNAs targeting every other position of 197 base regions in two genes.[8] The targets are human cyclophilin (M60857) and firefly Luciferase (U47298). On average, 55% of the duplexes identified by this method reduce expression by at least 50% or better. HEK-293 or HEK-293-Luc cells were transfected at 75% confluency with 100 nM siRNA complexed with Lipofectamine 2000. mRNA detection was carried out 24 h post transfection using a branched DNA assay (Genospectra). Protein activity was assessed 24 h post transfection using the SteadyGlo luciferase assay (Promega). Reprinted with permission from Reynolds et al.[8]

When averaged across several genes, over half of the randomly chosen siRNA sequences are functional (> F50), of which only 10% elicit silencing of > F95.

Because it is difficult to predict the ease with which genes are targeted, it is not unusual to test as many as 10 siRNAs or more in an effort to identify a functional duplex using random selection. This phenomenon is illustrated in Table 2.2, which

TABLE 2.2
Random Selection of siRNAs Reveals that Some Genes Are Easy to Silence While Other Genes Are Difficult to Silence; Functional siRNAs Are Those for Which mRNA Is Reduced by 70% or Greater (> F70) and Non-Functional siRNAs Are Those for Which Reduction Is Less than 70% (< F70)

Target	Functional (>F70)	Non-Functional (<F70)	% Functional (>F70)
Cyclophilin B	55	35	61
FLuc	58	32	64
GAPDH	41	48	46
DBI	27	63	30

lists the success rate for siRNAs that reduce expression of a set of four genes by 70% or more (≥ F70) (human cyclophilin, glyceraldehyde-3-phosphate dehydrogenase (GAPDH, NM_002046), diazepam binding inhibitor (DBI, NM_020548), and firefly luciferase (fLuc) (unpublished data). Finally, random selection implies a lack of sequence comparisons used to ensure that the duplex is unique. Sequence identity to unrelated targets will inevitably lead to downregulation of unintended targets or off-target silencing effects, ultimately making this strategy impractical as a functional tool.

2.2.2 Conventional Methods of siRNA Design

siRNA design strategies should assess attributes that contribute to duplex functionality and specificity and perform sequence comparisons, e.g., BLAST™ analyses, to minimize off-target effects. Currently one of the most commonly used strategies for selecting 19-21mer mRNA target sequences is based on early *in vitro* biochemical studies that assessed functional and structural requirements for siRNA duplexes using *Drosophila* and mammalian cell lysates.[9,10] Figure 2.2 illustrates the initial steps of this method.

The search begins in the coding sequence for unique regions that are least likely to be occupied by translational or regulatory proteins.[6] This typically means excluding the first approximately 100 bases downstream of the start codon and the UTRs from the search. Sequence motifs characterized by an AA (or NA) dinucleotide preceding a 19 base sequence with 30 to 70% G/C content are then selected. The resulting siRNA is comprised of complementary antisense and sense strands with characteristic 3' overhangs that mimic the naturally occurring silencing intermediates. The AA dinucleotide leader motif is not part of the final duplex, but defines the sequence composition of the 3' overhangs (UU or TT). This motif was initially chosen in part for practical reasons when producing siRNA by chemical strategies. Chemical synthesis proceeds in the 3' → 5' direction and U or T represent the most economical of the nucleosides to initiate synthesis for some chemistries. Early studies also suggested that paired

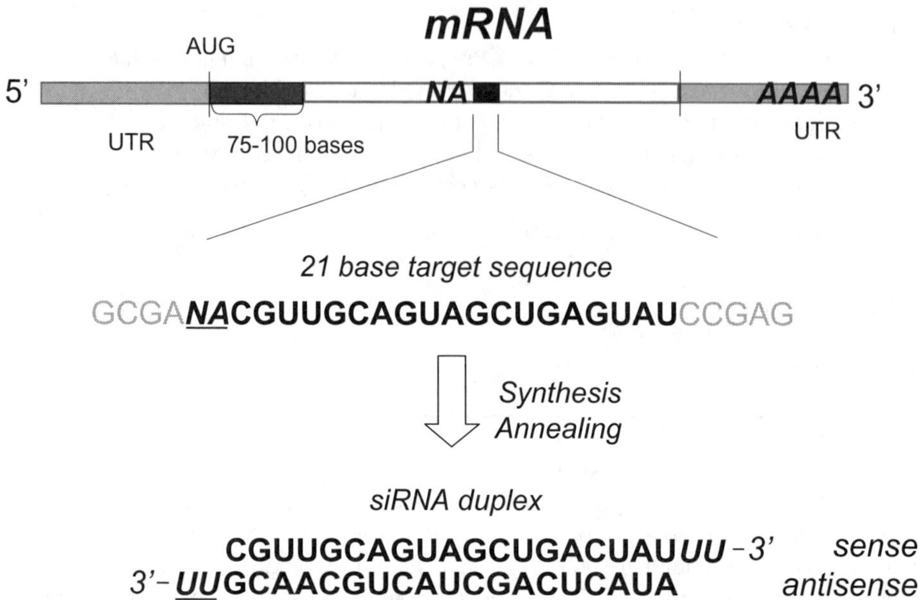

FIGURE 2.2 Selection of siRNAs by conventional methods. The coding region is evaluated for the $AA(N_{19})$ or $NA(N_{19})$ motif having a GC content of ~50%. The untranslated regions (UTRs) and the sequence following the start codon (~75 to 100 bases) are avoided to prevent targeting sites occupied by regulatory proteins or that are potentially polymorphic.[1] Reprinted with permission from Reynolds et al.[8]

deoxyribonucleotides might provide a degree of nuclease resistance, therefore duplexes are often designed to have TT overhangs.[1] While these same studies determined that the penultimate base in the overhang may contribute to target identification, more recent work suggests that the dinucleotide motif and the choice of 3' UU or TT overhangs have no functional consequence.

In general, $AA(N_{19})$ or $NA(N_{19})$ target motifs can also be found in the 5' or 3' UTRs of the transcript. Targeting the 3' UTR may be desirable for silencing and rescue strategies that employ expression constructs with heterologous UTRs to confirm sequence-specific silencing by a given siRNA.[11] However, because the UTRs tend to be more polymorphic or to contain conserved regulatory sequences, target sites in these regions should be selected carefully. Notably, the relative position of the target in the mRNA does not appear to follow a general trend, rather it is the sequence composition and biochemical properties of the duplex core that are key to functionality.

The final step in conventional design is to ensure that the selected siRNA is unique to prevent off-target effects. Typically this is achieved by performing sequence comparisons via BLAST™ to species-specific EST or annotated Unigene databases (see Table 2.1). The "EST" or "EST other" databases, which can be searched directly via BLAST™, are used in the absence of a species-specific collection e.g., when targeting virus, hamster, canine, etc.). Alignment parameters should be stringent with respect to allowable identity or mismatches. Typically, sequences with three mismatches or less to unrelated targets are eliminated as candidate target sites. Regardless of the database searched, the source of information populating the database should also be considered and those collections that include genomic, protein, and predicted sequence information (e.g., "nr" nucleotide database), should be used carefully for reasons mentioned previously).

On average, 60 to 70% of the duplexes identified by conventional design methods silence their target but with significant variability in silencing efficiencies. The consequence of this is the necessity to design three or more duplexes per target to ensure selection of a functional siRNA. For example, in a study to assess siRNA as a high throughput tool in a screen for regulators of apoptosis,[12] 62% (31/50) of a preliminary set of duplexes was found to exhibit silencing functionality of F70 or more. With this in mind, three siRNAs each were designed to target a panel of 510 genes. Genes selected for further analyses were then targeted by two additional duplexes to confirm sequence specific silencing (total of five). This strategy demonstrates the utility of siRNAs as a molecular tool but also highlights the inherent uncertainty in their design and functionality.

The variability in silencing is further illustrated for conventionally designed siRNAs targeting human GAPDH and DBI (see Figure 2.3). When selected by this

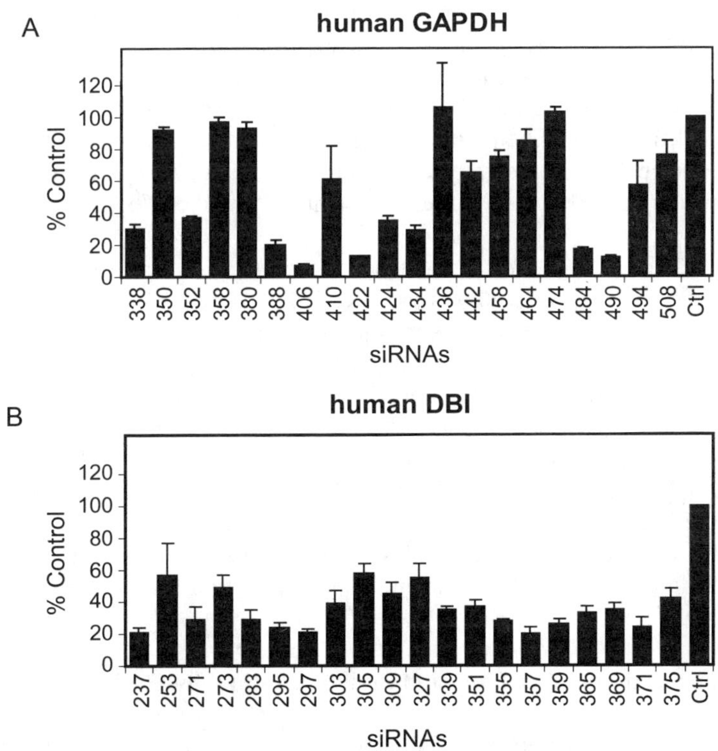

FIGURE 2.3 Silencing functionality of siRNAs designed using early conventional guidelines originated by Dr. Thomas Tuschl and colleagues.[6] The targets are human GAPDH (NM_002046) and DBI (NM_020548). (See Figure 2.1 legend for transfection conditions and functional assessments.) Reprinted with permission from Reynolds et al.[8]

strategy, the probability of identifying a duplex with a silencing functionality of ≥ F80 is approximately 25% (5/20) for human GAPDH and less than 5% (1/20) for human DBI. More importantly, the moderate silencing levels (F50–F80) induced by the majority of the conventionally designed duplexes may not be biologically significant, thus requiring additional screens to identify more potent siRNAs.

2.2.3 Rational Methods of siRNA Design

Conventional design methods are based on only a subset of features attributed to functional siRNAs. The inherent uncertainty of these selection strategies and unpredictable range of silencing efficiencies achieved with the resulting siRNAs add a costly and time-consuming screening step to RNAi. Thus, the development of sophisticated *in silico* procedures that accurately select functional duplexes and eliminate unnecessary screening are key to broadening the utility of RNAi as a tool for research and therapeutic applications.

A more effective method for identifying active siRNA target sites involves *in silico* rational design. Rather than randomly choosing target sites or identifying siRNA sequences using a limited number of parameters, rational design applies a multi-component algorithm to evaluate candidate sequences for optimal properties. The advantage of such a computational approach is the ability to adapt it to high throughput *in silico* selection of targets from tens of thousands of gene sequences in a relatively short period of time. Combined with extensive bioinformatic analyses, potential off-target effects can be minimized and the selected duplexes can be immediately sent into production via comparable high throughput synthetic methodologies (e.g., chemical synthesis).

One rational design algorithm evaluates the target sequences for attributes associated with functional siRNA duplexes and effectively eliminates non-functional siRNA from the pool of candidate sequences for optimal properties. An advanced version of this algorithm employs more than 30 parameters including (1) overall G/C content, (2) thermodynamic profiles of the duplex, (3) potential competing internal or fold-back structures, and (4) base preferences at select positions within the duplex. Each of these factors alone provides only a limited selection advantage; however when combined into a single algorithm, highly functional siRNAs can be identified from a population of candidate duplexes and all non-functional siRNAs are eliminated (see Figure 2.4). Notably, there are functional duplexes that are also eliminated based on these criteria, indicating that further refinements to the algorithm can be achieved.

2.2.3.1 Bioinformatically Controlled Pools of Rationally Designed siRNAs

The natural mechanism of RNAi is initiated by a long dsRNA that is processed by Dicer into a mixture of overlapping siRNAs.[9,13–15] One strategy that can be employed to further enhance gene silencing is to recapitulate the natural phenomenon of siRNA mixtures or pools, but employ bioinformatics and rational design algorithms to ensure

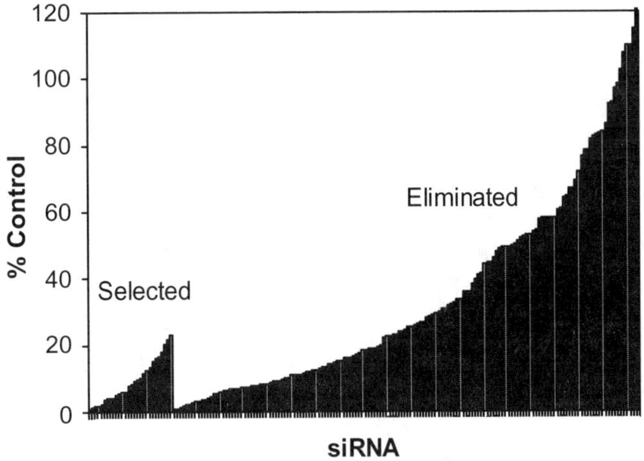

FIGURE 2.4 Application of rational criteria to evaluate sequences for optimal silencing efficiency. Application of multiple criteria significantly improves the probability of selecting a functional duplex. The "Selected" class is composed of only functional duplexes. Reprinted with permission from Reynolds et al.[8]

specificity and maximize potency. Previous studies suggested that multiple siRNAs directed against one or more target sequences could diminish gene-silencing effects,[5,16,17] possibly due to a mechanism that involves competition among the siRNAs for access to the silencing machinery (i.e., RISC). Interestingly, this phenomenon is not observed with mixtures of rationally designed siRNAs, which represent bioinformatically controlled correlates of naturally produced pools. Figure 2.5 demonstrates the advantage of rational design for the selection of functional siRNAs targeting two reporter genes, human secreted alkaline phosphatase (SEAP) and firefly luciferase (fLuc). For each gene, the silencing ability of randomly designed siRNA duplexes are compared to that of four individuals or one pool of four rationally designed siRNAs. Contrary to previous observations, pools perform with greater reliability, significantly raising the frequency of F95 silencing. Pooling rationally designed siRNA mimics the natural RNAi phenomenon while simultaneously maintaining high sequence specificity. This method for siRNA selection and pooling to achieve potent gene specific knockdown has been experimentally tested for over a thousand siRNA sequences against multiple gene targets and radically improves the probability of identifying functional siRNAs for any given target mRNA sequence.

2.2.3.2 Rational siRNA Design: Potency and Longevity

One attribute of rationally designed siRNAs is that on average, the potency and longevity of these duplexes is greater than that of conventionally designed molecules. Highly functional siRNAs e.g., ≥ F95) silence when used at sub-nanomolar concentrations (see Figure 2.6). In addition, these duplexes with a silencing efficiency of ≥ F90 have been shown to sustain gene knockdown for up to 7 days, while less potent siRNAs

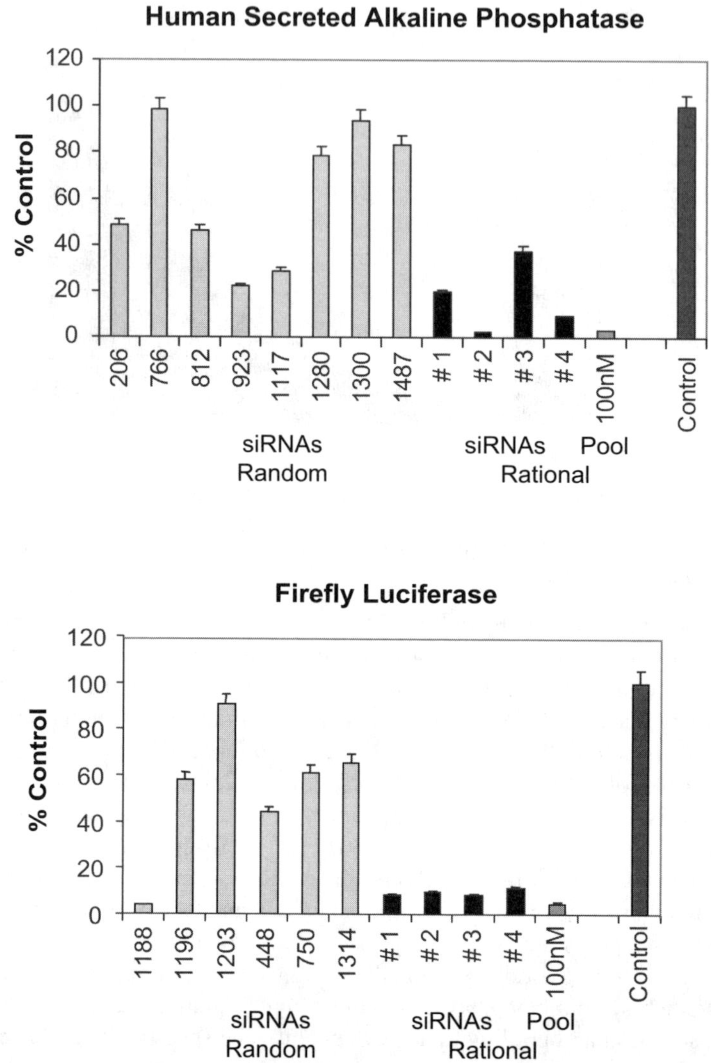

FIGURE 2.5 Silencing data comparing siRNAs designed by random or by an advanced rational selection method developed by the scientists at Dharmacon, Inc. The gene targets are human SEAP (NM_001632) and firefly Luciferase (U47298). Randomly selected siRNAs (light gray bars) are designated by their relative position in the coding sequence. Rationally selected siRNAs (black bars) and the pool of four rationally designed siRNAs (medium gray bars) illustrate significantly improved reduction in expression. siRNAs complexed with Lipofectamine 2000 were transfected at a total final concentrations of 100 nM into HEK293-fLuc or with a SEAP reporter into HEK293. Silencing was assessed 24 hrs (SEAP) or 72 hrs (fLuc) post-transfection by assaying for activity (Great EscAPe SEAP™ [Clontech]; SteadyGlo luciferase assay [Promega]). Reprinted with permission from Reynolds et al.[8]

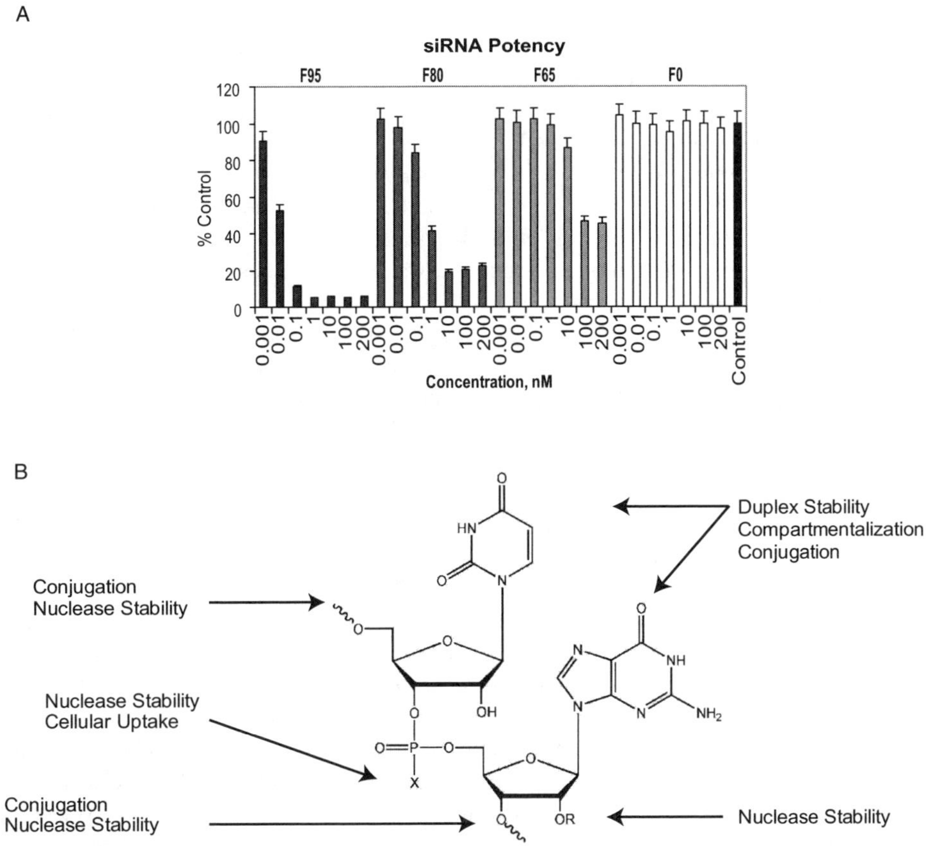

FIGURE 2.6 Representative data illustrating attributes of rationally designed siRNAs. (A) Highly functional (> F95) siRNAs are very potent, silencing at sub-nanomolar concentrations. (B) Chemical synthesis offers the flexibility of adding modified nucleosides or conjugates that may enhance siRNA potency without increasing toxicity.

e.g., < F70) maintain knockdown for shorter durations (data not shown). Potent duplexes such as these offer an alternative to expression systems designed to induce stable siRNA expression especially when knockdown periods of 5 to 7 days are sufficient. In addition, siRNAs can be chemically modified with functional groups that further enhance silencing and stability. Thus while conventional design methodologies frequently result in silencing reagents with variable functionality, rational design of highly potent siRNAs expands the utility of RNAi.

2.2.3.3 Rational siRNA Design: Specificity

Recent publications analyzing the genome-wide effects of siRNA via microarray analysis revealed the occurrence of duplex dependent off-target signatures. In one instance, data was derived from duplexes designed by conventional selection methods,[18] while in another study siRNAs were identified by a combination of conventional design guidelines plus additional sequence and biophysical criteria.[19] The latter group, which relied on conventional guidelines and additional selection criteria, observed few off-target signatures that were easily resolved by optimizing tranfection conditions (e.g., siRNA:lipid ratios, reducing final siRNA concentrations). In contrast, the group that relied solely on conventional design to select siRNAs targeting MAPK14 and IGF1R, identified down-regulation of multiple unrelated targets. The analysis of these off-target signatures revealed that the unintended targets had no direct or indirect functional relationship to the intended target and shared as little as 11 out of 19 bases of identity with the siRNA sequence.[18]

Upon closer examination of these off-target signatures,[18] the following observations can be made. First, the level of downregulation was relatively low for the unrelated targets, with 11 to 16 bases of identity to the siRNA (2- to 3-fold decrease relative to normal gene expression). Recent studies suggest that this low level of reduction may not be functionally significant.[20] The analysis of these off-target signatures revealed that the intended targets had no direct or indirect functional relationships to the intended target. While the off-target sequences shared as little as 11 out of 19 bases of identity with the siRNA sequence,[18] the level of down-regulation was relatively low for the unrelated targets (2- to 3-fold decrease relative to normal gene expression). Recent studies suggest that this low level of reduction may not be functionally significant.[20] Such analyses highlight the need for careful bioinformatic analyses to minimize off-target events associated with partial sequence identity followed by careful interpretation of the silencing experiments.

In addition to more stringent sequence comparisons, another approach for minimizing off-target signatures is based on current models for siRNA-RISC processing.[7,8,21] In these models the RISC interacts with the duplex by unwinding and then selecting one strand over another. The strand selection process can be predicted by the thermodynamic duplex stability profile of the siRNA. As chemical modifications and base pair mismatches can alter the thermodynamic profile and internal strand stability, it is possible to drive affinity of the RISC toward the strand complementary to the target (antisense), thus eliminating off-target effects directed by the sense strand. Furthermore, by modulating the affinity of RISC for the sense strand, this strategy may reduce competition between the two strands thus enhancing the potency of the siRNAs and reducing the effective concentration needed to induce silencing. Preliminary work using modified nucleosides in key positions of the duplex supports this hypothesis of biased strand selection by RISC (data not shown). Certain positional modifications in the sense strand had no effect on duplex silencing capacity yet effectively eliminated sense strand off-target effects. Moreover, these chemical modifications enhanced potency of moderately functional duplexes suggesting that as RISC–sense strand associations were eliminated, the probability of RISC–antisense strand interactions was improved.

2.3 SIRNA PRODUCTION TECHNOLOGIES

A variety of RNA production technologies are available for the assembly of candidate siRNA inhibitors. These include chemical synthesis, *in vitro* transcription,[22–24] and *in vivo* expression of siRNA duplexes or short hairpin RNA (shRNA) from PCR cassettes,[25] plasmids, or viral vectors.[26–29] *In vitro* digestion of unique dsRNA using recombinant Dicer[30,31] is an alternative method that does not require target site selection and mimics the natural intracellular RNAi process. Production methods should yield high quality RNA quickly, reliably, and economically. At the same time, production methods should be amenable to high-throughput formats and bulk syntheses, and the synthetic platform of choice should be flexible enough to permit the introduction of chemical modifications or conjugates that enhance silencing potency, longevity, and specificity. Among the methods mentioned above, duplexes generated by commercial suppliers using chemical synthesis methods deliver siRNAs that require the least amount of preparation time by the end user (see Protocol 2.1 to Protocol 2.5). The alternative production methods discussed in the adjoining chapters offer various cost-saving features but are associated with limitations regarding reliability, purity, scalability, and flexibility in design.

Chemically synthesized siRNAs are available through commercial suppliers who employ solid-phase synthesis strategies using automated synthesizers.[32] Sequences are generated as single strands on solid support by the sequential addition of protected nucleoside phosphoramidites. Extensive washes between each nucleotide addition effectively remove all reagents and reaction intermediates, ensuring sequence purity. Final yields are determined by UV spectrophotometry (260 nm), and molecular identity and purity are readily confirmed by mass spectrometry, strong anion exchange HPLC, or PAGE analyses.

Currently three RNA chemical synthesis platforms are commercially employed. These platforms are named for the protecting group that protects the 2' position of the ribose sugar during synthesis: 2'-tBDMS (2'-*O*-t-butyldimethylsilyl),[32,33] 2'-TOM (2'-*O*-triisopropylsilyloxymethyl),[34] and 2'-ACE® (2'-acetoxyethoxy)[35,36] (see Figure 2.7.a-c). As an example, the typical yield from a standard 25 mmole scale synthesis of a 21mer duplex (average molecular weight of 13,300 g/mole) employing 2'-ACE technology is 20 nmoles (approximately 266 μg). As a general guide, under optimized transfection conditions, when using a final concentration of 100 nM siRNA complexed with a lipid reagent, one nmole of siRNA is sufficient to transfect cells plated in 20 to 24 wells of a 24-well plate format.

Among the available RNA synthesis methods, 2'-ACE RNA chemistry offers several positive attributes.[37] Rather than being as adaptation of traditional DNA synthesis methods to RNA synthesis, the 2'-ACE method represents a fundamental redesign of protecting group strategies. The optimized protecting group scheme allows for synthesis of RNA with unprecedented crude product yield and purity. The approach replaces the traditional transient 5' acid labile protecting group with a fluoride ion labile protecting group. This alteration then allows for the implementation of the acid labile acetoxyethoxy orthester (ACE) at the 2' position. Implementation of the 2'-ACE allows for the RNA to remain in an intermediately protected form after the initial deprotection step. This offers several advantages in the ease of handling and also confers a blockage to the formation of higher order structures. The 2'-ACE protected RNA is water soluble, resistant to nuclease degradation, and stable during long-term storage. The protecting group is easily removed using mildly acidic conditions and can be carried

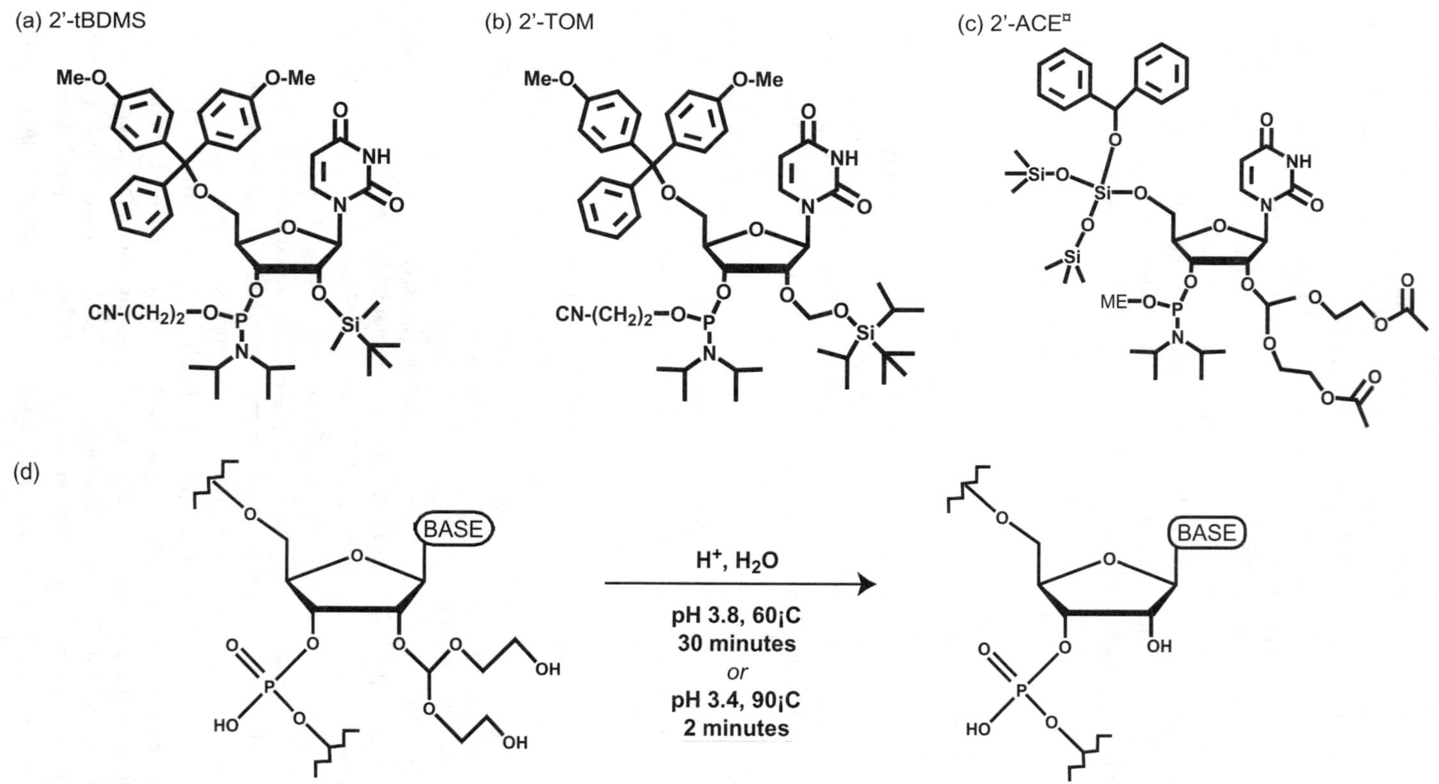

FIGURE 2.7 Chemical RNA synthetic strategies. Structure of protected phosphoramidite nucleoside monomers used in (a) 2'- tBDMS, (b) 2'- TOM and (c) 2'-ACE® RNA synthesis, respectively. (d) The 2'-deprotection reaction for 2'-ACE protected RNA oligos is achieved rapidly under mildly acidic conditions. The reaction is efficient, quantitative, and irreversible. Byproducts of this reaction are ethylene glycol and formic acid, both of which are volatile and removed under vacuum. Panels (a), (b), and (c) reprinted with permission from Boese et al., Strategies for Synthesizing Small Interfering RNA (siRNA), in *RNA Interference: Nuts and Bolts of siRNA Technology*, D. Engelke, Ed., DNA Press, 2004.

out at any point after receipt of the protected oligos (see Figure 2.7.b). High crude product yield and purity combined with the unique properties of the 2'-ACE intermediate make this technology particularly attractive for high-throughput production of siRNA.

2.4 CONCLUDING REMARKS

SiRNA-mediated gene knockdown is now widely accepted as a precision tool for sequence-specific attenuation of gene expression. This technology enables a variety of sophisticated studies to be performed including functional analyses,[38–42] pathway dissection,[20] gene mapping,[43] high throughput genome-wide surveys,[44] and drug target validation. Key challenges to successful implementation of this technology are the design and production of specific and potent siRNAs. There are several options for selection and synthesis, each characterized by distinct but important time, labor, and cost considerations. Rational design methods reduce the initial time and labor investments and ensure silencing with minimized off-target effects. When rational design approaches are coupled with chemical synthesis of siRNA production, a cost effective, high throughput platform can be established because of the ability to produce siRNAs quickly, reliably, and with any modification to further enhance stability and longevity for downstream investigations.

PROTOCOLS

The following protocols allow for processing chemically synthesized siRNA duplexes.

PROTOCOL 2.1 2′-ACE DEPROTECTION AND ANNEALING

Materials

All materials and solutions should be RNase-free.

- Complementary 2'-ACE protected oligonucleotides (Dharmacon, Inc.)
- Buffer for resuspending nucleic acids — commonly used buffers include:

    i. 1 × universal siRNA buffer (20 mM KCl, 6 mM HEPES-KOH, pH 7.5, 0.2 mM $MgCl_2$) (Dharmacon, Inc)
    ii. 1 × TE (10 mM Tris-HCl, pH 7.4-7.6, 1 mM EDTA)
    iii. 100 mM NaCl, 50 mM Tris-HCl, pH 8.0
    iv. 1 × HAM or HEPES-acetate-magnesium buffer (100 mM KOAc, 30 mM HEPES, pH 7.4, 2 mM MgOAc)

- RNase-free water (Sigma-Aldrich)
- Ethanol, 100% and 95% (v/v) (Sigma-Aldrich)
- 2'-Deprotection buffer (100 mM acetic acid, pH 3.8 [Dharmacon])
- 10m ammonium acetate (Sigma-Aldrich)

Experimental Procedures

1. Resuspend the 2'-ACE protected oligomers in 2'-deprotection buffer. If purchased as a duplex, re-suspend the duplex in 400 μl of 2'-deprotection buffer. If purchased as single strands, combine equimolar concentrations of the complementary oligomers into a single microcentrifuge tube.
2. Vortex and centrifuge.
3. Incubate at 60°C for 45 min to cleave the 2'-ACE protecting groups, converting the RNA into the free 2'-OH form (the acid-catalyzed hydrolytic byproducts of the 2'-deprotection step are ethylene glycol and formic acid that are volatile and readily removed upon drying and/or desalting).
4. Remove from heat and centrifuge in a microcentrifuge briefly (5 sec).
5. Cool at room temperature for 30 min to anneal the complementary sequences.
6. The duplexes may be (a) dried in a SpeedVac® and stored as a pellet or (b) desalted to remove any excess salts (see Protocol 2.2) and re-dried or resuspended in a buffered RNAse-free solution in small aliquots for storage.

Technical notes: Some modifications (e.g., Cy3™ or Cy5™) or sequences (e.g., homopolymers of Adenosine) require modified 2'-deprotection steps (e.g., prolonged 2'-deprotection of 2 h or less acidic conditions). However, the oligonucleotides or duplexes should not be incubated for longer than recommended and should not be stored in the 2'-deprotection buffer as this may lead to acid catalyzed depurination and degradation.

Protocol 2.2 De-Salting and Ethanol Precipitation of Duplexed RNA Material

Materials

All materials and solutions should be RNase-free.

- Complementary oligonucleotides or 21 to 23mer duplexes
- Buffer for resuspending nucleic acids — commonly used buffers (see Protocol 2.1)
- RNase-free water (Sigma-Aldrich)
- Ethanol, 100% and 95% (v/v) (Sigma-Aldrich)
- 10 M ammonium acetate (Sigma-Aldrich)

Experimental Procedures

1. Add 1/10 vol of 10 M ammonium acetate and 2.5 vol of 100% ethanol to one vol of siRNA duplex solution (i.e., add 40 μl of 10 M ammonium acetate and 1.5 μl 100% ethanol to 400 μl of duplex in RNase-free water or buffer).
2. Vortex the solution.
3. Place the solution at –20°C for approximately 16 h or at –70°C for 2 h.
4. Centrifuge the solution at 14000 × g for 30 min at 4°C.
5. Carefully pipette the supernatant away from the pellet.
6. Rinse the pellet with 200 μl of cold 95% ethanol.
7. Dry the sample under vacuum in a lyophilizer or SpeedVac®.

8. The dry pellet can be stored until use or re-suspended in an appropriately buffered solution for immediate use or aliquoted to small volumes for storage.

Protocol 2.3 Resuspension and Annealing of Complementary RNA Oligonucleotides To Form siRNA

Materials

All materials and solutions should be RNase-free.

- Complementary oligonucleotides or 21 to 23mer duplexes
- Buffer for resuspending nucleic acids — commonly used buffers (see Protocol 2.1)
- RNase-free water (Sigma-Aldrich)

Experimental Procedures

1. Resuspend pellets of oligonucleotides using RNase-free water or buffer to convenient stock concentrations of 20 μM to 100 μM (20 pmol/μl to 100 pmol/μl).
2. Combine equimolar amounts of complementary strands in a microcentrifuge tube.
3. Incubate the resuspended oligonucleotides at 90°C for 1 min to denature.
4. Remove from heat and centrifuge briefly (5 sec).
5. Cool at room temperature for 30 min to anneal the complementary sequences.

Protocol 2.4 Resuspension and Annealing of Preformed siRNAs

Materials

All materials and solutions should be RNase-free.

- Complementary oligonucleotides or 21 to 23mer duplexes
- Buffer for resuspending nucleic acids — commonly used buffers (see Protocol 2.1)
- RNase-free water (Sigma-Aldrich)

Experimental Procedures

1. Resuspend the dry duplexed material using RNase-free water or buffer to convenient stock concentrations of 20 μM to 100 μM (20 pmol/μl to 100 pmol/μl).
2. Vortex for 10 to 15 sec.
3. Centrifuge the solution briefly for (5 sec).
4. Aliquot to convenient small volumes and store at –20°C or –80°C or use immediately.

Technical notes. Resuspension in RNase-free water is sufficient for immediate use. For long-term storage at –20°C or –80°C, resuspending in small aliquots in an RNase-free buffered solution is recommended to preserve integrity of the duplex. Preformed duplexes typically do not require additional heating to melt and re-anneal. However when RNA has been dried completely, heating at 90°C for 1 min followed by slow cooling at room temperature may help re-suspend dried aggregate material.

Protocol 2.5 Quantitation of an siRNA Duplex

Materials

All materials and solutions should be RNase-free.

- Stock solution of 21 to 23mer duplexes (20 μM to 100 μM)
- RNase-free water (Sigma-Aldrich)
- 100 mM Tris-HCl, pH 7.5

Experimental Procedures

1. Prepare a 1/100 dilution by adding 10 μl sample from the stock solution to 990 μl of RNase-free water or 100 mM Tris-HCl, pH 7.5 at room temperature.
2. Blank the UV spectrophotometer at 260 nm with the appropriate solution used for diluting the stock solution (RNase-free water or 100 mM Tris-HCl pH 7.5).
3. Fill the cuvette with 1/100 dilution sample.
4. Record the absorbance at 260 nm. This value is the experimental A_{260} (A_{260}exp). Epsilon (ε) is the molar extinction coefficient (L/mol·cm). Assuming a 1 cm path-length, calculate the concentration (mol/L or μg/L) using Beers Law: $A_{260} = \varepsilon bc$, where ε = molar absorptivity ($M^{-1}cm^{-1}$), b = path length in cm (1), and c = concentration (M):

Equation 2.1: For calculating μmol/L (μM):

$$\frac{\left[A_{260}\,\text{exp}\right]\left[100\right]}{\varepsilon \text{ of the duplex}} = \Box\ \frac{\text{mole}}{\text{L}} \text{ or } \left[\Box\ \frac{\text{mole}}{\text{L}}\right]\left[\frac{1\ \mu\text{mole}}{1 \times 10^{-6}\ \text{mole}}\right] = \Box\ \mu\text{M}$$

Equation 2.2: For calculating μg/mL:

$$\left[\Box\ \frac{\text{mole}}{\text{L}}\right]\left[\text{M.W. in} \frac{\text{g}}{1\ \text{mole}}\right]\left[\frac{1 \times 10^{6}\ \mu\text{g}}{1\ \text{gram}}\right]\left[\frac{1\ \text{L}}{1 \times 10^{3}\ \text{mL}}\right] = \Box\ \frac{\mu\text{g}}{\text{mL}}$$

Protocol 2.6 Verifying Duplex Integrity by Polyacrylamide Gel Electrophoresis (PAGE)

Materials

- 20–100 μM stock solution of siRNA
- 20 μM stock solution of a 21mer single strand control RNA and a 21mer duplex control
- Polyacrylamide gel electrophoresis system (BioRad)
- Transilluminator for UV shadowing (Fotodyne)
- Polyacrylamide gel electrophoresis solutions (15%, w/v)

- TEMED (N,N,N',N'-tetramethyl ethylenediamine) (Sigma-Aldrich)
- 10% (w/v) Ammonium persulfate (APS) (Sigma-Aldrich)
- 5 × TBE (450 mM Tris Base, 450 mM boric acid, 10 mM Na_2 EDTA, pH 9.0) (Sigma-Aldrich)
- Loading buffer (1 × TBE: 90 mM Tris HCl, 90 mM boric acid, 2 mM Na_2 EDTA, pH 9.0) (Sigma-Aldrich)
- Bromophenol blue-xylene cyanole (BPB-XC) tracking dye mixture of 0.5% each (Sigma-Aldrich)

Experimental Procedures

1. Prepare a 15% (w/v) non-denaturing polyacrylamide gel appropriate for analyzing a small aliquot of siRNA.
2. Prepare aliquots of approximately 3 nmoles (0.5–1.0 ODU_{A260}) of siRNA and controls of single strand or duplexed material in 1 × loading buffer without tracking dye.
3. Prepare an aliquot of loading buffer plus BPB-XC dye for tracking progress through the gel.
4. Run the gel for approximately 1 h at room temperature.
5. Analyze the duplexes for integrity by UV shadowing on a transilluminator.

Technical note. UV shadowing is sufficient to determine the integrity of the duplex; however more sensitive detection methods such as ethidium bromide staining or radioactive labeling may also be employed.

REFERENCES

1. Elbashir, S.M. et al., Functional anatomy of siRNAs for mediating efficient RNAi in *Drosophila melanogaster* embryo lysate, *EMBO J.*, 20, 6877, 2001.
2. Harborth, J. et al., Identification of essential genes in cultured mammalian cells using small interfering RNAs, *J. Cell Sci.*, 114, 4557, 2001.
3. Spankuch-Schmitt, B. et al., Effect of RNA silencing of polo-like kinase-1 (PLK1) on apoptosis and spindle formation in human cancer cells, *J. Natl. Cancer Inst.*, 94, 1863, 2002.
4. Laposa, R.R., Feeney, L., and Cleaver, J.E., Recapitulation of the cellular xeroderma pigmentosum-variant phenotypes using short interfering RNA for DNA Polymerase H, *Cancer Res.*, 63, 3909, 2003.
5. Holen, T. et al., Positional effects of short interfering RNAs targeting the human coagulation trigger Tissue Factor, *Nucleic Acids Res.*, 30, 1757, 2002.
6. Elbashir, S.M. et al., Analysis of gene function in somatic mammalian cells using small interfering RNAs, *Methods*, 26, 199, 2002.
7. Khvorova, A., Reynolds, A., and Jayasena, S., Functional siRNAs and miRNAs exhibit strand bias, *Cell*, 115, 209, 2003.
8. Reynolds, A. et al., RNA intereference: mechanistic implications and rational siRNA design, *Nat. Biotechnol.*, 22, 326, 2004.
9. Elbashir, S.M., Lendeckel, W., and Tuschl, T., RNA interference is mediated by 21- and 22-nucleotide RNAs, *Genes Dev.*, 15, 188, 2001.

10. Martinez, J. et al. Single-stranded antisense siRNAs guide target RNA cleavage in RNAi, *Cell*, 110, 563, 2002.
11. Lassus, P., Rodriguez, J., and Lazebnik, Y., Confirming Specificity of RNAi in mammalian cells, *Science STKE*, 2002, PL13., 2002.
12. Aza-Blanc, P. et al., Identification of modulators of TRAIL-induced apoptosis via RNAi-based phenotypic screening, *Mol. Cell*, 12, 627, 2003.
13. Hammond, S.M. et al., Argonaute2, a link between genetic and biochemical analyses of RNAi, *Science*, 293, 1146, 2001.
14. Hammond, S.M. et al., An RNA-directed nuclease mediates post-transcriptional gene silencing in *Drosophila* cells, *Nature*, 404, 293, 2000.
15. Bernstein, E. et al., Role for a bidentate ribonuclease in the initiation step of RNA interference, *Nature*, 409, 363, 2001.
16. McManus, M.T. et al., Small interfering RNA-mediated gene silencing in T lymphocytes, *J. Immunol.*, 169, 5754, 2002.
17. Gonczy, P. et al., Functional genomic analysis of cell division in *C. elegans* using RNAi of genes on chromosome III, *Nature*, 408, 331, 2000.
18. Jackson, A.L. et al., Expression profiling reveals off-target gene regulation by RNAi, *Nat. Biotechnol.*, 21, 635, 2003.
19. Semizarov, D. et al., Specificity of short interfering RNA determined through gene expression signatures, *Proc. Natl. Acad. Sci. U.S.A.*, 100, 6347, 2003.
20. Huang, F. et al., Global analysis of the clathrin-dependent endocytosis by high efficiency RNA interference, *J. Biol. Chem.*, 279, 16657, 2004.
21. Schwarz, D.S. et al., Unexpected Asymmetry in the assembly of the RNAi Enzyme Complex, *Cell*, 115, 199, 2003.
22. Donze, O. and Picard, D., RNA interference in mammalian cells using siRNAs synthesized with T7 RNA polymerase, *Nucleic Acids Res.*, 30, e46., 2002.
23. Capodici, J., Kariko, K., and Weissman, D., Inhibition of HIV-1 Infection by Small Interfering RNA-Mediated RNA Interference, *J. Immunol.*, 169, 5196, 2002.
24. Sohail, M. et al. A simple and cost-effective method for producing small interfering RNAs with high efficacy, *Nucleic Acids Res*, 31, E38., 2003.
25. Castanotto, D., Li, H., and Rossi, J.J., Functional siRNA expression from transfected PCR products, *RNA*, 8, 1454, 2002.
26. Sui, G. et al., A DNA vector-based RNAi technology to suppress gene expression in mammalian cells, *Proc. Natl. Acad. Sci. U.S.A.*, 99, 5515, 2002.
27. Brummelkamp, T.R., Bernards, R., and Agami, R., A system for stable expression of short interfering RNAs in mammalian cells, *Science*, 296, 550, 2002.
28. Paul, C.P. et al., Effective expression of small interfering RNA in human cells, *Nat. Biotechnol.*, 20, 505, 2002.
29. Lee, N.S. et al., Expression of small interfering RNAs targeted against HIV-1 rev transcripts in human cells, *Nat. Biotechnol.*, 20, 500, 2002.
30. Yang, D. et al., Short RNA duplexes produced by hydrolysis with *Escherichia coli* RNase III mediate effective RNA interference in mammalian cells, *Proc. Natl. Acad. Sci. U.S.A.*, 99, 9942, 2002.
31. Calegari, F. et al., Tissue-specific RNA interference in postimplantation mouse embryos with endoribonuclease-prepared short interfering RNA, *Proc. Natl. Acad. Sci. U.S.A.*, 99, 14236, 2002.
32. Caruthers, M.H., Gene synthesis machines: DNA chemistry and its uses, *Science*, 230, 281, 1985.

33. Usman, N. et al., The automated chemical synthesis of long oligoribuncleotides using 2'-O-silylated ribonucleoside 3'-O-phosphoramidites on a controlled-pore glass support: synthesis of a 43-nucleotide sequence similar to the 3'-half molecule of an *Escherichia coli* formylmethionine tRNA, *J. Am. Chem. Soc.*, 109, 7845, 1987.
34. Wu, X. and Pitsch, S., Synthesis and pairing properties of oligoribonucleotide analogues containing a metal-binding site attached to beta-D-allofuranosyl cytosine, *Nucleic Acids Res.*, 26, 4315, 1998.
35. Scaringe, S.A., Advanced 5'-silyl-2'-orthoester approach to RNA oligonucleotide synthesis, *Methods Enzymol.*, 317, 3, 2000.
36. Scaringe, S.A., RNA oligonucleotide synthesis via 5'-silyl-2'-orthoester chemistry, *Methods*, 23, 206, 2001.
37. Hartsel, S.A. et al., RNA Oligonucleotide Synthesis Via 5'-Silyl-2'-Orthoester Chemistry, in *Oligonucleotide Synthesis: Methods and Applications*, P. Herdewijn, Ed.. Totowa, Humana Press Inc., 2003, in press.
38. Zhou, T. et al., A role for Plk1 phosphorylation of NudC in cytokinesis, *Dev. Cell*, 5, 127, 2003.
39. Zender, L. et al., Caspase 8 small interfering RNA prevents acute liver failure in mice, *Proc. Natl. Acad. Sci. U.S.A.*, 100, 7797, 2003.
40. Hsieh, A. et al., A library of siRNA duplexes targeting the phosphoinositide 3-kinase pathway: Determinants of gene silencing for use in cell-based screens, *Nucleic Acids Res.,* 32, 893, 2004.
41. Jiang, Z.Y. et al., Insulin signaling through Akt/protein kinase B analyzed by small interfering RNA-mediated gene silencing, *Proc. Natl. Acad. Sci. U.S.A.*, 100, 7569, 2003.
42. Kurisaki, K. et al., Nuclear Factor YY1 inhibits Transforming Growth Factor {beta}- and Bone Morphogenetic Protein-induced cell differentiation, *Mol. Cell. Biol.*, 23, 4494, 2003.
43. Li, T. et al., Identification of the gene for vitamin K epoxide reductase, *Nature*, 427, 541, 2004.
44. Dudley, N.R., Labbe, J.-C., and Goldstein, B., Using RNA interference to identify genes required for RNA interference, *Proc. Natl. Acad. Sci. U.S.A.*, 99, 4191, 2002.

3 Oligonucleotide Scanning Arrays in the Design of Small Interfering RNAs

Valentine M. Macaulay and Muhammad Sohail

CONTENTS

3.1 INTRODUCTION

Small interfering RNAs are powerful gene silencing tools. The most potent RNAi effectors in *Drosophila* are RNA duplexes with 19 bp of homology to the target gene and 2 nucleotide 3' overhangs.[1,2] Nearly two years ago these 21 to 23 nucleotide small interfering RNAs (siRNAs) were shown to effect potent and sequence-specific silencing of exogenous and endogenous genes in mammalian cells.[1]

In many cases, the gene silencing effect has proven to be more robust, and less variable than that induced by antisense, or ribozyme techniques.[3–5] As this approach became more widely used, it was apparent that only about 50% of siRNAs were effective, and the determinants of activity were unclear. For example, Elbashir and colleagues initially were unable to knockdown expression of vimentin, and postulated that this was due to the fact that vimentin was expressed at higher levels that other targeted genes.[1] However, in a later report, the same group successfully knocked down

0–8493–2141–7/05/$0.00 + $.50

expression of vimentin using an siRNA targeting a different region of the same gene.[4] The problem is that there is no way to predict in advance the efficacy of individual duplexes. Other research has shown that siRNA efficacy is subject to positional effects, with some reports that even a few base shifts can influence activity, suggesting possible effects of RNA secondary structure.[6–9]

We have been studying the effect of mRNA structure on the efficacy of antisense oligonucleotides. It is well recognized that the major determinant of efficacy is mRNA structure. Intramolecular folding of mRNAs renders all but 5 to 10% of most transcripts inaccessible to binding with complementary nucleic acids.[10] Several computational methods are available for prediction of RNA secondary structure, calculated from thermodynamic properties and nearest-neighbor interaction.[11,12] However, these programs are generally believed to be unreliable in predicting folding of large RNAs.[13,14]

We wished to study the folding of long mRNAs, and utilized the strategy of antisense oligonucleotide scanning arrays.[15] As our target gene, we selected the type 1 insulin-like growth factor receptor (IGF1R). This is a promising focus for anti-cancer drug development, being frequently overexpressed by tumors and mediating tumor growth, motility, and protection from apoptosis.[16] The IGF1R shows significant sequence homology to the insulin receptor.[17] It will be important for the potential clinical use of this approach that an IGF1R inhibitor should block only its intended target, without influencing the expression, or function of the insulin receptor. This chapter describes the methods we used to identify regions of *IGF1R* mRNA that are most effectively targeted by RNA interference, and the implications of this information for the design of effective siRNAs.

3.2 IDENTIFICATION OF ACCESSIBLE REGIONS IN MRNAS

3.2.1 RNase H Mapping

Mapping of RNA transcripts with RNase H in combination with an oligonucleotide library is an effective approach to broadly define accessible regions in mRNAs.[18] In principle, this method maps the regions of the transcript that are accessible to binding with complementary oligonucleotides. The assay uses a random 12mer deoxyribophosphodiester oligonucleotide library (dN_{12}), synthesized as described in Protocol 3.1, and an end-labeled transcript (Protocol 3.2). RNase H cleaves the transcript where oligonucleotides in the library have hybridized (Figure 3.1a), and the cleavage products are resolved by gel electrophoresis (Protocol 3.3). The observation that this assay generates distinct products rather than a smear is manifest proof of the concept that only limited regions of most mRNAs are accessible.[10]

We identified a region of IGF1R mRNA approximately 530 to 690 nt from the 5' end of the transcript where there were several distinct cleavage products, indicating accessibility to RNase H-mediated cleavage (region marked with asterisk in Figure 3.1b). This region contained mixed sequence without long stretches of individual nucleotides, and was selected for further study.

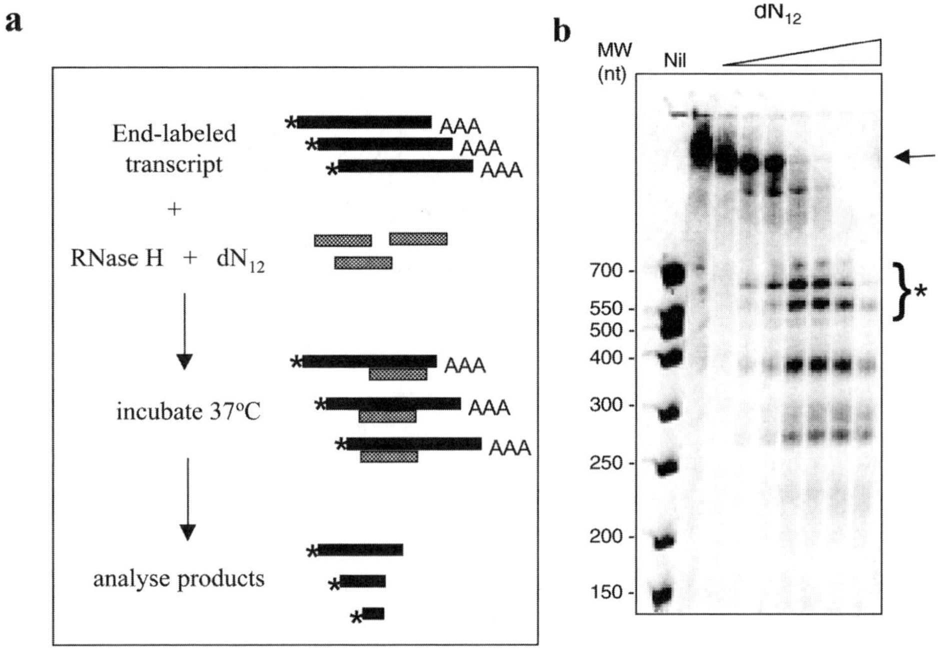

FIGURE 3.1 RNase H mapping to identify regions of IGF1R mRNA accessible to hybridize with IGF1R antisense oligonucleotides. (a) Principle of RNase mapping. Transcripts are end-labeled (asterisk) and incubated with RNase H and random 12mer oligonucleotide library (dN_{12}). Where allowed by transcript secondary structure, complementary 12mers hybridize and induce RNase H-mediated cleavage. (b) RNase mapping of accessible sites within IGF1R mRNA. End-labeled IGF1R mRNA (1-1581nt, 5fmol) was incubated with RNase H in the absence, or presence of 5, 10, 20, 50, 100 or 500pmol random 12mers (dN_{12}). Arrow: full-length (uncleaved) IGF1R1-1581 transcript. Asterisk: region selected for array screen (modified from Bohula et al. *J. Biol. Chem.*, 278, 15991, 2003; with permission).

3.2.2 Antisense Oligonucleotide Arrays

The precise sequence of the accessible sites can be determined by hybridization of radio-labeled transcript with an array of antisense oligonucleotides (ASOs).[15] Figure 3.2 describes the principle of array design, and Protocol 3.4 outlines the method of array fabrication. The synthesis is carried out using standard nucleotide-CE-phosphoramidites on an adapted ABI DNA synthesizer (Applied Biosystems). Reagents are delivered to the substrate surface using a diamond-shaped or a circular mask. The mask is sealed against the substrate to create a cell into which reagents for first base synthesis are delivered. After each nucleotide coupling, the mask is displaced by a certain step size (usually 1 to 2.5 mm) to create an array of overlapping sequences, ranging in size from monomers to a maximum of ~20 nt. The oligonucleotides are covalently bound to the surface by their 3' ends, and represent all possible complements of a region. See [15] for more details.

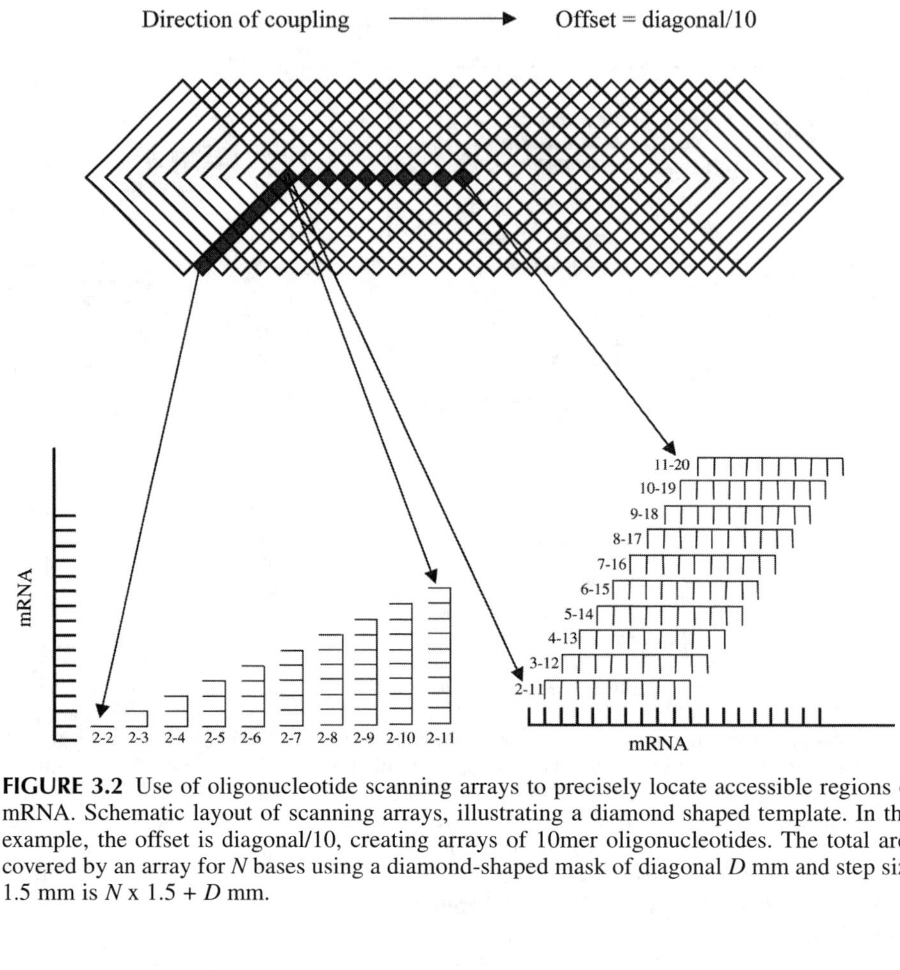

FIGURE 3.2 Use of oligonucleotide scanning arrays to precisely locate accessible regions of mRNA. Schematic layout of scanning arrays, illustrating a diamond shaped template. In this example, the offset is diagonal/10, creating arrays of 10mer oligonucleotides. The total area covered by an array for *N* bases using a diamond-shaped mask of diagonal *D* mm and step size 1.5 mm is *N* x 1.5 + *D* mm.

Arrays were hybridized to internally labeled IGF1R mRNA (Protocol 3.5). Our initial arrays were of antisense oligonucleotides complementary to nucleotides 530 to 650 of the IGF1R coding sequence. These appeared to show high-intensity hybridization between nucleotides 645 to 650 of the IGF1R mRNA, and antisense oligonucleotides at the extreme 3' end of the array (right side of image; not shown). To clarify the significance of this, we made further arrays complementary to nucleotides 536 to 685, i.e., extending ~20 nt further downstream than the intial array. Hybridization of radiolabeled IGF1R mRNA to these arrays showed that the apparent "hot spot" at 645 to 650 was much less intense (region of peak 5, Figure 3.3). This indicates that array hybridization can be subject to end-artefact. To avoid this phenomenon, we suggest that arrays should extend at least 15 to 20 nt up- and downstream of regions of interest.

Probing with IGF1R mRNA at 37°C (Figure 3.3) showed negligible binding to monomers at the edges of the array. Along the centerline, 20mers showed significant heteroduplex yield only in accessible regions of the transcript, with three peaks of hybridization in the region 590 to 668nt (peaks 1, 2, 4; Figure 3.3). Hybridizations at 25°C resulted in higher heteroduplex yield, with a similar hybridization pattern, and an additional focus of hybridization not present at 37°C (region 3, Figure 3.3). We also hybridized the array to human insulin receptor mRNA in order to identify antisense oligonucleotides that could affect expression of the insulin receptor, which has 60% sequence homology with the IGF1R.[17] This analysis revealed two regions of heteroduplex formation between insulin receptor mRNA and IGF1R antisense oligonucleotides; one of these regions was immediately upstream of peak 1 of hybridization with the IGF1R transcript (lower panel, Figure 3.3).

Following transfection into human tumor cells, IGF1R antisense oligonucleotides corresponding to peaks 1, 2, and 4 induced IGF1R downregulation (to 25–30% of control levels) that was proportional to the relative affinity of hybridization when tethered on the scanning array. See Protocols 3.6, 3.7, and 3.8 for details of cell lysis, SDS PAGE, and immunoblotting methods. We noted that ASO6, which showed negligible binding to IGF1R mRNA on the array, was completely ineffective within intact cells. Furthermore ASO3, which had hybridized intensely to IGF1R mRNA at 25°C but weakly at 37°C, was similarly lacking in antisense activity. This demonstrates the influence of temperature on RNA secondary structure, and reinforces the importance of conducting array hybridizations at physiological temperature. As predicted by the array, ASO1 caused minor downregulation (85 to 90%) of the insulin receptor, but ASOs 2, and 4 had no effect ([19] and data not shown). These results confirmed our previous study in which we showed correlation between array hybridization and effects on gene expression by assessing *in vitro* translation of B type cyclin transcripts following ASO microinjection into *Xenopus* oocytes.[18] However, despite this optimization of ASO targeting, we were unable to suppress IGF1R expression to below 20 to 30% of levels in control-transfected cells, as we and others have reported.[19–21]

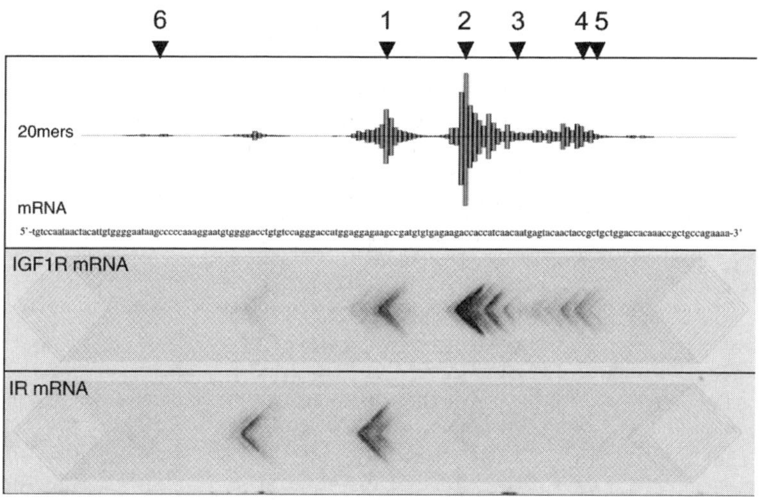

FIGURE 3.3 Hybridization of *IGF1R* or *IR* mRNA to scanning array of IGF1R antisense oligonucleotides. Middle panel: hybridization of *IGF1R* transcript to scanning array of oligonucleotides complementary to nucleotides 536 to 685 of the human *IGF1R* sequence. Histogram (upper panel) represents quantification of binding of 20mer antisense oligonucleotides to *IGF1R* mRNA. Lower panel: hybridization to *IR* transcript. Numbered arrowheads: oligonucleotides selected for further study (modified from Bohula et al., *J. Biol. Chem.*, 278, 15991, 2003; with permission).

3.3 ASSESSMENT OF ANTI-IGF1R siRNAS

3.3.1 DESIGN OF siRNAs

Aiming to improve the efficiency of IGF1R targeting, we elected to use the information on IGF1R mRNA folding to investigate the effects of target site accessibility on siRNA efficacy. We synthesized 21mer siRNA duplexes that were homologous to 20mer ASOs that had been identified on the array as showing intense (ASO 2, 4), or negligible (ASO 5, 6) binding to IGF1R mRNA at 37°C (Figure 3.3 and Table 3.1). These sequences were all immediately downstream of an AA doublet, and none were homologous to ASOs that had hybridized to insulin receptor mRNA on the array. We

TABLE 3.1
Sequences of siRNAs and Control Duplexes

Duplex	Sequence	*IGF1R* mRNA homology
siRNA 2	5'-GCCGAUGUGUGAGAAGACCTT-3 3'-TTCGGCUACACACUCUUCUGG-5'	612-630
siRNA 2 (18mer)	5'-GCCGAUGUGUGAGAAGTT-3' 3'-TTCGGCUACACACUCUUC-5'	612-627
siRNA 2 (24mer)	5'-GCCGAUGUGUGAGAAGACCACCTT-3' 3'-TTCGGCUACACACUCUUCUGGUGG-5'	612-633
siRNA 2 (Mouse)	5'-GCCCAUGUGUGAGAAGACCTT-3' 3'-TTCGGGUACACACUCUUCUGG-5'	1051-1069*
siRNA 2 (Mutant)	5'-GCCGAUGUGUG<u>U</u>GAAGACCTT-3' 3'-TTCGGCUACACAC<u>A</u>CUUCUGG-5'	612-630
siRNA 2 Inv	5'-CCAGAAGAGUGUGUAGCCGTT-3' 3'-TTGGUCUUCUCACACAUCGGC-5'	N/H
siRNA 4	5'-CAATGAGTACAACUACCGCTT-3' 3'-TTGUUACUCAUGUUGAUGGCG-5'	639-657, 1078-1096*
siRNA 4 Inv	5'-CGCCAUCAACAUGAGUAACTT-3' 3'-TTGCGGUAGUUGUACUCAUUG-5'	N/H
siRNA 5	5'-TGAGTACAACTACCGCUGCTT-3' 3'-TTACUCAUGUUGAUGGCGACG-5'	642-660, 1081-1099*
siRNA 5 Inv	5'-CGUCGCCAUCAACAUGAGUTT-3' 3'-TTGCAGCGGUAGUUGUACUCA-5'	N/H
siRNA 6	5'-UACAUUGUGGGGAAUAAGTT-3' 3'-TTGAUGUAACACCCCUUAUUC-5'	547-565
siRNA 6 (Mouse)	5'-CUACAUUGUGGGGAACAAGTT-3' 3'-TTGAUGUAACACCCCUUGUUC-5'	985-1003*
siRNA 6 Inv	5'-GAAUAAGGGGUGUUACAUCTT-3' 3'-TTCUUAUUCCCCACAAUGUAG-5'	N/H

Table shows sequences of siRNAs used in this study. All siRNAs were 21mer chimeric RNA/DNA duplexes, with the exception of 18 to 27mer variants of siRNA R2. The sequence of duplex Mut2 was as for R2 with the exception of a single bp mutation (underlined). N/H = No Homology, *Homology to murine *IGF1R* cDNA sequence. Human and mouse *IGF1R* mRNA sequences obtained from GenBank, accession NM_000875.2 and XM_133508.2, respectively.[17] The siRNAs R2 and R4 sequences are based on strongly hybridizing ASOs in array experiments (Figure 3.3). The siRNA R6 sequence corresponds to an antisense oligonucleotide that failed to hybridize to the same array and R5 is homologous to a weakly hybridizing antisense oligonucleotide.

designed RNA/DNA chimeric oligonucleotides as described:[1,2] the 5' base of each sense strand was immediately downstream of the AA doublet, and each strand incorporated 19 bases of RNA with two 3' deoxythymidines in the 3' overhang. An inverted sequence control duplex was made for each siRNA. The region of peak 1 had also shown strong hybridization to the IGF1R transcript. However, as previously described, this sequence had shown itself capable of influencing insulin receptor expression, both on the array and within intact tumor cells, and furthermore, it lacked an upstream AA motif. ASO3, which had shown weak hybridization to IGF1R mRNA on the array, also lacked an upstream AA doublet. Therefore, neither ASO1 nor ASO3 were evaluated as siRNA.

Unlike single-stranded RNA, which is highly unstable in serum, chemically synthesized siRNAs appear to have stability comparable to that of phosphorothioate DNA (Figure 3.4). This was attributed to the fact that the siRNAs were double-stranded and incorporated two bases of DNA in the 3' overhang of each strand (Figure 3.4).

3.3.2 Assessment of *IGF1R* Gene Silencing in Human and Murine Tumor Cells

The effects of siRNAs R2, R4, R5, and R6 were assessed using transfection conditions that resulted in uptake of fluorescently-tagged ASO into 90 to 95% of cells, 5 to 24 h after transfection (data not shown; Protocol 3.6). Transfections are performed in serum-free conditions in Optimem (Gibco BRL), using the lipid transfection reagent Oligofectamine (InVitrogen). Transfection conditions given here achieve 100-fold dilution of 20 μM stock duplex, giving a final concentration of 200 nM in the transfection. Many effective duplexes will cause significant knockdown at concentrations as low as 0.5 nM. To test siRNA concentrations lower than 200 nM, dilute 20 μM stock duplex with 1 × annealing buffer, or Optimem, prior to transfection.

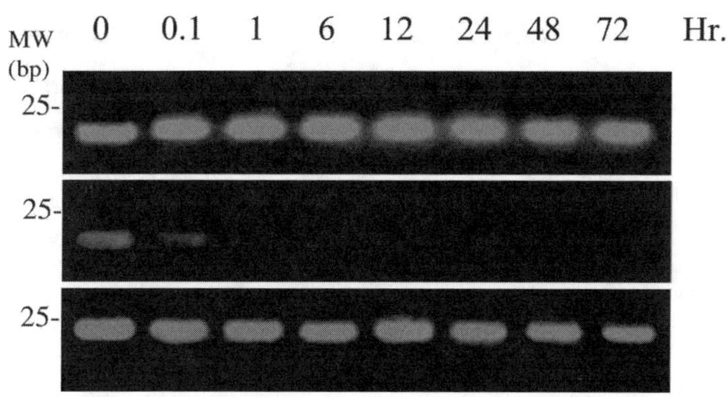

FIGURE 3.4 Stability of siRNA. Aliquots of phosphorothioate antisense oligonucleotide ASO2 (40 pmol), siRNA R2 (40 pmol), or R2 sense strand (100 pmol) were incubated at 37°C in fetal calf serum for the indicated times and analyzed by electrophoresis through a 5% (w/v) low melting point agarose/TBE gel. PS ASO, phosphorothioate antisense oligonucleotide; ssRNA, single stranded RNA.

First we evaluated the effects of duplex R2, which corresponded to the most intensely hybridizing ASO from the scanning array (Figure 3.3). This siRNA induced profound sequence-specific *IGF1R* gene silencing to ~1% of levels in cells treated with an inverted control duplex (Figure 3.5a). This suggests that the inability to suppress IGF1R expression by greater than 80% using antisense was unlikely to be due to poor transfection efficiency, but was more likely a fundamental limitation of the

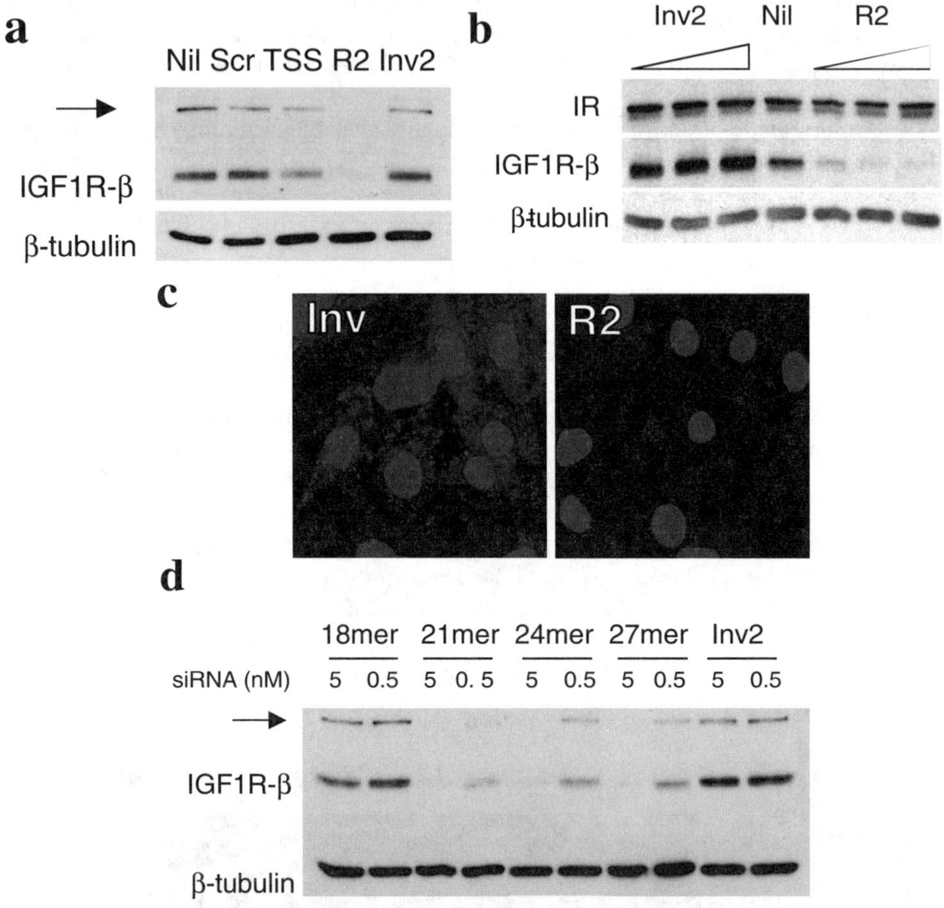

FIGURE 3.5 *IGF1R* gene silencing induced by siRNA equivalent to IGF1R ASO that hybridizes intensely to *IGF1R* mRNA. (a) Human ME melanoma cells were transfected with oligofectamine alone or with 200nM TSS scrambled sequence control (Scr) or TSS ASO, or with siRNA R2 or inverted sequence duplex (Inv2). Cells were lysed 48 h after transfection and analyzed by immunoblotting for IGF1R. Arrow, ~220kDa IGF1R proreceptor. (b) Sequence-specificity of *IGF1R* gene silencing in ME melanoma cells transfected with oligofectamine alone, or with 5, 50 or 500nM Inv2 or R2. Cell lysates were analyzed by immunoblotting for IGF1R and IR. Similar results were obtained in a second set of independently-prepared ME cell lysates, in MDA-MB231 breast cancer and UC101 ovarian cancer cells. (c) Immunofluorescent staining of MDA-MB-231 cells 48 h after transfection with 100 nM Inv2 or R2. Inv2 transfectants (left panel) showed detectable IGF1R staining, which was almost completely abolished by siRNA R2 transfection (right panel). (d) MDA-MB-231 cells were transfected with 0.5 or 5 nM R2, or with 18 to 27mer variants of the same duplex (see Table 3.1). IGF1R protein levels were analyzed after 48 h. Arrow, ~220 kDa IGF1R proreceptor (modified from Bohula et al., *J. Biol. Chem.*, 278, 15991, 2003; with permission).

antisense approach. Duplex R2 had no effect on the expression of the insulin receptor (Figure 3.5b), confirming that the array screen had enabled identification of siRNAs that induce profound sequence-specific *IGF1R* gene silencing.

We synthesized additional duplexes of 18 to 27 bp based on 3' deletions or extensions of duplex R2 (Table 3.1). An18mer R2 duplex with a 3 bp 3' deletion was less effective than the 21bp R2, while comparable IGF1R downregulation was induced by 24mer and 27mer duplexes representing 3 and 6 bp 3' extensions of R2 (Figure 3.5c). This is in contrast to duplex length requirements for RNAi in *Drosophila*[2] but consistent with the characteristics of RNAi in mammalian cytoplasmic lysate.[22] It was notable that in this system siRNA activity was reduced, but not abolished, by a single base pair mutation in the center of 21mer duplex R2 (Figures 3.6a and b). In contrast, others have found that gene silencing can be abolished by a single point mutation, suggesting that this approach may be able to specifically target mutant oncogenes.[23]

The effects of R2 were compared with R6, a duplex corresponding to an ASO that failed to hybridize to IGF1R mRNA on the array (Figure 3.3). In MDA-MB-231 breast cancer cells, duplex R2 induced more profound dose-dependent inhibition of IGF1R protein levels than R6 ($p < 0.05$ at 0.5 and 50 nM, $p < 0.01$ at 5 nM (Figure

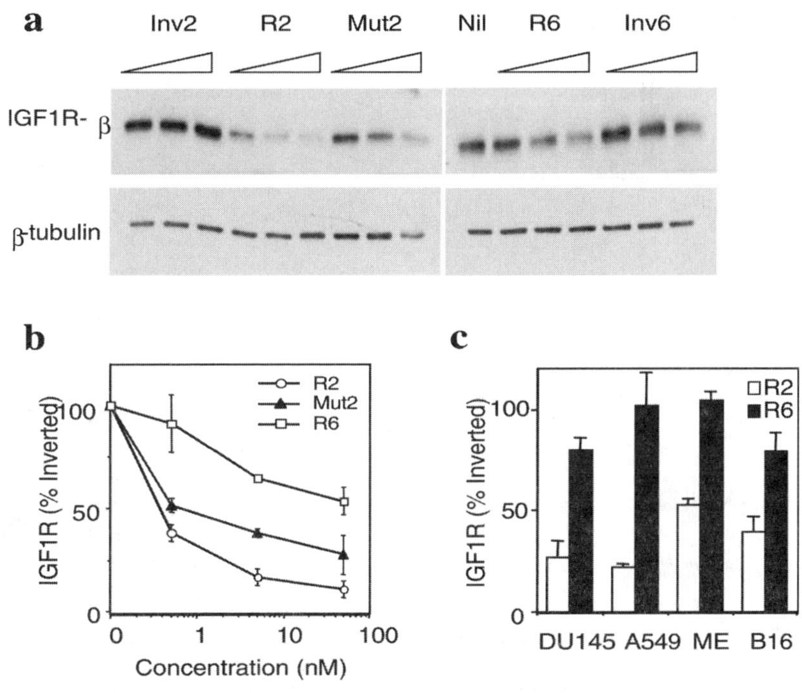

FIGURE 3.6 Influence of mRNA secondary structure on siRNA efficacy. (a) Immunoblot for IGF1R in MDA-MB-231 cells 48 h after transfection with R2, Mut2 (sequence as R2 with a single bp mutation in the center of the duplex), or R6 at 0.5, 5 or 50 nM. (b) Analysis of IGF1R protein levels in siRNA-transfected MDA-MB-231 cells. Points represent mean (± range of values from two independent experiments) of IGF1R levels, expressed as percent value in cells treated with equivalent concentration of inverted control duplex. (c) Analysis of IGF1R levels (mean ± sem of triplicate experiments) in human DU145 prostate cancer, A549 non-small cell lung cancer, ME melanoma and murine B16.F1 melanoma cells transfected with 10nM R2 or R6 (modified from Bohula et al., *J. Biol. Chem.*, 278, 15991, 2003; with permission).

3.6a and b). Indeed, the difference between these two duplexes was greater than that between R2 and the mutant R2 duplex (Figure 3.6a and b). R2 and R6 were also compared in a range of human and murine cell lines, and in all cases R2 caused significantly more profound IGF1R downregulation (DU145 ($p < 0.01$), A549 ($p < 0.05$), ME ($p < 0.01$), B16 ($p < 0.05$); Figure 3.6c). This indicates that intra-molecular structure in the transcript has a major effect on siRNA efficacy, and suggests that the structural features dictating access are robust, and conserved between different cell lines and species.

We then assessed the efficacy of a second pair of duplexes based on the sequence around peak 4 of hybridization between *IGF1R* mRNA and the IGF1R ASO scanning array (Figure 3.7a). Duplexes R4 and R5 were designed to target 19mer sequences

a

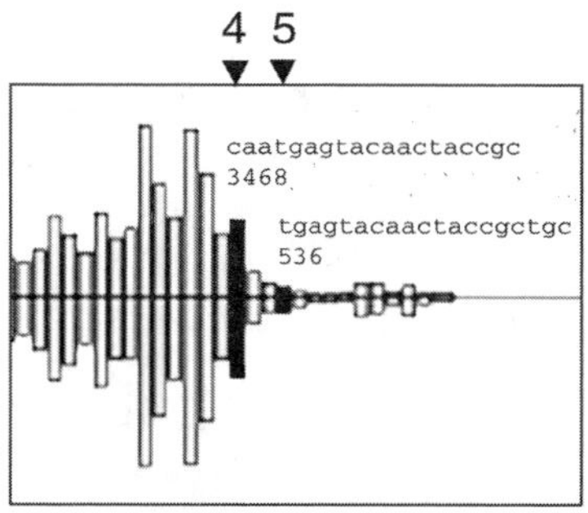

b

siRNA	Hybridization intensity
R2	6749
R4	3468
R5	536
R6	263

c **d**

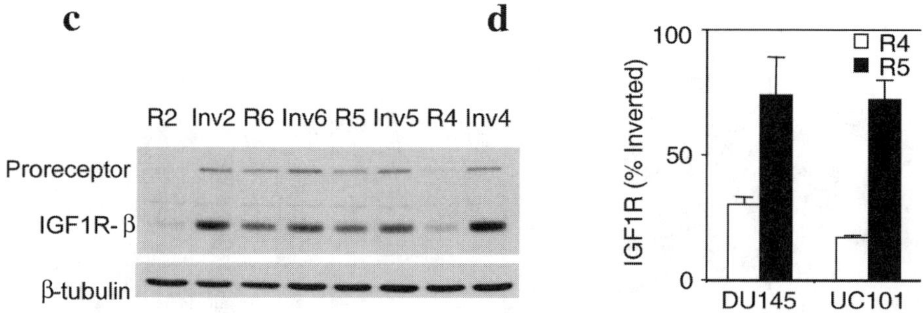

FIGURE 3.7 Small sequence shift has major effect on hybridization intensity and siRNA efficacy. (a) Phosphorimager analysis of peak 4 region of hybridization of IGF1R mRNA to 19mer IGF1R ASOs on the scanning array. ASOs equivalent to duplexes R4 and R5 are marked by arrowheads, and by black bars in the histogram. (b) Table shows hybridization intensity (arbitrary units, from scanning array) of 19mer ASOs equivalent to duplexes R2, R4, R5, and R6. (c) Representative immunoblot showing effects of 10 nM siRNAs, or inverted controls on IGF1R levels in human DU145 prostate cancer cells. (d) Results (mean ± sem) of triplicate analyses of IGF1R protein levels in DU145 prostate cancer, and UC101 ovarian cancer cells transfected with 10 nM R4 or R5 (modified from Bohula et al., *J. Biol. Chem.*, 278, 15991, 2003; with permission).

immediately downstream of AA motifs at bases 636 and 639, respectively, of the IGF1R sequence.[17] Data from the scanning array indicated that there was a ~6.5-fold difference in hybridization intensity between the equivalent 19mer oligonucletides (Figure 3.7b). Both duplexes induced IGF1R downregulation in human ovarian and prostate cancer cells, but R4 was significantly more potent ($p < 0.05$) in both cell lines (Figure 3.7c and d). Thus, a 3 BP shift had major effects on siRNA efficacy, paralleling differences in heteroduplex yield on the scanning array.

These results confirmed that intensity of array hybridization at 37°C successfully predicted for target downregulation within intact tumor cells. This is despite the fact that in several respects array conditions were non-physiological: ASOs were tethered at the 3' ends, and the composition of the hybridization buffer did not reflect the protein-rich cytosolic microenvironment. Furthermore, hybridizations were performed in 1 M NaCl, since there was insignificant heteroduplex formation at near-physiological (150 mM) salt concentration (not shown).

3.3.3 Correlation of *IGF1R* Gene Silencing with Biological Efficacy

Finally, it was important to assess whether the improvement in potency of gene silencing translated into any difference in biological efficacy. We compared effects on colony-forming efficiency and post-irradiation survival (Protocol 3.10) of transfection with 100 nM siRNA duplexes R4 and R5, equivalent to strongly and weakly hybridizing antisense oligonucleotides, respectively. These experiments also utilized ASO4, a 20mer phosphorothioate antisense oligonucleotide homologous to the same target as R4. This oligonucleotide caused inhibition of overall survival, and survival after 2Gy irradiation (SF2), but this effect was entirely nonspecific, with similar results in cultures treated with the scrambled sequence oligonucleotide (Scr4, Figure 3.8). Transfection with siRNAs R4 or R5 inhibited survival of MDA-MB-231 ($p < 0.001$ for comparison of each siRNA with inverted control duplex), with a significantly greater effect in R4-treated cells ($p < 0.01$ for comparison with R5; Figure 3.8b). In cultures treated with the more potent siRNA duplex R4 there was significant inhibition of survival after 2Gy irradiation (SF2; $p < 0.01$ for comparisons with Inv4 and R5; Figure 3.8c).

3.4 CONCLUSIONS

These results suggest that the structure of target transcripts can influence siRNA efficacy. In addition, the findings have two key implications. First they shed light on the mode of action of siRNA at the molecular level. The need for access comparable to that required for ASO binding supports the concept, as originally proposed when RNAi was first recognized,[24,25] of direct interaction by base-pairing between the transcript and component(s) of the duplex. This is consistent with the demonstration that antisense strands can mediate RNAi in mammalian cytoplasmic lysate.[22,26]

Second, the influence of mRNA structure has clear implications for the design of effective siRNAs. The availability of an empirical tool to measure RNA folding has allowed us to identify IGF1R siRNAs capable of inducing profound IGF1R knock-

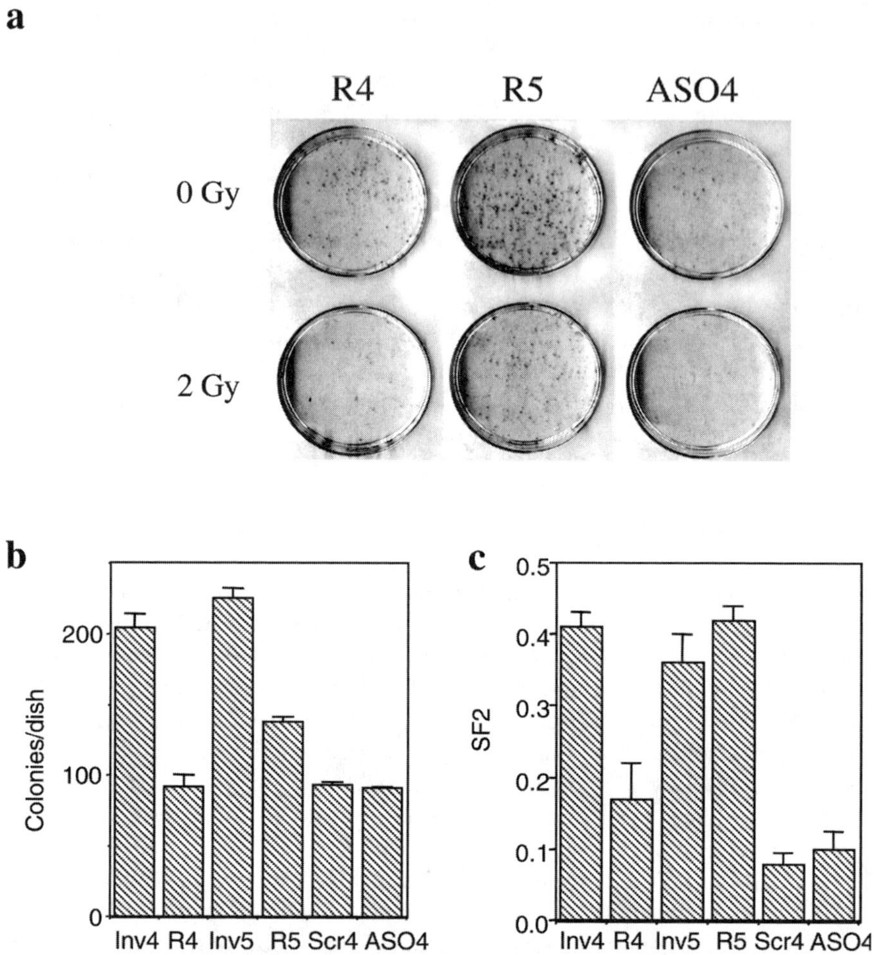

FIGURE 3.8 siRNAs corresponding to intensely-hybridizing ASOs block survival and enhance radiosensitivity of human breast cancer cells. (a) Clonogenic survival of MDA-MB-231 cells following transfection with 100nM siRNAs 4, or 5, or ASO4. Dishes show representative survival of unirradiated cells (upper row) or of cells following 2 Gy ionizing radiation (lower row). (b) Clonogenic survival of MDA-MB-231 cells following transfection with 100 nM siRNA, inverted control duplexes (Inv), ASO4 or scrambled control (Scr4). Results represent mean ± sem of colony counts in triplicate dishes. Similar results were obtained in two further clonogenic assays. (c) Survival of transfected MDA-MB-231 cells that were unirradiated, or irradiated, at 2Gy. The fraction surviving 2Gy irradiation (SF2) is shown as mean ± sem of three independent assays (modified from Bohula et al., *J. Biol. Chem.*, 278, 15991, 2003; with permission).

down. The potency of these siRNAs in blocking IGF signaling, tumor cell survival, and the stability of siRNAs, at least *in vitro*, suggests that IGF1R-targeting agents incorporating this mode of action have potential as anticancer therapy. Encouragingly, chemically-synthesized and plasmid-based siRNAs are now being used *in vivo*,[27–32] and are showing objective anticancer activity.[33] It seems clear that RNA interference is not only a powerful research tool for studying gene function, but also shows genuine therapeutic potential.

PROTOCOLS

Protocol 3.1 Synthesis of an Oligonucleotide Library (dN_{12})

Materials

- Standard DNA synthesis reagents (Cruachem, Applied Biosystems, or Glen)
- Ammonia solution (~30%) (BDH)
- DNA synthesis columns (Cruachem or Applied Biosystems)
- NAP10 Column (Amersham Biosciences)
- DNA synthesizer (Applied Biosystems)

Experimental Procedures

1. Pool contents of dA, dG, dC and dT columns into a single column.
2. Program the synthesizer to add all four amidites at each coupling step.
3. Carry out synthesis using the standard 1 μmol synthesis cycle.
4. Elute with 33% ammonia and de-protect at 55°C for 8 h.
5. Dry in a SpeedVac and resuspend in 1 ml nuclease-free water.
6. De-salt through a NAP10 Sephadex G25 Column and store at –20°C.

Protocol 3.2 Preparation of End-labeled Transcripts by *In Vitro* Transcription

Since the first nucleotide incorporated by T7, T3 or SP6 RNA polymerase is a G, this protocol makes use of [γ^{32}P] GTP for 5'end-labeling. See Protocol 5.4 for internal labeling of transcripts.

Materials

- DNA template (linearized plasmid or PCR product) with an appropriate viral promoter sequence (T7, T3 or SP6)
- [γ^{32}P] GTP (> 5000Ci/mmol, Amersham Biosciences)
- MicroSpin G25 Columns (Amersham Biosciences)
- *In vitro* transcription reagents (Ambion, Promega)
- RNase inhibitor (Ambion, Promega)

Experimental Procedures

1. To a microfuge tube on ice, add reagents in this order:

10 × transcription buffer	2 μl
RNase-free H_20 to	20 μl
100 mM DTT	2 μl (optional with the type of kit)
RNasin	20 U (optional with the type of kit)
10 mM ATP, CTP and UTP	1.5 μl each (final conc. 750 μM)
1 mM GTP	1.5 μl (final conc. 75 μM)
Linearized plasmid or PCR product	2–3 μg
T7, T3 or SP6 RNA polymerase	20 U
[γ^{32}P] GTP	5 μl (50 μCi)

2. Mix and incubate at 37°C for 1 h.
3. Remove unincorporated ribonucleotides by Sephadex G25 column chromatography (e.g., MicroSpin G-25 columns).
4. Check the integrity of the transcript by electrophoresis through a 6% (w/v) denaturing polyacrylamide gel.

Protocol 3.3 RNase H/dN$_{12}$ Mapping of Transcripts

Materials

- 10 × RNase H buffer (500 mM Tris-Cl pH 7.4, 200 mM KCl, 100 mM $MgCl_2$)
- Oligonucleotide library (dN$_{12}$)
- RNase H (Promega, Ambion, Sigma)
- RNase inhibitor (Promega, Ambion)
- 100 mM DTT (Promega)
- End-labeled transcript

Experimental Procedures

1. Assemble on ice a mix of reaction components, making sufficient for n + 1 reactions:

Buffer × 10	1 µl
RNase inhibitor	0.5 U
100 mM DTT	1 µl
RNase H	1 µl (0.5 U)
5' end-labeled transcript	5 fmol
Nuclease-free H_2O to	9 µl

2. Dilute dN$_{12}$ library to, for example, 20, 50, 100, 250, 500 pmol/ml.
3. To individual eppendorf tubes on ice and add 1 µl of each dN$_{12}$ dilution or nuclease-free H_2O.
4. Add 9 µl of the RNase H reaction mix.
5. Mix and incubate at 30°C or 37°C for 30–40 min.
6. Terminate reactions by adding 10 µl gel loading buffer (30% (v/v) formamide, 0.2 M EDTA).
7. Analyze 5 µl of each reaction on a 6% denaturing polyacrylamide gel. Map fragment sizes by comparing with radiolabeled MW markers, e.g., Gibco 100 bp ladder, and with transcripts of known size, e.g., 200 nt, 600 nt, 1000 nt. (*N.B.* DNA markers will not give 100% accurate sizing for cleavage fragments since the mobility of DNA and RNA are not equivalent)
8. Analyze the image by autoradiography.

Protocol 3.4 Fabrication and Processing of Oligonucleotide Scanning Arrays

Materials

- Synthesis Mask (machined to specification, e.g., JFC Engineering Ltd, High Wycombe, UK)

- Array Maker[15]
- Aminated polypropylene (Beckman Instruments) or derivatized glass[15]
- DNA synthesizer (Applied Biosystems)
- Standard DNA synthesis reagents (Cruachem, Applied Biosystems, or Glen)
- Ammonia solution (~30%) (BDH)
- Duran glass bottle or a high-density polyethylene chamber[15]

Experimental Procedures

1. Cut derivatized polypropylene or glass to the correct size (see Figure 3.2). Add 4 to 5 mm to margins for ease of handling.
2. Mount polypropylene on a glass plate, e.g., 3-mm thick soda glass.
3. Fix the array maker to the front of a DNA synthesizer.
4. Connect inlet and outlet holes of the mask via standard 19 SWG (1.1 mm) syringe needles to the synthesizer's reagent supply.
5. Program the DNA synthesizer with an appropriate synthesis cycle (see [15]).
6. Enter the sequence of the antisense strand in the 5' → 3' direction.
7. Place the substrate polypropylene or glass on the array maker assembly in the desired starting position.
8. Mark the first footprint of the mask on the support by cutting notches into the polypropylene or with a diamond scriber on glass.
9. Tighten the mask against the substrate to produce a seal, using sufficient pressure to prevent leakage but not enough to indent the polypropylene.
10. Start the DNA synthesizer to go through the preprogrammed cycle to couple the first nucleotide, i.e., the first base at the 3' end of the antisense sequence.
11. After completion of the first synthesis step, slacken the pressure clamp and advance the plate one increment.
12. Tighten the pressure clamp and start the synthesizer for the second nucleotide in the sequence. Continue until the full sequence is synthesized.
13. Detach the polypropylene array from its glass support.
14. Place the array in a Duran bottle or another suitable chamber.[15]
15. Add 30% ammonia solution to cover the array completely and seal the chamber.
16. Incubate at 55°C in a water bath inside a fume hood for 16 to 18 h.
17. Cool to 4°C before opening. The array is now ready for hybridization.

Protocol 3.5 Hybridization of Transcript to Antisense Arrays and Image Analysis

Materials

- Hybridization buffer (1 M NaCl, 10 mM Tris-Cl pH 7.4, 1 mM EDTA, 0.01% SDS), or (150 mM NaCl, 2 mM $MgCl_2$, 10 mM Tris-Cl pH 7.4, 1 mM EDTA, 0.01% (v/v) SDS)
- 50 to 100 fmol internally radio-labeled transcript (see Protocol 5.4)
- Hybridization tube and oven (Techne or Hybaid)
- Storage Phosphor screen (Kodak)
- PhosphorImager or STORM (Molecular Dynamics)

Experimental Procedures

1. Place the array in the hybridization tube, coiling it in a spiral.
2. Make hybridization mix by diluting 50 to 60 fmol internally labeled transcript in 10 to 20 ml hybridization buffer, using sufficient volume to cover the array along the length of the tube.
3. Place hybridization tube and mix separately into hybridization oven to pre-warm to desired temperature for 30 min. Also put ~100 ml hybridization buffer into oven, for later use in step 5.
4. Pour the hybridization mix into the tube and hybridize for 3 to 4 h.
5. Tip off the hybridization mix, and wash the array with pre-warmed buffer.
6. Dry in layers of Whatman paper, cover with cling film, and expose to a phosphor storage screen for 16 to 20 h.
7. Scan the phosphor screen in a PhosphorImager (STORM: Molecular Dynamics).
8. Analyze the image using the computer program xvseq.[15]
9. Superimpose template grid onto image, using notches for alignment (see Protocol 3.3).
10. Display as histograms the results of integrated pixel values for oligonucleotides of any desired length. Generally the most useful information derives from the longest sequences in the array (usually 15 to 20 mers), and there is often negligible signal from shorter oligonucleotides at the edges of the array.
11. Click on a bar in the histogram to highlight the equivalent cell (oligonucleotide) in the array, and the corresponding region of mRNA sequence, or click on a cell in the array to highlight the corresponding region of sequence, the equivalent bar in the histogram, and the integrated pixel value.
12. Identify regions of mRNA sequence accessible to hybridization with antisense oligonucleotides for further study.

Protocol 3.6 Transfection of Mammalian Cells with siRNAs

This protocol is for transfection of monolayer cultures of mammalian cells. We found that this method gives high efficiency of transfection of a variety of human and murine cells.

Materials

- siRNA duplex (synthesized, e.g., from MWG-Biotech, Qiagen, or enzymatically-produced, see Chapters 4 and 5).
- Store aliquots at –20°C; avoid > 4 freeze–thaw cycles
- Standard cell culture facility, cell media, and disposable plastic consumables
- Oligofectamine (InVitrogen)
- Optimem (Gibco)

Experimental Procedures

1. On the day prior to transfection, seed cells to achieve ~30 to 40% confluence on day of transfection. Transfection efficiency may be compromised if cultures are > 50% confluent. For many cell lines, a seeding density of ~3×10^5 cells per

6 cm dish will give approximately the correct degree of confluence for transfection the following day. For rapidly growing cultures, the seeding density should be reduced to 0.7-1 × 10^5 per 6 cm dish, but this should be determined empirically for each cell line.

2. Prepare complexes in sterile eppendorf tubes or wells (e.g., of a 24 well plate) as below: Tube 1. **a)** 20 µM duplex plus **b)** Optimem (see below)
   Tube 2. **c)** Oligofectamine (OF) plus **d)** Optimem

	a) Duplex (µl)	**b)** Optimem (µl)	**c)** OF (µl)	**d)** Optimem (µl)
6-well plate	5	250	2	68
6-cm dish	10	550.5	4.5	150
10-cm dish	29	1500	12	410

3. Incubate tubes 1 and 2 separately at room temperature for 10 min.
4. Combine contents of tubes 1 and 2.
5. Incubate at room temperature for 25 min to allow complexes to form.
6. Wash cell monolayer with **e)** ml Optimem (see below); remove wash completely.
7. Add to dish **f)** µl Optimem and then **g)** µl complexes from step 5.

	e) Wash (µl)	**f)** Optimem (µl)	**g)** Complexes (µl)	**h)** FCS (µl)	**j)** Medium (µl)
6-well plate	1	175	325	50	3
6-cm dish	2	285	715	100	4
10-cm dish	3	950	1951	300	7

8. Incubate at 37°C for 4 h.
9. Top up with **h)** µl FCS and **j)** µl usual growth medium (e.g., RPMI and 10% (v/v) FCS).
10. Assay after 48 to 72 h.

Protocol 3.7 Preparation of Cell Lysates

Materials

- Phosphate-buffered saline, pH 7.4
- Cell lysis buffer[34] (50 mM HEPES, pH 7.5, 100 mM NaCl, 10 mM EDTA, 1% (v/v) Triton X-100, 4 mM Na pyrophosphate, 2 mM Na vanadate, 10 mM NaF, 2 µg/ml each leupepton and aprotinin, 1 mM PMSF)
- BCA protein assay kit (Pierce)
- Refrigerated bench-top microfuge

Experimental Procedures

1. Wash cells with ice-cold PBS.
2. Remove cells from plate by scraping into 1 ml ice-cold PBS.
3. Transfer to an eppendorf tube, and spin in microfuge at 4°C for 30 seconds at maximum speed.
4. Remove PBS wash completely.
5. Re-suspend cell pellet in cell lysis buffer (at ~100 µl per 10^6 cells).
6. Lyse on ice for 30 min.
7. Spin for 15 min at 4°C at maximum speed in microfuge, and transfer the supernatant to fresh tube. Keep on ice.

8. Assay lysates for total protein using BCA assay.
9. Determine volume of lysate for 20 μg total protein, equalize volumes of all samples with IGF1R lysis buffer, add 0.5 × volume 3 × SDS PAGE sample buffer.
10. Store remaining lysate at –20 or –80°C.

Protocol 3.8 SDS-Polyacrylamide Gel Electrophoresis

Materials

- For gels: Protogel 30% (w/v) acrylamide, 0.8% (w/v) bis-acrylamide stock solution (37.5:1; National Diagnostics); 1.5 M Tris-Cl pH 8.8, 0.5 M Tris-Cl pH 6.8, TEMED, 10% (w/v) ammonium persulphate.
- Alternative: Precast PAGE gels (Criterion gels, BioRad)
- Rainbow marker (Amersham Biosciences RPN 756)
- SDS-PAGE sample buffer (3 ×; 150 mM Tris-Cl pH 6.8, 15% (v/v) 2-mercaptoethanol, 9% (w/v) SDS, 30% (v/v) glycerol, 0.3 mg/ml bromphenol blue)
- Electrophoresis buffer (3g Tris-Cl, 14.4g glycine, 1g SDS per liter).
- Electrophoresis apparatus: Mini Protean III (BioRad)

Experimental Procedures

1. Make standard denaturing SDS-PAGE gel: 7.5% resolving gel, 4% stacking gel, or use pre-cast gels.
2. Mix protein samples (and molecular weight markers) with 0.5 × volume of 3 × SDS-PAGE sample buffer.
3. Denature samples at 95°C for 3 min. Centrifuge for 3 min.
4. Load samples onto gel and run at ~100 to 125 V until the dye front reaches bottom of the gel (~1 to 1.5 h).

Protocol 3.9 Immunoblotting for IGF1R

This method is for simultaneous determination of the IGF1R (130kD α-subunit or 98kD β-subunit), and β-tubulin as loading control.

Materials

- SemiPhor semi dry transfer unit (Hoefer)
- Transfer buffer, pH 9.2 (48 mM Tris-Cl, 39 mM glycine, 1.3 mM SDS, 20% (v/v) methanol)
- Hybond C Extra membrane (Amersham Biosciences)
- Blocking buffer (PBS with 0.05 (v/v)% Tween-20 (PBS/T), plus 5% (w/v) fat-free milk)
- Anti-IGF1R antibody (Santa Cruz, β-subunit sc713, or α-subunit, sc712)
- Anti-β-tubulin antibody (Sigma)
- Anti-rabbit and anti-mouse secondary antibodies conjugated to HRP (Dako)
- ECL Plus Detection Kit (Amersham Biosciences)

- Rocking platform
- X-ray film (Kodak)

Experimental Procedures

1. Transfer proteins to nitrocellulose membrane by electro-blotting at 0.8 mA/cm^2 (i. e., ~35 mA per minigel) for 90 to 120 min. Use 180 min for high MW proteins (> 200 kD).
2. Incubate membrane in blocking buffer at room temperature for 1 h, or in blocking buffer without Tween–20 for 2 h to overnight at 4°C.
3. Cut membrane in half horizontally through 66 kD (pink) marker band (*N.B.* β-tubulin runs with 46 kD (yellow) marker and IGF1R β-subunit with 98 kD (brown) marker of the Rainbow Markers). Orientate filters, e.g., by cutting off lower right corner.
4. Prepare primary antibody mixes, 1 ml per half filter as follows:
   Anti-IGF1R antibody at 1:100-1:500 in blocking buffer
   Anti-β-tubulin antibody at 1:500 (1:300 to detect mouse protein) in blocking buffer
5. Pipette each 1 ml mix in square area onto clean plastic surface and overlay membrane protein side down. Incubate for 120 min at room temperature.
6. Wash 3 × 10-15 min in PBS/T on rocker.
7. Prepare 30 ml secondary antibody of each mix in blocking buffer as follows:
   For IGF1R detection: 1:10,000 anti-rabbit HRP
   For β-tubulin detection 1:10,000 anti-mouse HRP
8. Incubate for 60 min on rocker.
9. Wash 3 × 10-15 min in PBS/T on rocker.
10. Detect using ECL Plus Kit. Briefly, blot off excess moisture from membranes. Place protein side up on clean plastic surface and pipette mixed detection solution (2 ml solution A plus 50 µl solution B per gel) onto membrane. Incubate for 5 min at room temperature then drain off excess detection fluid. Place in clean plastic folder, smooth out air bubbles, and expose to x-ray film for between 1 min-overnight and develop the film. A storage phosphor screen can also be used, scanned in a PhosphorImager (STORM, Molecular Dynamics).

Protocol 3.10 Survival Assay

This method measures clonogenic survival of cells seeded as monolayer cultures in wells or dishes. If using siRNA-transfected cells, the survival assay should be set up 24 to 48 h after transfection. This method can be adapted for measuring sensitivity to drugs or radiation. By seeding cells prior to these treatments, one can ensure that all dishes are identical prior to drug application. For details of assay for clonogenic survival in agarose, see [35].

Materials

- PBS/EDTA buffer (3 mM EDTA in 1 x PBS)
- Growth medium (e.g., RPMI 1640 or Dulbecco's modification of Eagle's medium plus 10% (v/v) FCS) (Sigma or InVitrogen)

- Hemocytometer (Improved Neubauer, VWR International)
- Fixative (Methanol:Acetic acid 1:3 (v/v))
- Stain (400 μg/ml crystal violet in 2% ethanol)

Experimental Procedures

1. Disaggregate cells in PBS/EDTA. We avoid the use of trypsin in assays measuring the function of the IGF1R or other cell surface molecules.
2. Wash cells in usual growth medium and resuspend in fresh growth medium.
3. Count cells using a hemocytometer.
4. Reseed cells in wells or dishes. For many human tumor cells, seed 2000 cells per 6 cm dish. For rapidly growing cells, use 1000 cells per 6 cm dish. Scale up for 10-cm dishes and down for 6-well plate. Seed sufficient dishes for triplicate data points in subsequent assay.
5. Allow cells to adhere for 4 to 16 h (to be determined empirically).
6. Treat cells, e.g., with irradiation or cytotoxic drugs. For the latter, remove growth medium and replace with fresh medium plus drug at desired final concentration. Treat control cultures with equivalent volume of diluent in fresh growth medium.
7. Incubate at 37°C until colonies of > 50 cells are visible. This may take 6 to 7 days for rapidly growing cultures and 2 to 3 weeks for slower growing cells.
8. To fix colonies pour off growth medium and carefully add fixative.
9. Incubate for 30 min to 2 h at room temperature. Pour off and discard fixative.
10. Stain colonies by careful addition of crystal violet stain. Incubate at room temperature for 2 to 16 h.
11. Pour off stain (which can be reused). Carefully wash dishes in tap water, invert over paper towels and leave to dry overnight.
12. Count colonies and express result as percent survival in untreated dishes.

ABBREVIATIONS

ASO antisense oligonucleotide
IGF insulin-like growth factor
IGF1R type 1 IGF receptor
IR insulin receptor
siRNA small interfering RNA
TSS translation start site.

ACKNOWLEDGMENTS

Work in the laboratories of the authors is supported by Cancer Research UK, the Medical Research Council, the Rhodes Trust, the Royal College of Physicians, and Health Foundation.

REFERENCES

1. Elbashir, S.M. et al., Duplexes of 21-nucleotide RNAs mediate RNA interference in cultured mammalian cells, *Nature,* 411, 494, 2001.
2. Elbashir, S.M. et al., Functional anatomy of siRNAs for mediating efficient RNAi in Drosophila melanogaster embryo lysate, *EMBO J.,* 20, 687, 2001.
3. Dudley, N.R., Labbe, J.C. and Goldstein B., Using RNA interference to identify genes required for RNA interference, *Proc. Natl. Acad. Sci. U.S.A.,* 99, 4191, 2002.
4. Harborth, J. et al., Identification of essential genes in cultured mammalian cells using small interfering RNAs, *J. Cell Sci.,* 114, 4557, 2001.
5. Ohta, T. et al., Characterization of Cep135, a novel coiled-coil centrosomal protein involved in microtubule organization in mammalian cells, *J. Cell Biol.,* 156, 87, 2002.
6. Holen, T. et al., Positional effects of short interfering RNAs targeting the human coagulation trigger Tissue Factor, *Nucleic Acids Res.,* 30, 1757, 2002.
7. McManus, M.T. et al., Gene silencing using micro-RNA designed hairpins, *RNA,* 8, 842, 2002.
8. Harborth, J. et al. Sequence, Chemical, and Structural Variation of Small Interfering RNAs and Short Hairpin RNAs and the Effect on Mammalian Gene Silencing, *Antisense Nucleic Acid Drug Dev.,* 13, 83, 2003.
9. Miyagishi, M., Hayashi, M. and Taira, K., Comparison of the suppressive effects of antisense oligonucleotides and siRNAs directed against the same targets in mammalian cells, *Antisense Nucleic Acid Drug Dev.,* 13, 1, 2003.
10. Stein, C.A., The experimental use of antisense oligonucleotides: a guide for the perplexed, *J. Clin. Invest.,* 108, 641, 2001.
11. Zuker, M., Computer prediction of RNA structure, *Methods Enzymol.,* 180, 262, 1989.
12. Dumas, J.P. and Ninio, J., Efficient algorithms for folding and comparing nucleic acid sequences, *Nucleic Acids Res.,* 10, 197, 1982.
13. Ho, S.P. et al., Mapping of RNA accessible sites for antisense experiments with oligonucleotide libraries, *Nat. Biotechnol.,* 16, 59, 1998.
14. Sohail, M., Akhtar, S. and Southern, E.M., The folding of large RNAs studied by hybridization to arrays of complementary oligonucleotides, *RNA,* 5, 646, 1999.
15. Sohail, M. and Southern, E.M., Using oligonucleotide scanning arrays to find effective antisense reagents, *Methods Mol. Biol.,* 170, 181, 2001.
16. Bohula, E., Playford, M. and Macaulay, V., Targeting the type 1 insulin-like growth factor receptor as anticancer treatment. *Anti-Cancer Drugs,* 14, 669, 2003.
17. Ullrich, A. et al., Insulin-like growth factor I receptor primary structure: comparison with insulin receptor suggests structural determinants that define functional specificity, *EMBO J.,* 5, 2503, 1986.
18. Sohail, M. et al., Antisense oligonucleotides selected by hybridisation to scanning arrays are effective reagents *in vivo*, *Nucleic Acids Res.,* 29, 2041, 2001.
19. Bohula, E.A. et al., The efficacy of small interfering RNAs targeted to the type 1 IGF receptor is influenced by secondary structure in the IGF1R transcript, *J. Biol. Chem.,* 278, 15991, 2003.
20. Wraight, C.J. et al., Reversal of epidermal hyperproliferation in psoriasis by insulin-like growth factor I receptor antisense oligonucleotides, *Nat. Biotechnol.,* 18, 521, 2000.
21. Hellawell, G.O. et al., Chemosensitization of human prostate cancer using antisense agents targeting the type 1 insulin-like growth factor receptor, *BJU Int.,* 91, 271, 2003.
22. Martinez, J. et al., Single-Stranded Antisense siRNAs Guide Target RNA Cleavage in RNAi, *Cell,* 110, 563, 2002.

23. Borkhardt, A., Blocking oncogenes in malignant cells by RNA interference-New hope for a highly specific cancer treatment?, *Cancer Cell,* 2, 167, 2002.
24. Hammond S.M., Caudy A.A. and Hannon G.J., Post-transcriptional gene silencing by double-stranded RNA, *Nat. Rev. Genet.,* 2, 110, 2001.
25. Fire, A. et al., Potent and specific genetic interference by double-stranded RNA in Caenorhabditis elegans, *Nature,* 391, 806, 1998.
26. Holen, T. et al., Similar behaviour of single-strand and double-strand siRNAs suggests they act through a common RNAi pathway, *Nucleic Acids Res.,* 31, 2401, 2003.
27. McCaffrey, A.P. et al., RNA interference in adult mice, *Nature,* 418, 38, 2002.
28. Xia, H. et al., siRNA-mediated gene silencing *in vitro* and *in vivo*, *Nat. Biotechnol.,* 20, 1006, 2002.
29. Brummelkamp, T.R., Bernards, R. and Agami, R., Stable suppression of tumorigenicity by virus-mediated RNA interference, *Cancer Cell,* 2, 243, 2002.
30. Wilda, M. et al., Killing of leukemic cells with a BCR/ABL fusion gene by RNA interference (RNAi), *Oncogene,* 21, 5716, 2002.
31. Jacque, J.M., Triques, K. and Stevenson, M., Modulation of HIV-1 replication by RNA interference, *Nature,* 418, 435, 2002.
32. Novina, C.D. et al., siRNA-directed inhibition of HIV-1 infection, *Nat. Med.,* 8, 681, 2002.
33. Filleur, S. et al., SiRNA-mediated inhibition of vascular endothelial growth factor severely limits tumor resistance to antiangiogenic thrombospondin-1 and slows tumor vascularization and growth, *Cancer Res.,* 63, 3919, 2003.
34. Blakesley, V.A. et al., Tumorigenic and mitogenic capacities are reduced in transfected fibroblasts expressing mutant insulin-like growth factor (IGF)-I receptors. The role of tyrosine residues 1250, 1251, and 1316 in the carboxy-terminus of the IGF-I receptor, *Endocrinology,* 137, 410, 1996.
35. Macaulay, V.M. et al., Downregulation of the type 1 insulin-like growth factor receptor in mouse melanoma cells is associated with enhanced radiosensitivity and impaired activation of Atm kinase, *Oncogene,* 20, 4029, 2001.

4 siRNA Production by *In Vitro* Transcription

Olivier Donzé, Peter Dudek, and Didier Picard

CONTENTS

4.1 INTRODUCTION

RNA interference (RNAi) is a process in which double-stranded RNA (dsRNA) induces the posttranscriptional degradation of homologous transcripts. This phenomenon has been observed in a variety of organisms, including plants, fungi, insects, protozoans, and mammals.[1] Introduction of dsRNA into cells leads to the sequence-specific destruction of endogenous RNAs that are complementary to the dsRNA.[2] According to the prevailing model, long dsRNA molecules are cleaved into short interfering dsRNA species (siRNAs) of about 21 to 24 nt.[3] In mammalian cells, long dsRNA (more than 30 nt) molecules cannot be used since they elicit a nonspecific interferon response.[4] This problem can be avoided with siRNAs of 21 nt. These siRNAs efficiently inhibit target gene expression in a sequence-specific manner[5] but are too short (< 30 bp) to trigger the nonspecific dsRNA responses. Since RNAi technology has begun to be applied to mammalian cells, many researchers have performed it with expensive chemically synthesized siRNAs. However, we have recently reported[6] a technically simple and inexpensive alternative method for the production of siRNAs. It makes use of *in vitro* transcription with T7 RNA polymerase from DNA oligonucleotides. This chapter provides a more comprehensive description of this method with detailed experimental protocols.

0-8493-2141-7/05/$0.00 + $.50

4.2 STRATEGY TO GENERATE siRNAs BY USING *IN VITRO* TRANSCRIPTION

To generate siRNAs *in vitro*, we designed the strategy presented in Figure 4.1. The transcribed sequences are 20 nt plus 2 nt, which can be any nucleotides in the case of the sense RNA (see comments about T7 transcription) but must be complementary

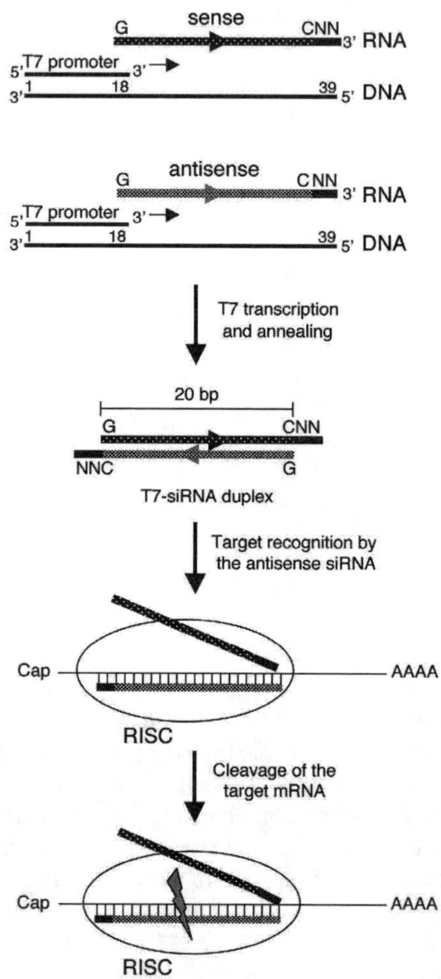

FIGURE 4.1 Strategy for generating T7-siRNAs. The T7-siRNA sense and antisense strands are shown in dark and light gray, respectively. The two nucleotides (NN) in black represent the two protruding nucleotides present at the 3' ends of siRNAs. RISC stands for the RNA-induced silencing complex that targets the mRNA for cleavage. See text for further details. This figure is a modified version of Figure 1 from Donzé and Picard, RNA interference in mammalian cells using siRNAs synthesized with T7 RNA polymerase, *Nucleic Acids Res.*, 2002, vol. 30, e46, by permission of Oxford University Press.

nucleotides in the antisense RNA since in the siRNA duplex, it is the antisense strand that guides target recognition.[7] During the elaboration of this method, we noticed that siRNAs of 22 nt are more efficient than those of 21 nt. For this reason, we chose 22 nt as the default length for the current T7-siRNA protocol (Protocols 4.1 and 4.2).

T7 RNA polymerase is routinely used to synthesize RNAs longer than 10 or 20 nucleotides. It can initiate transcription from a partially single-stranded DNA oligonucleotide template, in which only the region encompassing the 18 nt of the T7 promoter is double-stranded (Figures 4.1 and 4.2).[8] The core promoter sequence is commonly considered to be an 18-bp sequence from position –17 to –1, with the start site of transcription at the G at position +1 (Figure 4.3).[9] Certain sequences may be problematic for T7 transcription, such as strings of eight or more A's or U's. Moreover, too many A or U bases in the first six nucleotides can substantially reduce the yield. With some low frequency, T7 RNA polymerase can add one or more nucleotides at the 3' end of runoff transcripts.[10] Indeed, with some T7-siRNA preparations we did notice a small fraction of products that were longer by 1 nt (data not shown). However, we have not seen any obvious deleterious effects of this population of molecules on RNAi (data not shown) either because these molecules represented only a

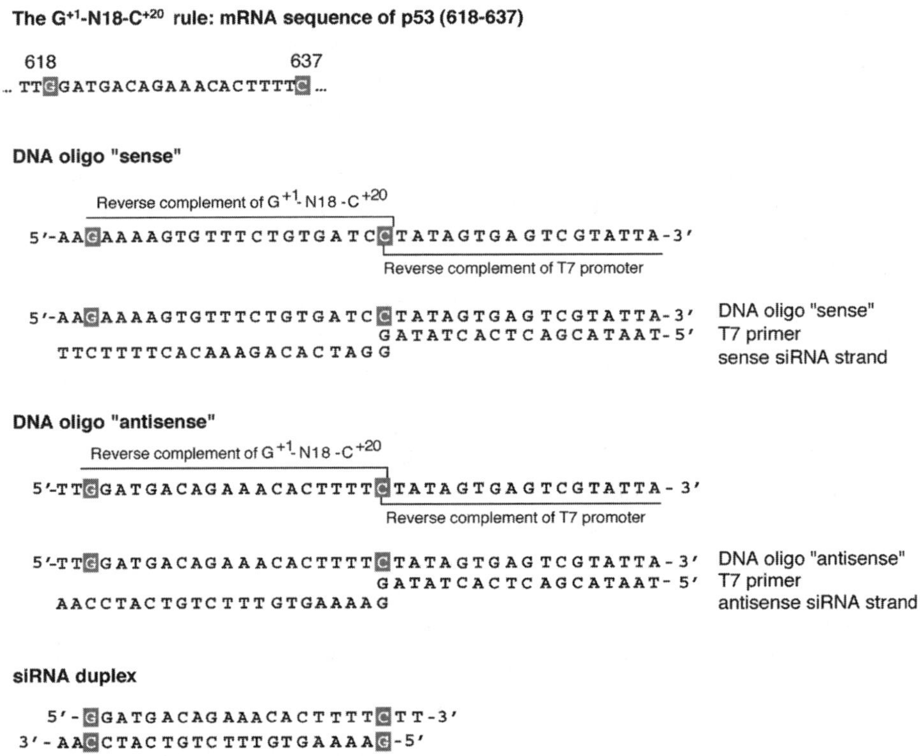

FIGURE 4.2 Design of DNA oligonucleotides for generating a particular T7-siRNA duplex. Details shown here are for the human p53 mRNA (position 618) as an example. For simplicity, RNA sequences are shown with I's instead of U's.

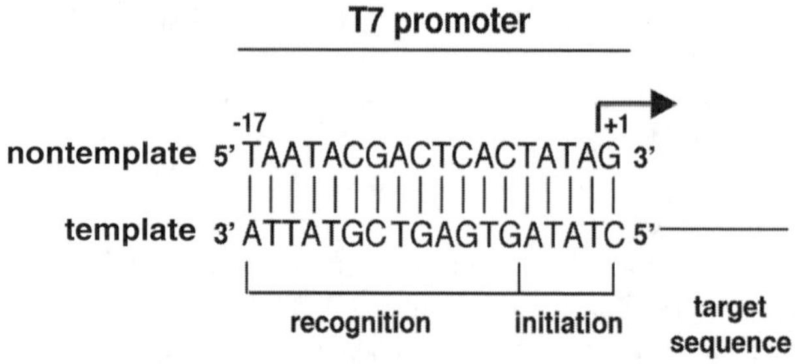

FIGURE 4.3 Consensus promoter elements of T7 RNA polymerase. The recognition domain encompasses base pairs –17 through –5. The initiation domain includes base pairs –4 to +1. Transcription initiates at position +1 (arrow), giving rise to an RNA transcript that is complementary to the template strand.

small fraction of the total T7-siRNA preparation or because an additional nucleotide at the 3' overhang may be tolerated.[11] T7-siRNAs differ from normal Dicer products (and from chemically synthesized siRNAs) in that they contain a 5'-triphosphate group, but clearly, despite this difference, they are biologically functional.[6,12]

T7 RNA polymerase has a strong preference for a G at the start site and, thus, the last G of the T7 promoter sequence is the first ribonucleotide that is incorporated into the RNA. Therefore, all siRNAs generated by this method must start with a G. Thus, a prerequisite for the synthesis of functional siRNA by T7 transcription is a G at +1 and a C at position +20 to allow annealing with the complementary siRNA, which inevitably also starts with a G (see Figure 4.2). The resulting siRNA duplex is double-stranded for 20 nt and contains protruding 3' ends of 2 nt on either side. The G^{+1}-N18-C^{+20} rule is not a serious constraint since this sequence module is frequently found in any gene (about six times in a random sequence of 100 bp). As illustrated in Figures 4.1 and 4.2, the T7 promoter sequence is identical for the synthesis of both sense and antisense RNAs so that only two DNA oligonucleotides of 39 nt corresponding to the sense and antisense templates are required for any new target.

4.3 DESIGN OF DNA OLIGONUCLEOTIDES FOR PRODUCTION OF T7-siRNAs

4.3.1 Criteria

To date, relatively few criteria have been established that provide guidance in the design of siRNAs,[13,14] and they are discussed in other chapters of this book. In addition to the G^{+1}-N18-C^{+20} rule, we suggest respecting the following ones: (a) a relatively low GC content (30 to 50%), (b) an AT-rich dinucleotide at the 3' end of the antisense RNA strand that is complementary to the dinucleotide present in the mRNA preceding the 5' G of the sense strand, and (c) avoidance of regions with long stretches of a single nucleotide, especially G.[14]

4.3.2 Facilitated Design with the Web Application TROD

While we have designed numerous pairs of DNA oligonucleotides that respect the above-mentioned rules by scanning target sequences by eye, we recently developed a Web application to facilitate this step. The "T7 RNAi DNA Oligo Designer" (TROD) can be used online at http://www.cellbio.unige.ch/RNAi.html (updates to the method are also posted there). Sequences can be entered directly, from a local file, or with a GenBank accession code. The output of TROD is a table listing both sense and antisense DNA oligonucleotides and the position of the target sequence in the query sequence. Oligonucleotides are ordered by GC content, and those yielding favored terminal dinucleotides in the antisense RNA strand are highlighted. Direct links allow one to blast selected oligonucleotides. Compared to other RNAi-related Web applications, TROD has the advantage of yielding the DNA oligonucleotide sequences exactly as they need to be ordered.

4.4 INHIBITION OF HUMAN p53 EXPRESSION AS AN EXAMPLE

We previously reported that expression of both the transfected GFP and the endogenous gene encoding PKR could be knocked down with T7-siRNAs.[6] To illustrate the methodology described in this chapter, we present the inhibition by RNAi of another endogenous gene, the gene encoding the human protein p53.[17]

Two sequences within the human p53 mRNA sequence were selected as targets (Table 4.1). The first is the published sequence of a small hairpin RNA (shRNA) selected against human p53, starting at the G at position 775 relative to the first nucleotide of the start codon and ending with the C at position 794.[18] We added two Ts downstream at the 3' end to create the 2-nt overhang that is required to form functional

TABLE 4.1
List of the T7-siRNAs Used for the Experiments in Figure 4.4

(1) **T7/p53 (618-637)**[1]

- sense RNA strand: 5'-GGATGACAGAAACACTTTTCTT-3'
- antisense RNA strand: 5'-GAAAAGTGTTTCTGTCATCCAA-3'

(2) **T7p53 (775-794)**[1]

- sense RNA strand: 5'-GACTCCAGTGGTAATCTACTT-3'
- antisense RNA strand: 5'-GTAGATTACCACTGGAGTCTT-3'

(3) **T7/PKR (1201-1220)**[1]

- sense RNA strand: 5'-GGGGTGGATTATATACATTCAA-3'
- antisense RNA strand: 5'-GAATGTATATAATCCACCCCTT-3'

[1]The sequence numbers of each gene do not include the first two nucleotides 5' of the G^{+1} that hybridize to the antisense RNA strand.

TABLE 4.2
Pipetting Scheme for the Ca-P Transfection Experiment Presented in Figure 4.4A

Wells/lanes	pEGFP-C1t	pSuper /siRNAs	water	8 × Ca	2 × HBS
1. pSuper	2 µg	2 µg	86 µl	30 µl	120 µl
2. pSuper/p53	2 µg	2 µg	86 µl	30 µl	120 µl
3. T7/p53^{775}	2 µg	2 µl	86 µl	30 µl	120 µl
4. T7/p53^{618}	2 µg	2 µl	86 µl	30 µl	120 µl
5. T7/PKR	2 µg	2 µl	86 µl	30 µl	120 µl
6. Control	2 µg	—	88 µl	30 µl	120 µl

siRNA duplexes, resembling Dicer-generated products.[5] This particular T7-siRNA duplex is 21 nt long. The second T7-siRNA is targeted against the sequence from 618 to 637 and strictly follows the G^{+1}-N18-C^{+20} rule recommended above, to give a 22-nt T7-siRNA (Figure 4.2). As a control, we designed a new T7-siRNA targeted against human PKR (1204 to 1223) (Table 4.1). After synthesis of the siRNAs and hybridization to form siRNA duplexes, both siRNAs as well as the control T7-siRNA against PKR were transfected into HEK293T cells using the Ca-P procedure (Tables 4.2 and 4.3; see also Protocol 4.4). It is worth noting that the cell line used here (HEK293T) contains high levels of endogenous p53. The cells were seeded at 50% confluency into 6-well dishes. Transfection efficiency was routinely checked using GFP fluorescence and was generally about 70%.

Figure 4.4A presents the result of a typical RNAi experiment, illustrating that siRNAs synthesized *in vitro* with T7 RNA polymerase are functional in tissue culture cells. 293T cells were transfected as described in detail in Table 4.2 and lysed 70 h later. Equivalent amounts of cell extracts were loaded onto a 10% SDS-polyacrylamide gel and probed for either PKR or the molecular chaperone Hsp90 as an internal loading control. The slight difference between the two p53-directed T7-siRNAs might

TABLE 4.3
Pipetting Scheme for the Ca-P Transfection Experiment Presented in Figure 4.4B

Wells/lanes	pEGFP-C1	T7-siRNAs	water	8 × Ca	2 × HBS
1. Control	2 µg	—	88 µl	30 µl	120 µl
2. T7/p53^{618}	2 µg	0.5 µl	87.5 µl	30 µl	120 µl
3. T7/p53^{618}	2 µg	1 µl	86 µl	30 µl	120 µl
4. T7/p53^{618}	2 µg	2 µl	86 µl	30 µl	120 µl
5. T7/PKR	2 µg	1 µl	87 µl	30 µl	120 µl
6. T7/PKR	2 µg	2 µl	86 µl	30 µl	120 µl

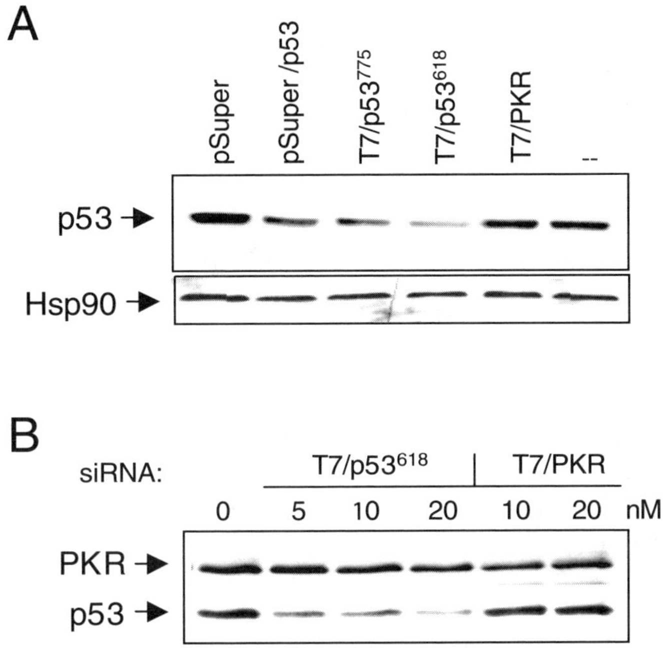

FIGURE 4.4 T7-siRNAs inhibit expression of human p53 in HEK293T cells. (A) Comparison of RNAi by different siRNAs directed against human p53. siRNAs were either generated *in vitro* by T7 transcription or *in vivo* from plasmid pSuper. Note that the T7/p53^{775} siRNA targets the exact same mRNA sequence as the shRNA generated by pSuper/p53. 293T cells were transfected with the calcium-phosphate co-precipitation protocol. The protein p53 was revealed by immunoblotting with a monoclonal antibody (Santa-Cruz, sc-126). Hsp90 was used as a loading control. T7-siRNA directed against PKR was transfected as a specificity control. (B) Dose response analysis of RNAi with the T7/p53^{618} siRNA in 293T cells. PKR and p53 were revealed by immunoblotting using specific anti-p53 and anti-PKR (Santa-Cruz, sc-707) antibodies, respectively. Note that this particular T7/PKR siRNA not only, as expected, has no effect on p53 levels, but it is also inefficient against PKR itself (see text).

be due either to their different sizes, since the T7/p53^{775} siRNA is 21 nt long, or to their sequences, since it has been reported that the efficiency of RNAi is sequence-dependent for reasons that are still unclear.[19] Interestingly, we observed that our best T7-siRNA against p53 was more efficient than the plasmid-generated shRNA targeted against p53,[18] although it is difficult to compare the two systems due to a lack of information concerning the levels of shRNA expression. The control treatments, including transfection of the empty vector pSuper, of the T7-siRNA T7/PKR or a mock transfection did not affect p53 levels, indicating that the T7/p53 or the pSuper/p53 specifically targets p53.

As shown in Figure 4.4B, a dose-response experiment was performed with the more potent T7-siRNA T7/p53^{618}. 293T cells were transfected as detailed in Table 4.3 and processed for immunoblotting 70 h later. PKR levels varied only slightly, confirming the specificity of the T7/p53 siRNA observed in Figure 4.4A. p53 levels began to drop in the presence of 5 nM $T7/p53^{618}$, the strongest effect being reached at 20 nM. This suggests that T7-siRNA works to the same extent as or even better than, chemically synthesized siRNAs.[20] Note that this experiment also indicates that our new PKR-directed T7-siRNA (22 nt long) is inefficient. As the exact rules for efficient RNAi have yet to be discovered, a certain percentage of selected sequences fail to work. It is, therefore, advisable to design and test several T7-siRNAs in parallel.

4.5 ADVANTAGES AND DISADVANTAGES OF THE METHOD

4.5.1 Advantages

1. Low cost, allowing the test of several siRNAs targeted against the same gene at a reasonable price
2. Rapid and easy
3. High yield of siRNAs
4. No risk of production of longer dsRNA molecules that could trigger the interferon response
5. Efficient at relatively low concentrations, thereby minimizing the risk of off-target effects

4.5.2 Disadvantages

1. Cannot be used for long term RNAi (inhibitory effects last for 4 to 5 days, as observed for siRNA generated by other methods described in other chapters of this book)

PROTOCOLS

Protocol 4.1 Preparation of the dsDNA Template

Materials

- H_2O: distilled water made RNase-free, for example, by autoclaving
- 5 × T7 transcription buffer (200 mM Tris–HCl pH 7.9, 30 mM $MgCl_2$, 50 mM DTT, 150 mM NaCl, and 10 mM spermidine) (comes with the high concentration T7 RNA polymerase, e.g., from Fermentas)

Experimental Procedures

1. Set up two tubes: *Tube A : "sense" dsDNA*
   Tube B: "antisense" dsDNA

Products	*Volume*	*Final Concentration*
Oligo "T7" (100 μM)	10 μl	20 μM
Oligo (100 μM) sense (Tube A) or antisense (Tube B)	10 μl	20 μM

5 × T7 transcription buffer	10 µl	1 ×
RNAse-free water	20 µl	
Final volume	50 µl	

2. Heat to 92°C for 2 min using a heating block, turn off the heating block and let it cool down for 30 min, followed by 15 min at 37°C, and finally place on ice to obtain annealed dsDNA "sense" and "antisense" templates.

PROTOCOL 4.2 T7-siRNA SYNTHESIS

Materials

- H_2O and 5 × T7 transcription buffer; see Protocol 4.1.
- rNTPs: ribonucleoside triphosphate set (4 × 200 µl) from Roche, order #1277057.
- T7 RNA polymerase, high concentration, 10,000 units, MBI Fermentas, order #EPO113.
- Pyrophosphatase, inorganic (yeast), 100 µg lyophilized (1000 U/mg) (Sigma, order #I1891). Add 200 µl RNase-free water to obtain 0.5 U/µl and store at –20°C.
- RNaseOUT inhibitor: 5000 U (GIBCO - Life Technologies, order #10777019).
- RNase-free DNase: 1000 U (Promega, order #M6101).

Experimental Procedures

1. Set up two tubes: *Tube 1: "sense" siRNA with template from tube A*
Tube 2: "antisense" siRNA with template from tube B

Products	*Volume*	*Final Concentration*
5 × T7 transcription buffer	10 µl	1 ×
5 µM rNTPs	10 µl	1 µM
Pyrophosphatase (0.5 U/µl in water)	1 µl	0.5 U
RNaseOUT (40 U/µl)	1 µl	40 U
T7 RNA polymerase (200 U/µl)	1 µl	200 U
dsDNA template (tubes A or B)	5 µl	2 µM
RNase-free water	22 µl	
Final volume	50 µl	

2. After incubation at 37°C for 2 hr, add 1 µl RNase-free DNase (1 U/µl) to tubes 1 and 2 and incubate for 15 min at 37°C.
3. Sense and antisense RNAs generated in tubes 1 and 2, respectively, are then annealed by mixing contents of tubes 1 and 2, heating at 95°C for 5 min followed by 1 hr at 37°C to obtain 100 µl of T7-siRNA.
4. *Purification.* This T7-siRNA preparation can be used as is or further purified with commercial RNA purification columns or simply by ethanol precipitation (adjust the 100-µl T7-siRNA preparation to 0.2 M sodium acetate pH 5.2 and precipitate with 2.5-vol ethanol; after centrifugation, wash the pellet once with 70% (v/v) ethanol, dry, and resuspend in 100 µl of RNase-free water).
5. *Yield. In vitro* transcription with highly concentrated T7 RNA polymerase is very efficient. Spectroscopic quantification is possible following purification of dsRNA

over a column. Alternatively, T7-siRNA can be displayed by gel electrophoresis (see Protocol 4.3) to estimate the yield visually by comparison to a known amount of control dsRNA. As an average, we obtain between 15 to 20 μg T7-siRNA duplex in the final mix of 100 μl, which means that we synthesize in one set of reactions about 10 to 15 μM of 22-nt T7-siRNA. After dilution in the medium, we obtain a good RNAi effect at about 5 nM with a peak at 20 nM (see above and Figure 4.4B).

6. *Storage*. The T7-siRNAs are stored at –20°C and can be freeze-thawed several times without loss of activity.

Protocol 4.3 Gel Analysis of T7-siRNAs

To check the integrity as well as to estimate the quantity of T7-siRNA, an electrophoretic analysis on a nondenaturing 12% (w/v) polyacrylamide gel can be used.

Materials

- 10 × TBE (Tris-Borate-EDTA) (0.9 M Tris base, 0.9 M Boric acid, 20 mM EDTA, pH 8.0)
- 12% (w/v) Polyacrylamide gel; prepared by mixing 1.5 μl 10 × TBE, 6 μl 30% (w/v) acrylamide/bisacrylamide (19:1), 7.5 ml distilled water, 150 μl 10% (w/v) ammonium persulfate, and 15 μl TEMED; mix briefly, and pour the gel immediately (gel size: 10 cm × 8 cm × 0.75 mm)
- Loading buffer (30% (v/v) glycerol with 0.1% (w/v) bromophenol blue)

Experimental Procedures

1. Mix 2 μl of T7-siRNA with 2 μl of loading buffer (total volume may be increased with RNase-free water).
2. Load the sample on a nondenaturing 12% (w/v) polyacrylamide gel with TBE as running buffer, and run at 200 V.
3. Stop the gel when the bromophenol blue dye has migrated two thirds of the way down (about 30 to 45 min).
4. Stain the gel for 5 min in water containing ethidium bromide (1 μg/ml).
5. Soak the gel for 5 min in water.
6. Visualize the T7-siRNA on a UV transilluminator.

Technical notes. The T7-siRNA duplex should migrate as a 22 bp band slightly behind the bromophenol blue dye front. If the single-stranded siRNA has to be analyzed, it is preferable to heat the sample for 5 min at 65°C in order to disrupt any stable secondary structures that hinder the migration of the RNA.

Protocol 4.4 Transfection of T7-siRNAs

The T7-siRNA preparation can be transfected into tissue culture cells by a number of different methods. Both calcium phosphate (Ca-P) and commercial transfection reagents such as lipofectamine can be used. The optimal amounts must be determined experimentally. The following are useful guidelines to begin with:

A. 1 to 2 µl of T7-siRNA preparation per transfection in 35-mm plates.
B. 2 to 4 µl of T7-siRNA preparation per transfection in 60-mm plates.

As we reported originally on the T7-siRNA method,[15] Ca-P transfection can be used for RNAi for a range of cell lines. While commercial transfection reagents are expensive, the price of Ca-P reagents is negligible. Our protocol is adapted from Jordan et al.,[16] who modified the original Ca-P protocol in several important details; notably, the salt concentration of the 2 × HBS has been reduced to half.

The key to success is good cells and the pH of the Ca-P mix. Transfection efficiency with HeLa, COS, and 293T can reach more than 80%. If you observe a decrease in your transfection efficiency over time, change your cells and take a fresh tube. Serum is also crucial. Over time, cells grown in calf serum or newborn calf serum lose their high transfection potential compared to cells growing in fetal calf serum (FCS). The pH depends in part on the pH of the tissue culture medium, which in turn depends, among other things, on the HCO^3 concentration in the medium and on the CO_2 concentration in the incubator.

Where possible, grow cells at 37°C in Dulbecco's modified Eagle's medium (DMEM) supplemented with 5% FCS, L-glutamine, penicillin and streptomycin. For best results, cells should be growing vigorously; i.e., they should be split about twice a week and not overgrown.

Materials

- 8 × Ca (1 M CaC^{l2})
- 2 × HBS (50 mM Hepes, 1.5 mM Na^2HP0^4. 7 H_2O, 140 mM NaCl), adjust pH to 6.92 to 7.04 with NaOH for 5% CO_2 in incubator
- Filter sterile both solutions and store at 4°C
- TBS (10 mM Tris-HCl pH 7.4, 154 mM NaCl)

Experimental Procedures

Procedure for 6-well plates (or 35-mm plates)

1. Trypsinize cells and place in wells of a 6-well plate in 2.5 ml medium, 1 to 3 h prior to transfection. Replace medium once again ≥ 15' prior to transfection to remove traces of trypsin-EDTA, leaving enough time to equilibrate in the incubator.
2. Prepare a microfuge tube with T7-siRNA and a total of 2 µg of DNA. Keep total amount of DNA to 2 µg (up to 5 µg) with a carrier DNA (for example, an empty vector or the vector encoding GFP, as shown in the example presented in paragraph 4.4). Adjust total volume to 90 µl with sterile distilled water.
3. Add 30 µl of 1 M $CaCl_2$. Mix well and leave ≥ 5' to equilibrate.
4. Add the calcium-DNA solution quickly to 120 µl of 2 × HBS in another microfuge tube. Mix by gently pushing air bubbles through a pipette tip. Note that the order of addition of the solutions as described is important.
5. Add precipitate to cells after 1 to 2 min.
6. For many standard cell lines, overnight (12 to 18 h) incubation is best. The day after, wash with TBS (not PBS) in order to dissolve the remaining calcium-phosphate precipitate.

7. Add 2.5 ml of fresh medium containing 5% (v/v) FCS and let the cells grow for 1 to 2 days, depending on the half-life of the target mRNA and/or protein.

Procedure for 60-mm plates

The process is as above, except that the cells are seeded in 5 ml and the volumes are doubled, i.e., 4 μg of DNA (up to 10 μg) and water in a volume of 180 μl, 60 μl of 1 M $CaCl_2$, to be added to 240 μl of 2 × HBS.

Technical notes.

1. Trypsinization of cells just before transfection improves the DNA uptake. For HeLa or COS cells, it can be done 1 h before transfection, while for other cells attachment can be a bit slower (2 to 3 h).
2. For this protocol, adding a "neutral" carrier DNA to a certain amount of total nucleic acids is important because making a Ca-P mix is pure chemistry, and, thus, the nucleic acid concentration and purity are some of the critical parameters. Highly purified DNA is probably the best.
3. The pH of the phosphate solution (2 × HBS) must be optimized. We suggest making five solutions with a pH ranging from 6.94 to 7.04. Check for transfection efficiency using GFP as a marker.
4. Everything is mixed in sterile tubes (e.g., microfuge tubes) at room temperature (i.e., 8 × Ca and 2 × HBS should be equilibrated at room temperature); it can be done on a regular lab bench (antibiotics in tissue culture medium during transfection are sufficient).

APPENDIX 4.1 TROUBLESHOOTING GUIDE

1. *Low yield of T7-siRNA.* The transcription reactions routinely yield at least 15 μg of siRNA. A lower yield could result from inefficient transcription through the target sequence (see following comments). We suggest performing a positive control reaction with the p53 sequences described in this chapter to check the different components. If the problem persists (very rare with this protocol), choose another sequence.
2. *No reduction of target levels.*
    - 2.1. To check the inhibition of expression, an immunoblot analysis with a specific antibody is the easiest approach. However, it depends on the half-life of your protein, and it implies that you have to wait for a period of time long enough to allow protein turnover. If you don't know the half-life of your protein, harvest the transfected cells at least 2 or 3 days posttransfection.
    - 2.2. The efficacy of a particular T7-siRNA can easily be assessed by testing its ability to block the expression of a tagged version of the same target protein expressed by cotransfection. An immunoblot analysis is performed 24 to 36 h after transfection to check the levels of the tagged protein. This is particularly useful when (1) you do not want to optimize the transfection efficiency (the vast majority of the transfected cells take up both the T7-siRNA and the expression vector), (2) transfection efficiency is low, (3) you do not have an antibody against the endogenous protein, or when (4) you do not know the half-life of your protein.

2.3. Many researchers who are using RNAi have observed that some siRNAs simply do not inhibit expression. Currently, it is difficult to predict the efficiency of a particular siRNA. If the T7-siRNA preparation looks normal on a gel and the yield is as expected, and if transfection efficiency has been optimized (checked with GFP as control), simply redesign your siRNA with another sequence of your favorite gene.
2.4. Increase the levels of the T7-siRNA (but be aware that at higher concentrations, siRNAs can have some nonspecific effects).[21]
2.5. The transfection efficiency is a crucial parameter for a successful RNAi experiment. Below 60 to 70%, silencing becomes more difficult to observe unless you have a single-cell assay. If you face this problem, see comments in Protocol 4.4.

ACKNOWLEDGMENTS

We thank our colleagues, notably Philippe Parone, for their input and feedback during the development of our method. We are grateful to Dr. Agami for providing plasmid pSuper. OD was supported by the Fondation Medic. DP's laboratory was supported by the Canton de Genève, the Swiss National Science Foundation, Recherche Suisse contre le Cancer, and the Fondation Medic.

REFERENCES

1. Moss, E.G., RNA interference: it's a small RNA world, *Curr. Biol.,* 11, R772, 2001.
2. Bernstein, E., Denli, A.M., and Hannon, G.J., The rest is silence, *RNA,* 7, 1509, 2001.
3. Zamore, P.D., Tuschl, T., Sharp, P.A., and Bartel, D.P., RNAi: double-stranded RNA directs the ATP-dependent cleavage of mRNA at 21 to 23 nucleotide intervals, *Cell,* 101, 25, 2000.
4. Robertson, H.D. and Mathews, M.B., The regulation of the protein kinase PKR by RNA, *Biochimie,* 78, 909, 1996.
5. Elbashir, S.M. et al., Duplexes of 21-nucleotide RNAs mediate RNA interference in cultured mammalian cells, *Nature,* 411, 494, 2001.
6. Donzé, O. and Picard, D., RNA interference in mammalian cells using siRNAs synthesized with T7 RNA polymerase, *Nucleic Acids Res.,* 30, e46, 2002.
7. Lipardi, C., Wei, Q., and Paterson, B.M., RNAi as random degradative PCR: siRNA primers convert mRNA into dsRNAs that are degraded to generate new siRNAs, *Cell,* 107, 297, 2001.
8. Milligan, J.F. and Uhlenbeck, O.C., Synthesis of small RNAs using T7 RNA polymerase, *Methods Enzymol.,* 180, 51, 1989.
9. McGinness, K.E. and Joyce, G.F., Substitution of ribonucleotides in the T7 RNA polymerase promoter element, *J. Biol. Chem.,* 277, 2987, 2002.
10. Milligan, J.F. et al., Oligoribonucleotide synthesis using T7 RNA polymerase and synthetic DNA templates, *Nucleic Acids Res.,* 15, 8783, 1987.
11. Chiu, Y.L. and Rana, T.M., RNAi in human cells: basic structural and functional features of small interfering RNA, *Mol. Cell,* 10, 549, 2002.

12. Yu, J.Y., DeRuiter, S.L., and Turner, D.L., RNA interference by expression of short-interfering RNAs and hairpin RNAs in mammalian cells, *Proc. Natl. Acad. Sci. U.S.A.,* 99, 6047, 2002.
13. Elbashir, S.M. et al., Analysis of gene function in somatic mammalian cells using small interfering RNAs, *Methods,* 26, 199, 2002.
14. Dykxhoorn, D.M., Novina, C.D., and Sharp, P.A., Killing the messenger: short RNAs that silence gene expression, *Nat. Rev. Mol. Cell. Biol.,* 4, 457, 2003.
15. Donzé, O., Abbas-Terki, T., and Picard, D., The Hsp90 chaperone complex is both a facilitator and a repressor of the dsRNA-dependent kinase PKR, *EMBO J.,* 20, 3771, 2001.
16. Jordan, M., Schallhorn, A., and Wurm, F.M., Transfecting mammalian cells: optimization of critical parameters affecting calcium-phosphate precipitate formation, *Nucleic Acids Res.,* 24, 596, 1996.
17. Sharpless, N.E. and DePinho, R.A., p53: good cop/bad cop, *Cell,* 110, 9, 2002.
18. Brummelkamp, T.R., Bernards, R., and Agami, R., A system for stable expression of short interfering RNAs in mammalian cells, *Science,* 296, 550, 2002.
19. Aza-Blanc, P. et al., Identification of Modulators of TRAIL-Induced Apoptosis via RNAi-Based Phenotypic Screening, *Mol. Cell,* 12, 627, 2003.
20. Harborth, J. et al., Identification of essential genes in cultured mammalian cells using small interfering RNAs, *J. Cell Sci.,* 114, 4557, 2001.
21. Semizarov, D. et al., Specificity of short interfering RNA determined through gene expression signatures, *Proc. Natl. Acad. Sci. U.S.A.,* 100, 6347, 2003.

5 Production of siRNAs with the Application of Deoxyribozymes

Graeme Doran and Muhammad Sohail

CONTENTS

5.1 INTRODUCTION

Exogenous application of siRNAs is becoming a widely used approach to achieve gene knockdown.[1] However, siRNAs targeted to different regions of a gene exhibit different efficacies,[2–4] and often several reagents may have to be screened for optimal activity. Chemical synthesis of RNA remains expensive, making this approach uneconomical. Therefore, this is highly desirable to develop alternative methods of producing siRNAs.

A cheaper alternative to chemical synthesis of siRNAs is to produce siRNAs by *in vitro* transcription using DNA oligonucleotides as templates.[5] These oligonucleotides direct the transcription of small RNA transcripts by T7 RNA polymerase.[6] Sense and antisense strands of siRNAs are produced in separate transcription reactions,

0-8493-2141-7/05/$0.00 + $.50

oligonucleotide templates are removed by DNase digestion, and the two RNA strands are annealed to produce siRNA duplex. The siRNAs may then be purified by ethanol precipitation, by Sephadex column chromatography, or by gel electrophoresis before use in transfections. The reactions can be repeated many times over, lessening the reliance on a single precious RNA stock.

However, this useful method is limited by specific sequence requirements of the viral RNA polymerases; all siRNAs made with this method start with a 5'G residue and require a C3' residue at position 19 (i.e., 5'G-N17-C3') to allow annealing with the complementary RNA, which correspondingly must also start with a 5'G residue. Given the dramatic variations in the efficacy of siRNAs targeted to different regions of a gene, such sequence requirements greatly reduce the number of potential target sites and are, thus, disadvantageous in the identification of optimally effective siRNAs.

We describe the use of deoxyribozymes (Figure 5.1) to overcome these sequence limitations and to allow the synthesis of RNAs of any sequence by the addition of a simple enzymatic step.[7] Deoxyribozymes are DNA oligonucleotides with RNA-cleaving activity.[8–10] The deoxyribozyme "guide arms" have sequences complementary to the substrate RNA molecule and, on hybridization to a target site, cleave the RNA at a deoxyribozyme-specific dinucleotide sequence. The different dinucleotide sequence requirement for different deoxyribozyme groups makes them flexible tools for digesting a variety of sequences. For example, 10 to 23 type deoxyribozyme can digest an RNA strand containing the sequences 5'<u>A</u>T, 5'<u>A</u>C, 5'<u>G</u>C, and 5'<u>G</u>T; 8-17 type deoxyribozymes cleave at 5'<u>A</u>G, and Bipartite II deoxyribozymes optimally cleave at 5'<u>AA</u>. The underlined nucleotides are unpaired and the second nucleotides pair with a complementary nucleotide in the deoxyribozyme. In optimally active Bipartite II deoxyribozymes, five unpaired nucleotides separate the binding arms, including those at the catalytic site.[10]

By designing precursor RNAs that are targets for deoxyribozyme cleavage, it is, in principle, possible to produce siRNA sequences starting and ending with any nucle-

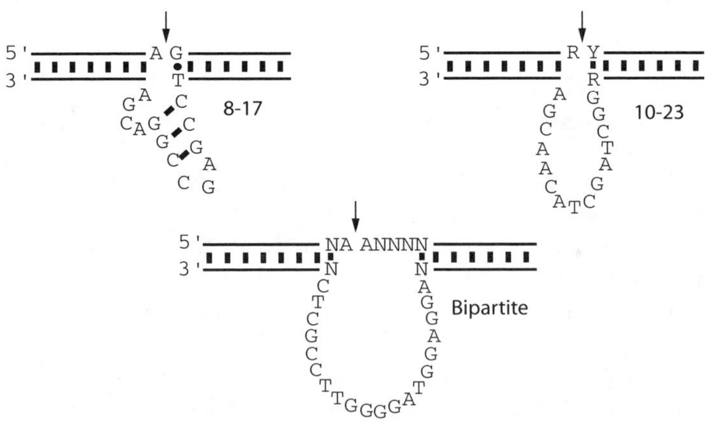

FIGURE 5.1 Three types of RNA-cleaving domains of deoxyribozyme (8-17 and 10-23 domains drawn after Santoro and Joyce[9] and Bipartite after Feldman and Sen.[10] R is A or G and Y is U or C. N is for any nucleotide.

otide. This overcomes the disadvantage of sequence limitation for *in vitro* transcription methods, allowing synthesis of siRNAs to be directed to any site in the target mRNA.

5.2 DESIGN OF TEMPLATE OLIGONUCLEOTIDES FOR siRNA PRODUCTION

T7 RNA polymerase recognizes a 17-bp promoter sequence (5'TAATACGACTCACTATA) to initiate transcription and requires the promoter to be double stranded. The promoter sequence is followed by the leader sequence of one (5'G) or more residues that appears in the transcripts. The first nucleotide of the leader sequence is 5'G, a hexanucleotide sequence 5'GGGCGA or 5'GGGAGA being the optimal leader sequence. Once the RNA polymerase has initiated transcription, a single-stranded template is sufficient for further RNA production.

Overhang sequences, for example, 5'TT, is added after the target gene sequence to provide a 5'UU overhang in the siRNA (other overhangs may also be used). Therefore, when the two siRNA strands are annealed, the resultant duplex has a 19-bp double-stranded region with a two-nucleotide overhang at both 3' ends.

There are three principle strategies for RNA production by *in vitro* transcription using synthetic oligonucleotides as templates.

5.2.1 PARTIALLY DOUBLE-STRANDED OLIGONUCLEOTIDE TEMPLATE

This rationale takes advantage of the processivity of the T7 RNA polymerase along single-stranded DNA. A short DNA oligonucleotide corresponding to the T7 promoter/leader sequence is hybridized with a longer oligonucleotide comprising the T7 promoter/leader sequence plus the target sequence (Figure 5.2A). The polymerse initiates transcription at the double-stranded region and continues along the single-stranded region of the partial duplex. Processivity of the polymerase is dependent on the sequence of the single stranded region, and production of aborted transcripts 1 to 2 nt less than the required length is common. The presence of such products would result in less than perfect siRNAs, but overall this may not have considerable effect on their activity.

5.2.2 PARTIALLY DOUBLE-STRANDED TEMPLATE COUPLED WITH KLENOW FILL-IN REACTION

In this method, a partial oligonucleotide duplex is extended to produce a fully double-stranded template with a Klenow fill-in reaction (Protocol 5.1).[11] Generally, the shorter oligonucleotide in the partial duplex corresponds to the T7 promoter/leader sequence, and the longer oligonucleotide contains a 3'end-truncated promoter/leader sequence plus an additional target sequence (Figure 5.2B). This method is advantageous if one prefers to use a fully double-stranded template, since only one strand corresponding to

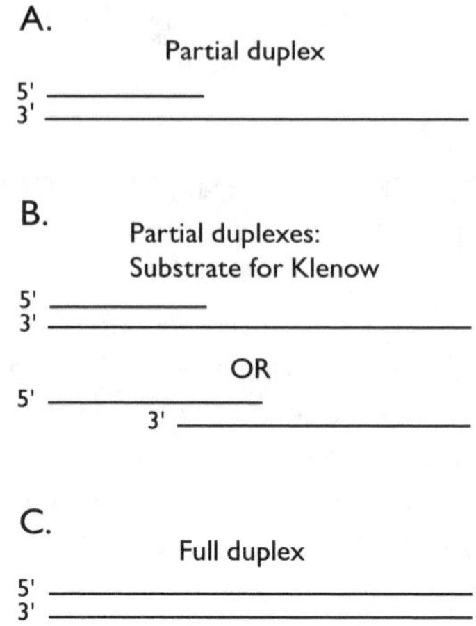

FIGURE 5.2 The various possible types of substrates for use in *in vitro* transcription to produce siRNAs. Substrates in **B** can be converted into fully double-stranded duplexes by Klenow fill-in reaction.

the target sequence needs to be synthesized, and the shorter T7 promoter/leader oligonucleotide remains constant. However, the enzyme reaction may not always be 100% efficient and may result in a partially double-stranded template; also, the additional enzymatic step is time consuming. A similar approach has also been used in the SiRNA Construction Kit™ (Ambion).

5.2.3 Fully Double-Stranded Template

Due to the ready availability of cheap DNA oligonucleotides from commercial sources, full-length double-stranded templates may also be used for RNA transcription (Figure 5.2C). This method, requiring the design of two DNA oligonucleotides per one strand of RNA, is straightforward, as it does not require additional enzymatic steps.

5.3 DESIGN OF DEOXYRIBOZYMES AND MODIFICATION OF THE T7 LEADER

The commonly used hexanucleotide leader sequence of the T7 promoter, 5'GGGAGA or 5'GGGCGA, appears in the transcripts. Thus, the 3'arm of the deoxyribozyme is designed to bind to this sequence. The 5'arm is complementary to the target gene sequence cassette in the transcript. The deoxyribozyme cleaves at the dinucleotide position between the leader and the siRNA target sequence. However, deoxyribozymes have been shown to work optimally with 7 to 8 nt binding arms[9]; therefore, the T7 leader sequence is extended by ~2 to 3 nt to provide a longer annealing site for the 3'arm of the deoxyribozyme. One of the preferable extension sequences is 5'AT (5'AU in RNA). This sequence offers two advantages; first, catalytic sites surrounded by A or U appear to be a better substrate,[9] and second, 5'T (of the 5'AT extension) is useful as we can trace the 5'cleavage product with the use of either [α-^{32}P] UTP or [α-^{33}P] UTP in *in vitro* transcription reactions. A third nucleotide is also added and, with the first nucleotide of the target sequence, forms the catalytic site for the deoxyribozyme. Therefore, the sequence of the 3'binding arm of the deoxyribozyme and that of the catalytic domain remain constant for one type of deoxyribozyme and the only variable is the sequence of the 5'binding arm that changes with the target sequence (Figure 5.3).

5.4 PRODUCTION OF SMALL INTERFERING RNAS

The production of small RNAs may be achieved through a variety of *in vitro* transcription solutions, though the MegaScript and MegaShortScript kits (Ambion) are specifically optimized for high yields of small RNAs.

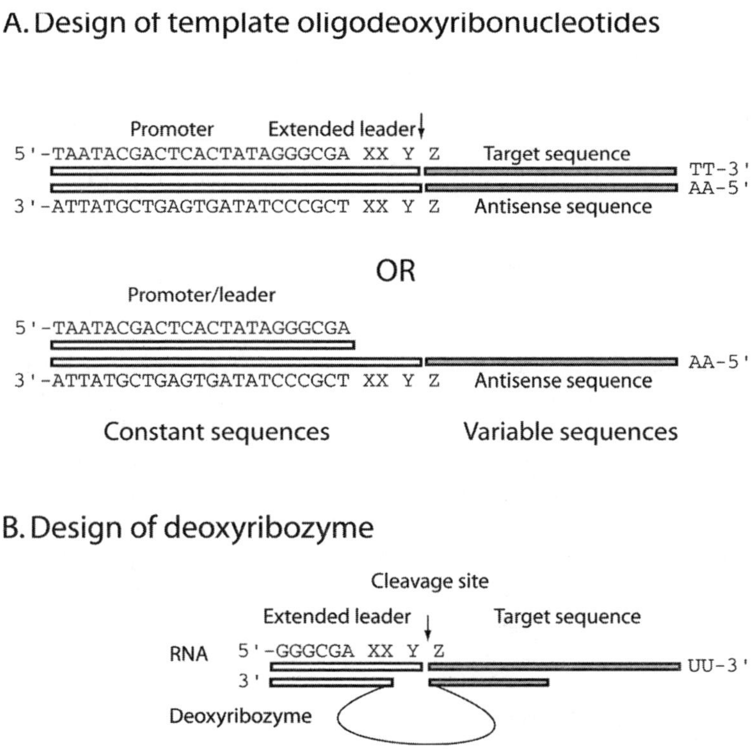

FIGURE 5.3 Design of oligonucleotides (**A**) and deoxyribozymes (**B**). 5'XX represents the dinucleotide extension of leader and 5'Y forms the unpaired base that, with the first base of the target sequence (5'Z), makes the catalytic site for the deoxyribozymes (8-17 and 10-23 types).

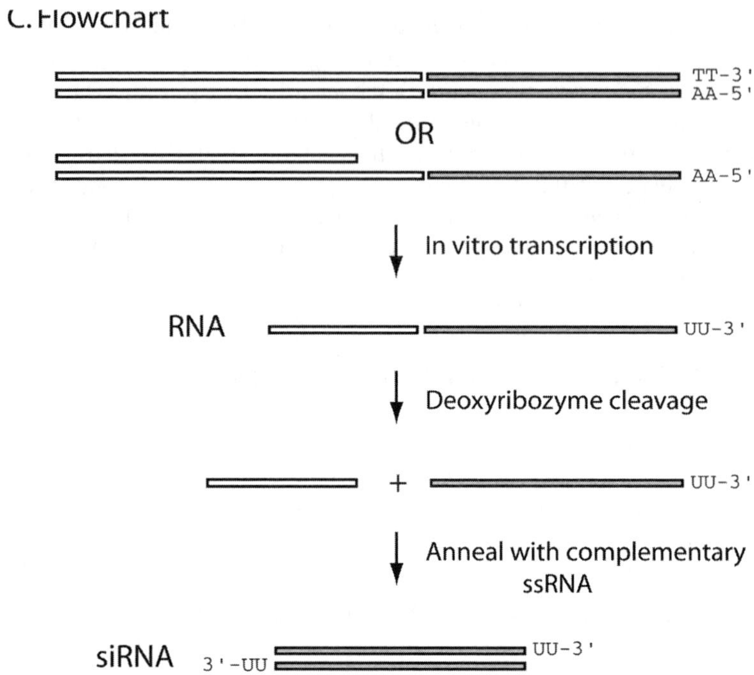

FIGURE 5.3 (*continued*) C. Flowchart. Two oligodeoxynucleotides are annealed to form a template for *in vitro* transcription. The two siRNA strands are prepared in separate reactions. The transcripts are treated with deoxyribozymes and are annealed to produce intact siRNA.

5.4.1 Template Oligonucleotides: Assessment and Purification

The primary concern in producing superior transcripts is the quality of the DNA oligonucleotide template. Commercial oligonucleotide suppliers offer desalted, HPLC purified oligonucleotides that are suitable for transcription, yet these may still contain variable quantities of aborted products produced during oligonucleotide synthesis. The short products may interfere with the transcription by acting as templates for the production of smaller-than-desired RNA transcripts, and these transcripts may further inhibit the deoxyribozyme cleavage reaction, lowering total yield of siRNA. The quality of oligonucleotides can be assessed by 5'end labeling with T4 polynucleotide kinase and [γ-^{33}P] ATP. The end-labelled products are analyzed by electrophoresis in a 12 to 15% (w/v) polyacrylamide gel.

5.4.2 *In Vitro* Transcription

Even when using gel-purified oligonucleotides for transcription, the addition of products smaller than intended may occur due to premature termination of transcription. This may be a result of reaction conditions (particularly salt concentration of the reaction buffer), RNase degradation of transcripts, exhaustion of NTPs, or poorly active RNA polymerase stock. Premature termination of transcription may be limited by using fully complementary templates or by manipulating buffer conditions to favor polymerase processivity. The elimination of buffer components that stabilize odd secondary structures, such as KCl, or the addition of cosolvents such as DMSO and glycerol are common approaches to this problem. Additives such as betaine (N,N,N-trimethylglycine) and TMANO (trimethylamine N-oxide) are commonly used to maintain processivity in PCR reactions and may be of similar use in cases of transcription, though their method of action is not clear. RNase inhibitors provide protection against most RNases likely to be contaminating preparations, with the exception of RNase H. Increasing the concentration of T7 polymerase may also improve low yields of transcript.

5.4.3 Deoxyribozyme Digestion and Preparation of Transfection-Ready siRNAs

Although the deoxyribozyme cleavage reaction is quite robust (Figure 5.4), there may be some differences in cleavage activity between different deoxyribozymes. For example, we found the deoxyribozyme activity to vary with buffer/salt conditions. Therefore, to obtain optimal results, more than one buffer may have to be tried. Some of the buffers we used are given in Protocol 5.5.

The deoxyribozyme digestion can be carried out at a fixed temperature or in a thermal cycler with cycling temperatures. The cycling reaction is designed to promote annealing by disrupting the secondary structure that may exist in the DNA/RNA oligonucleotides through temperature elevation to 50°C, followed by a cooling and annealing of the target to deoxyribozyme and a final catalytic step at a temperature optimal for the short RNA target. The cyclic nature is likely to promote dissociation of

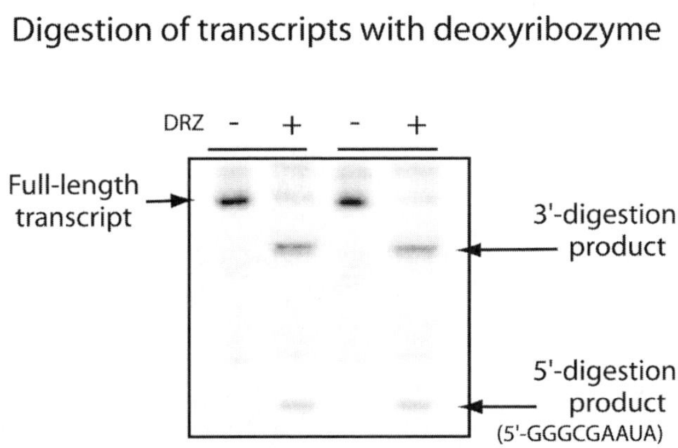

FIGURE 5.4 Examples of short RNA transcripts digested with deoxyribozyme to remove unwanted sequences. *In vitro* transcriptions were carried out in the presence of [α-^{33}P] UTP using two fully complementary oligonucleotides as templates.

deoxyribozymes from the cleaved product, allowing for lower concentrations of deoxyribozyme to be used.

5.5 CONCLUDING REMARKS

The deoxyribozyme processed in *in vitro* transcription siRNA synthesis still provides considerable cost savings over chemical synthesis in the production of RNA duplexes, even when taking into account the need to provide template DNA oligonucleotides, deoxyribozymes, and *in vitro* transcription reagents. The technical advantages provided by this method are not limited to the production of siRNAs but are extended to production of any single- or double-stranded RNA of defined length and sequence; also, the method allows RNA engineering. This method may also be used to produce hairpin RNA of a desired sequence and to produce long RNAs that are difficult to make chemically or for which the cost of chemical synthesis is high. This method is also suited to production of RNAs of specific sequences at large scale (e.g., as pharmaceutical agents) or to the production of libraries of RNA molecules of varying lengths and sequences (e.g., libraries of slightly varying siRNAs required for optimization of RNAi to any given gene target) that might otherwise be prohibitively costly by chemical synthesis.

PROTOCOLS

PROTOCOL 5.1 KLENOW REACTION

Materials

- Oligonucleotides (MWG-biotech, Qiagen)
- Klenow fragment (New England Biolabs or Cambio)
- 10 × Klenow buffer (100 mM Tris-HCl, pH 7.6, 100 mM $MgCl_2$)
- All four dNTPs (Promega, Roche, or Amersham Biosciences)

Experimental Procedures

1. Anneal the partially complementary oligonucleotides in 1 × Klenow buffer by heating equimolar concentrations to 95°C and slowly cooling to room temperature.
2. Add calculated amount of all four dNTPs required to achieve complete synthesis of the duplex.
3. Add Klenow enzyme at ~1U per 10 nmol of the duplex.
4. Incubate at 37°C for 30 to 40 min. The duplex can be directly used in *in vitro* transcription or stored at –20°C.

PROTOCOL 5.2 ASSESSING OLIGONUCLEOTIDE AND DEOXYRIBOZYME QUALITY

Materials

- Oligonucleotide/deoxyribozyme
- T4 polynucleotide kinase and reaction buffer (New England Biolabs)
- [γ-^{33}P] ATP (1000 to 3000 Ci/mmol) (Amersham)

Experimental Procedures

1. Set up a 5-μl reaction by mixing the following:

Oligonucleotide	5–10 pmol
10 × reaction buffer	0.5 μl
T4 polynucleotide kinase	0.5 U
[γ-^{33}P] ATP	0.5 μl
Nuclease-free water	to 5 μl

2. Incubate at 37°C for 15–20 min.
3. Analyze in a 12 to 15% (w/v) denaturing polyacrylamide gel.

Technical notes. Full-length oligonucleotides may be identified by autoradiography, excised from the gel, and purified for use in transcription by ethanol precipitation.[11] An alternative, labor-saving method is to visualize unlabelled oligonucleotides on a 12% denaturing polyacrylamide gel by UV shadowing. As the major product should still be of the required full length, excising and purifying this discrete band can remove the majority of aborted products.

PROTOCOL 5.3 GEL PURIFICATION OF OLIGONUCLEOTIDES

The following protocol can also be used to purify siRNA duplexes.

Materials

- Polyacrylamide gel electrophoresis system (Flowgen)
- Polyacrylamide gel solution (Ambion)
- TEMED (Sigma)
- 10% (w/v) APS
- X-ray film (Kodak) and film-developing facility
- Formamide loading buffer (95% (v/v) formamide, bromphenol blue/xylene cyanol 0.01% v/v: all from Sigma)
- Gel elution buffer (0.5 M ammonium acetate, 1 mM EDTA pH 8.0, 0.2% SDS, w/v)
- UV lamp

Experimental Procedures

1. Pour a standard preparatory polyacrylamide gel at desired concentration according to general protocol.[11] For single-stranded nucleic acids, a denaturing polyacylamide gel (12 to 20%, w/v) is preferred, and for duplex siRNAs, a nondenaturing polyacrylamide gel (6 to 10%, w/v) is used.
2. Load oligonucleotide sample in standard formamide gel loading buffer, and load siRNAs in a nondenaturing buffer (e.g., 40% (v/v) sucrose or glycerol)
3. After sufficient separation on the gel, locate the sample by exposing to an x-ray film (if radioactive sample is used) or by UV-shadowing. DNA is more readily observed by UV-shadowing than RNA.
4. Excise sample from the gel with a sharp disposable scalpel (nuclease-free) and transfer into a microfuge tube. Add 400 μl gel elution buffer. Carefully fragment the gel with a small clean implement to increase the surface area for elution.

5. Mix by vortexing and incubate for 4 hours or overnight at room temperature or 37°C.
6. Centrifuge for 30 to 60 sec to pellet the gel and transfer the supernatant containing the nucleic acid to a fresh microfuge tube.
7. To precipitate nucleic acid, add 0.1 volume of 3 M sodium acetate or 10 M ammonium acetate, mix gently, and add 2.5 to 3.0 volumes of absolute ethanol. Incubate at –20°C (–80°C for siRNAs) for 30 to 60 min.
8. Spin at 13,000 rpm for 20 minutes (spin at 4°C is preferred for siRNAs). Discard the supernatant.
9. Wash the pellet in cold 90% ethanol, and dry in a speed-vac or air-dry. Resuspend in appropriate volume of water or TE buffer (10 mM Tris pH 7.5, 1 mM EDTA).
10. Store at –20°C or –80°C.

Protocol 5.4 *In Vitro* Transcription of Precursor RNA

Materials

- Template oligonucleotides (MWG-Biotech or Qiagen Ltd)
- *In vitro* transcription reagents (e.g., MegaScript™ or MegaShortScript™, Ambion)
- RNase-free DNase (Ambion)
- [α-^{32}P] UTP (3000 Ci/mmol) or [α-^{33}P] UTP (2500 Ci/mmol)(Amersham)
- Sephadex G25 Microspin™ columns (Amersham)
- RNase inhibitor (Promega or Ambion)
- Spectrophotometer (Beckman)

Experimental Procedures

1. Mix equal concentrations of complementary oligonucleotides, heat to 95°C for 5 min, and slowly cool to room temperature.
2. Set up one 20-µl transcription reaction for each strand of siRNA duplex according to the kit supplier's instructions (the following recipe refers to one of the above Ambion's kits).

10 × transcription buffer	2.0 µl
75 mM ATP	3.0 µl
75 mM CTP	3.0 µl
75 mM GTP	3.0 µl
75 mM UTP	3.0 µl
Annealed template DNA	100 pmol
T7 RNA polymerase	2.0 µl
[α-^{32}P] UTP or [α-^{33}P] UTP	1.0 µl (optional)
Nuclease-free water	to 20 µl

3. Incubate at 37°C for 3 to 4 h, preferably in a cabinet incubator to prevent evaporation.
4. Add one unit of RNase-free DNase; mix and incubate for a further 15 min at 37°C.
5. Dilute the reaction to 40 µl with nuclease-free water and de-salt through a Sephadex G25 Microspin™ column according to the manufacturer's instructions. Elute RNA into a microfuge tube containing 20 to 40 units of RNase inhibitor. Mix gently.

6. A small aliquot of the filtered RNA may be taken to assess transcript quality by polyacrylamide gel electrophoresis (15–20%, w/v).

Technical notes. The optimal concentration of oligonucleotide template may vary depending on the nature of the RNA polymerase and the kit used, but 100 pmol in a 20-µl transcription reaction was found to be adequate when using the Ambion kits. T7 polymerase is known to add a variable number (usually one) of nucleotides that are not encoded by template DNA to the 3' termini of the RNA transcript. As such, siRNAs may eventually be produced with 3-nt 3'overhangs or greater. However, this does not detectably reduce their silencing activity. An alternative to the initial purification of DNA oligonucleotides is to purify the annealed siRNAs on a native polyacrylamide gel once synthesis is complete. In the case of the deoxyribozyme cleavage method, this approach may be preferable, as it will eliminate any RNA transcript that remains undigested, should the deoxyribozyme reaction not proceed to completion.

Protocol 5.5 Digestion of Precursor Transcripts with Deoxyribozymes

Materials

- Precursor transcripts
- Deoxyribozyme oligonucleotides
- 5 × deoxyribozyme reaction buffer (**A.** 500 mM NaCl, 150 mM $MgCl_2$, 100 mM Tris-HCl, pH 7.2. **B.** 1.2 M NaCl, 500 mM $MgCl_2$, 100 mM Tris-HCl pH 7.2. **C.** 750 mM NaCl, 10 mM $MgCl_2$, 500 mM Tris-HCL pH 7.5)
- RNase inhibitor (Promega or Ambion)
- Sephadex G25 Microspin Column (Amersham)
- Thermal cycler (MJ Research) (optional)

Experimental Procedures

1. Set up a 40-ml reaction as follows (a 20-ml reaction can also be set up):

5 × cleavage buffer	8.0 µl
RNA transcript	20 µl (~100 pmol)
Deoxyribozyme	200–400 pmol
RNase inhibitor	40 U
Nuclease-free water	to 40 µl

2. Incubate for 6 h to overnight at 37°C or in a thermal cycler (cycling at 50°C for 1 min, 20°C for 2 min and 37°C for 10 min).
3. Add 1 U of RNase-free DNase and incubate for a further 15 min at 37°C.
4. Dilute to 100 µl with nuclease-free water and desalt through a Sephadex G25 Microspin™ column as per manufacturer's instructions. Elute the sample into a microfuge tube containing 50 U of RNase inhibitor. Mix gently.

Protocol 5.6 Evaluation of Digested Products, Quantification, and Preparation of siRNAs

Materials

- Polyacrylamide gel electrophoresis system (Flowgen)
- Polyacrylamide gel solution (Ambion)

- TEMED (Sigma)
- 10% (w/v) APS (Sigma)
- Agarose (Flowgen)
- 10 × TBE electrophoresis buffer (homemade or from Ambion)
- Horizontal gel electrophoresis system (Anachem)
- 1000 × Ethidium bromide solution (10 mg/ml) (Sigma)
- X-ray film (Kodak) and x-ray film-developing facility, or a storage phosphor screen and PhosphorImager (Molecular Dynamics)
- Formamide loading buffer (95% (v/v) formamide, bromphenol blue/xylene cyanol 0.01% v/v: all from Sigma)
- Spectrophotometer (Beckman)

Experimental Procedures

1. Analyze a small aliquot in a 15 to 20% (w/v) polyacrylamide gel if the samples were radio labelled, or in a 3% (w/v) agarose gel if samples were unlabelled.
2. Expose radioactive samples to an x-ray film or a storage phosphor screen and analyze as appropriate (by developing the film or by scanning the screen on a PhosphorImager). Visualize samples in agarose gel by staining with ethidium bromide (0.5 μg/ml final concentration). (Note: Ethidium bromide is highly toxic. Handle with care.)
3. To quantify, measure OD_{260} of an appropriate dilution of the digested transcripts, e.g., 1:50). A fairly accurate estimate is shown below.
   A_{260} × dilution factor × ssRNA/dsRNA extinction coeficient = μg/ml
   [For single-stranded RNA, A_{260} = 1 OD unit = 40 μg/ml]
4. Mix equimolar concentrations of the sense and antisense strands.
5. Heat to 65°C for 2 min and cool to room temperature to obtain siRNA duplexes. At this stage siRNA solution can be concentrated (e.g., in a SpeedVac) or diluted as required to obtain required concentration. A convenient concentration is 20 μM.
6. Prepare small aliquots and store at –80°C. Avoid repeated freeze-thaw cycles.

Technical notes. Although polyacrylamide gels provide the most sensitive method for confirmation of cleavage, agarose gels allow for more simple identification of small RNAs and avoid the need for radioactive preparations. "Cold" RNA products may be easily visualized, as some manufacturers now offer agarose gels that provide discrimination of sufficient size to prove cleavage has occurred successfully (e.g., Nusieve).

Removal of deoxyribozymes by DNase digestion is recommended to prevent contamination of siRNA with DNA. Such contamination might result in unwanted and unpredictable "side-effects," such as RNase H-induced cleavage of RNA sequences that deoxyribozymes may bind with.

ACKNOWLEDGMENTS

Research in the authors' laboratory is supported by the Medical Research Council, UK.

REFERENCES

1. Dykxhoorn, D.M., Novina, C.D., and Sharp, P.A., Killing the messenger: short RNAs that silence gene expression, *Nat. Rev. Cell Biol.*, 4, 457, 2003.
2. Holen, T. et al., Positional effects of short interfering RNAs targeting the human coagulation trigger tissue factor, *Nucleic Acids Res.*, 30, 1757, 2002.
3. Vickers, T.A. et al., Efficient reduction of target RNAs by small interfering RNA and RNase H-dependent antisense agents: a comparative analysis, *J. Biol. Chem.*, 278, 7108, 2003.
4. Donze, O. and Picard, D., RNA interference in mammalian cells using siRNAs synthesized with T7 RNA polymerase, *Nucleic Acids Res.*, 30, e46, 2002.
5. Bohula, E.A. et al., The efficacy of small interfering RNAs targeted to the type 1 insulin-like growth factor receptor (IGF1R) is influenced by secondary structure in the IGF1R transcript, *J. Biol. Chem.*, 278, 15991, 2003.
6. Milligan, J.F. et al., Oligoribonucleotide synthesis using T7 RNA polymerase and synthetic DNA templates, *Nucleic Acids Res.*, 15, 8783, 1987.
7. Sohail, M. et al., A simple and cost-effective method for producing small interfering RNAs with high efficacy, *Nucleic Acids Res.*, 31, e38, 2003.
8. Breaker, R.R., DNA enzymes, *Nat. Biotech.*, 15, 427, 1997.
9. Santoro, S.W. and Joyce, G.F., A general purpose RNA-cleaving DNA enzyme, *Proc. Natl. Acad. Sci. U.S.A.*, 94, 4262, 1997.
10. Feldman, A.R. and Sen, D., A new and efficient DNA enzyme for the sequence-specific cleavage of RNA, *J. Mol. Biol.*, 313, 283, 2001.
11. Sambrook, J. and Russell, D.W., *Molecular Cloning: A Laboratory Manual*, 3rd ed., Cold Spring Harbor Laboratory Press, New York, 2001.

6 Production of siRNA *In Vitro* by Enzymatic Digestion of Double-Stranded RNA

Frank Buchholz, David Drechsel, Martine Ruer, and Ralf Kittler

CONTENTS

6.1 INTRODUCTION

RNAi is an ancient defense and regulatory mechanism to silence genes in a variety of organisms, including plants, insects and vertebrates (reviewed in references[1,2]). The discovery of this mechanism has enabled researchers to utilize it as a tool for gene function analyses in model organisms. Its relative ease has generated a wealth of new information, and for some organisms whole genome RNAi analyses have been accomplished.[3–5] In many species, long dsRNA can be used to trigger RNAi when introduced into the cytoplasm of a cell. However, long dsRNAs have a nonspecific negative effect on cell proliferation in many vertebrate cells because they trigger an interferon response (reviewed inreference [6]). Interferon production activates two pathways in the cell that negatively affect cell proliferation. First, interferons cause the activation of dsRNA-dependent protein kinase (PKR), leading to the general inhibition of protein production by phosphorylation of the translation factor eIF-2α. Second, 2'-5' oligoadenylate synthetase (2'-5'-OAS) is turned on, leading to mRNA degradation via activation of RNase L. As a consequence, long dsRNA is not useful for specific gene silencing in most mammalian cells.

Detailed analyses of the RNAi pathway revealed that long dsRNA is processed into ca. 20 to 25 bp fragments within cells by an RNase III-like endoribonuclease called Dicer.[7–10] These short dsRNA fragments are termed short interfering RNA (siRNA) and contain 2-nt 3'-end overhangs, the signature of all RNase III enzymes.[11] Interestingly, siRNA does not trigger a strong interferon response in mammalian cells

0-8493-2141-7/05/$0.00 + $.50

while efficiently and specifically acting as a silencing trigger for a corresponding gene.[12]

Shortly after the discovery of siRNAs as the effective molecules in RNAi, several groups demonstrated that either chemically synthesized, *in vitro* transcribed siRNAs or short hairpin RNAs (shRNAs) can be used for gene knockdown studies in mammalian cells (reviewed in reference [13]). While these methods for gene silencing are very useful, they all have major limitations, especially for functional genomic studies. First, these methods of generating siRNA reagents are expensive and labor intensive. Second, each siRNA or shRNA has a different potential to silence the mRNA of a corresponding gene. Some rules have been established that help to design siRNA or shRNA molecules.[14] However, we still do not understand enough about the silencing mechanism to predict effective silencing molecules at a very high rate. Therefore, each siRNA or shRNA has to be tested for its effectiveness in knocking down the gene of interest, and usually more than one siRNA or shRNA has to be produced and compared.

These drawbacks can be circumvented by using siRNA hydrolyzed *in vitro* from a long dsRNA with purified Dicer[15,16] or other dsRNA-specific endoribonucleases, including RNase III from *E. coli.*[17,18] Both methods are cheap, easy, fast to perform, and can be carried out in any standard molecular biology laboratory. Using this strategy, a pool of different siRNAs are generated that all have the potential to silence the gene of interest. The complex mixture of different siRNAs ensures that at least some effective molecules will be present, thus eliminating the requirement to screen for an active siRNA for each individual gene. We refer to these siRNAs as endoribonuclease prepared siRNAs or, in short, esiRNAs.

While both Dicer and RNase III can be used to generate esiRNAs, both enzymes produce different mixtures of siRNA molecules. Digestion with Dicer starts at the ends of the dsRNA and proceeds in a sequential manner.[12,19,20] As a consequence, this approach produces a set of nonoverlapping siRNAs. In contrast, RNase III from *E. coli* cleaves long dsRNA in a more or less random fashion,[17,21] leading to a more complex mixture of siRNAs. A larger number of different siRNAs may increase efficiency and specificity. A further disadvantage of the Dicer protocol is that the purification of the enzyme is considerably more difficult, and the activity of the enzyme is much lower than the activity of purified RNase III. Due to the high activity and processivity of *E. coli* RNase III, large quantities of esiRNA can be produced with very little enzyme. Much more of the purified Dicer has to be added to achieve the same yield of esiRNA. For both approaches, the digested dsRNAs of the correct size have to be purified away from residual longer dsRNA fragments that would otherwise trigger the interferon response in transfected mammalian cells.

A potential drawback of esiRNA may be the cross-silencing of highly homologous genes. However, previously published data[9] and unpublished data from our own laboratory argue that cross silencing of homologous genes is typically not observed.

A disadvantage of using *E. coli* RNase III for the production of esiRNA is that complete digestion produces dsRNA fragments of 12 to 15 bp, and these fragment sizes are too short to trigger efficient RNAi in mammalian cells. However, conditions can be developed in which the vast majority of the fragments are of a size range that allows efficient gene knockdown in mammalian cells (see also Figure 6.3).

In this chapter we present a protocol for esiRNA preparation from long dsRNA using an *E. coli* GST-RNase III fusion protein. The protocol can be divided into three basic steps that are summarized in Figure 6.1. Some of the materials and reagents are toxic and should be handled with care, taking the appropriate safety measures. Essentially, the three synthesis steps can be conveniently performed within two days. Day one involves preparation of *in vitro* transcription templates by PCR and large-scale synthesis of dsRNA. Limited digestion of dsRNA with GST-RNase III and purification of digestion products can be performed on day two.

6.2 CLONING, EXPRESSION, AND PURIFICATION OF *E. COLI* RNASE III

The coding sequence of *E. coli* RNase III can be amplified from bacterial genomic DNA and cloned into expression vectors. Because it is an endogenous *E. coli* gene, it expresses well and over-expression in its natural host does not produce any obvious

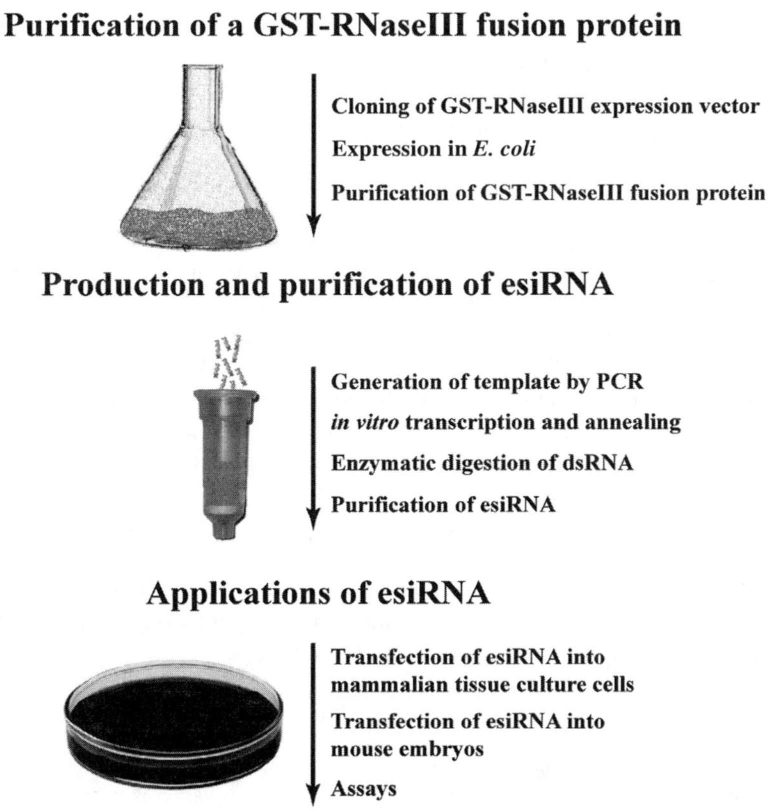

FIGURE 6.1 Overview of RNAi experiments utilizing esiRNAs. Distinct steps in the process are outlined and each step is subdivided into experimental procedures.

phenotype in the bacteria. We express *E. coli* RNase III as a GST-fusion protein to allow simple purification of the product. The GST-fusion protein is active without removal of the GST-moiety. In fact, the GST-RNase III fusion protein may alter the cleavage properties toward the generation of longer fragments. This property may serve the efficient production of esiRNAs for use in gene silencing in mammalian cells. Several companies sell purified *E. coli* RNase III. To our knowledge these proteins do not carry a GST-tag. We do not know whether the reaction conditions have to be altered using these proteins for esiRNA production.

PROTOCOLS

Protocol 6.1 Cloning of *E. coli* RNase III

Materials

- *E. coli* genomic DNA (QIAamp DNA Kit, Qiagen)
- Primer RNase III-Bam:
  5'-CGCGGATCCAACCCCATCGTAATTAATCGGCTTCA
- Primer RNase III-SMA: 5'-GACGTCCGACGATGGCAAT
- Taq DNA polymerase and buffers (Bioline)
- Thermal cycler (MJ Research)
- *Bam* HI and *Sma* I (NEB)
- Plasmid pGEX-2T (Amersham Biosciences)
- PCR Purification Kit (Qiagen)

Experimental Procedures

1. Amplify the coding sequence of *E. coli* RNaseIII from bacterial genomic DNA using the primers RNase III-Bam and RNase III-SMA (94°C/3 min; 30 cycles of 94°C/30 sec, 55°C/30 sec, and 72°C/60 sec; final elongation step, 72°C/10 min).
2. Purify the amplified fragment using the PCR purification kit.
3. Digest with *Bam* HI and *Sma* I and clone the fragment into the vector pGEX-2T digested with the same enzymes to produce pGEX-2T-RNase III.
4. Transform into a standard laboratory strain such as DH5α)

Protocol 6.2 Expression and Purification of GST-RNase III

RNase III is naturally present in *E. coli,* and its over-expression does not cause an obvious phenotype. Therefore, large amounts of highly pure enzyme can be expressed and purified in one simple purification step (Figure 6.2).

Materials

- Emulsiflex homogenizer (Avestin)
- Competent *E. coli* BL21 (Novagen)
- 5 × Bradford protein assay reagent (100 mg Coomassie Brilliant Blue **G** (Merck) 1.02082.0005, dissolve in 50 ml of 95% ethanol; add 100 ml 85.7% (v/v) phos-

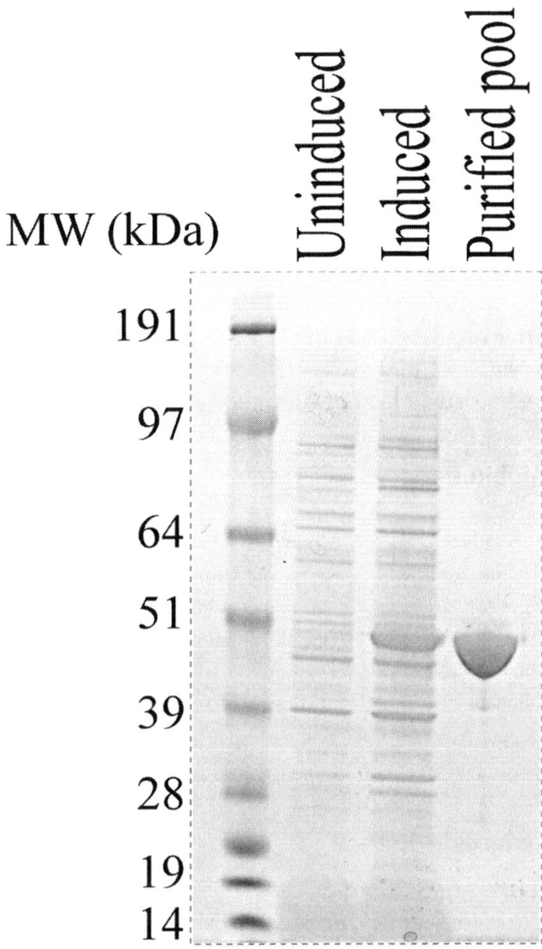

FIGURE 6.2 Purification of the GST-RNase III fusion protein. Samples taken at indicated steps during expression and purification are analyzed by SDS-PAGE stained with Coomasie blue. Uninduced, total cell extract prior to IPTG induction; induced, total cell extract at the end of induction period; purified pool, the final preparation.

phoric acid; mix thoroughly; add water to 200 ml for a 5 × stock; filter through Whatman #1 paper; dilute with H_2O to 1 × solution, which can be stored at 4°C for at least 2 weeks)

- Terrific Broth (TB) (Dissolve in 900 ml deionized H_2O, 12 g tryptone, 24 g yeast extract, and 4 ml glycerol; autoclave and cool to <60°C; add100 ml of 0.17 M KH_2PO_4/0.72 M K_2HPO_4 buffer)
- Luria-Bertani medium (LB) (Dissolve in 900 ml deionized H_2O, 10 g tryptone, 5 g yeast extract, and 10 g NaCl; adjust pH to 7.0 and volume to 1 l with H_2O; autoclave
- 20% (w/v) glucose (20 × stock)

- 100 mg/ml ampicillin (1000 × stock)
- IPTG (Diagnostic Chemicals)
- Lysis buffer (1 × PBS, 0.5% (v/v) Tween-20, 1 mM EDTA, 1 × Protease inhibitors to be added before lysing the cells)
- Column buffer with Tween-20 (1 × PBS, 0.25 M M NaCl, 0.05% (v/v) Tween-20, 0.5 mM DTT)
- Column buffer without Tween-20 (1 × PBS, 0.25 M M NaCl, 0.5 mM DTT)
- Elution buffer (50 mM HEPES, pH 8.0, 300 mM NaCl, 0.5 mM DTT, 10 mM reduced Glutathione; check pH and adjust to 8.0 with NaOH after adding reduced Glutathione from a 250 mM stock in H_2O)
- 2 × Storage buffer (40 mM Tris-Cl, pH 7.9, 1 mM EDTA, 10 mM $MgCl_2$, 2 mM DTT, 280 mM NaCl, 14 mM KCl)
- Glutathione-Sepharose 4 Fast Flow (Amersham, #17-5132-01)

1000 × Protease Inhibitors

Protease inhibitor	Supplier	Specificity	[Final](mg/ml)	Mode of action
Chymostatin	Calbiochem	Cysteine and serine	6	Reversible
Leupeptin	BioMol	Serine and cysteine	0.5	Reversible
antipain-HCl	Sigma-Aldrich	Serine and cysteine	10	Reversible
Aprotinin	Sigma-Aldrich	Serine	2	Reversible
Pepstatin	BioMol	Aspartic	0.7	Reversible
APMSF	Sigma-Aldrich	Serine	10	Irreversible

Prepare a 1000 × stock in dry DMSO (Sigma-Aldrich). Store 200-µl aliquots at –20°C.

Experimental Procedures

Induction of protein overexpression

1. Inoculate a single colony of *E. coli* BL21, freshly transformed with the pGEX-2T RNase III expression plasmid, in 50 ml of LB broth containing 100 µg/ml ampicillin and 1% (v/v) glucose in a 250-ml flask to prepare a preculture.
2. Incubate overnight (12 to 16 hr) at 37°C, with shaking at 180 rpm.
3. Transfer the overnight culture to a 50-ml conical tube, pellet (4000 rpm, 5 min at RT, Heraeus Multifuge 3SR), and resuspend in 50 ml fresh media.
4. Inoculate 20 ml of this washed culture in 1000 ml TB containing 100 µg/ml ampicillin.
5. Incubate at 37°C for 2.5 h and then induce protein expression by adding 0.2 mM IPTG.
6. Incubate with shaking at 37°C for 3 h.

Extract preparation

7. Harvest cells (5000 rpm, 10 min, 4°C, Beckman JLA 8.1000) and resuspend in ice-cold lysis buffer (5 ml lysis buffer/g bacterial pellet) containing 1 × protease inhibitors.
8. Prepare extracts immediately by passing the cell suspension once through an ice-cold Emulsiflex homogenizer set at 17,000 psi.

9. Centrifuge the lysate to remove cell debris (12000 rpm, 30 min, 4°C, Beckman JA12) prior to affinity chromatography.

Affinity chromatography

Note. Perform all subsequent steps at 4°C or on ice.

10. Preequilibrate a 3-ml (bed volume) Glutathione Sepharose column with 20 ml column buffer containing Tween-20.
11. Bind the GST-fusion protein to the column by passing the cleared extract once through the column and save the first flow-through fraction for further processing.
12. Wash the column with 40 ml of column buffer with Tween-20 and then with 20 ml of column buffer without Tween-20.
13. Elute bound protein by applying the elution buffer in 12 × 1 ml increments, and by collecting individual fractions.
14. To increase the total yield of purified GST-fusion protein repeat this affinity chromatography procedure but this time apply the first flow-through fraction to a re-equilibrated column.
15. Determine peak fractions by protein estimation (e.g., with Bradford assay) from both runs, pool the peak fractions, and then desalt into 2 × storage buffer.
16. Measure the absorbance at 280 nm, add an equal volume of glycerol, and store aliquots of the preparation at –20°C.

Notes. The final yields based on the absorbance of the sample in storage buffer (ε=1.032 for a 0.1% (v/v) solution at 280 nm) are of the order of 15 mg/l bacterial culture. While the amount of GST-RNase III for the limited digestion of dsRNA is fixed, we recommend performing a pilot digestion with every new preparation to determine the optimal ratio of dsRNA and GST-RNase III (see Figure 6.3).

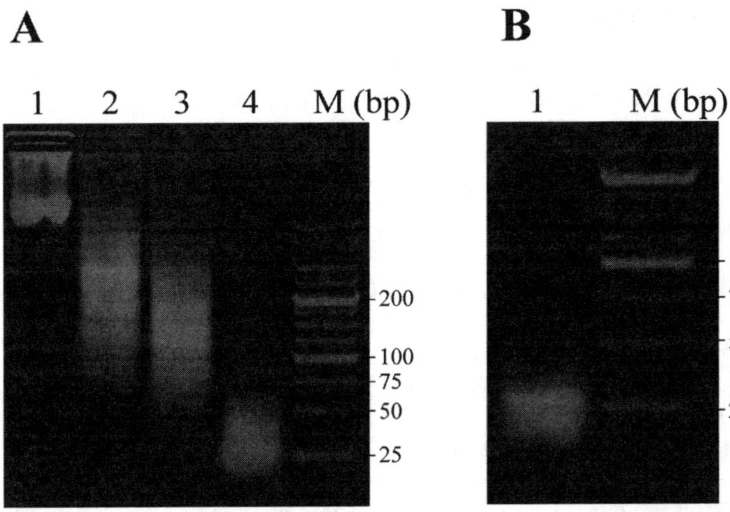

FIGURE 6.3 Titration experiment to determine the optimal amount of RNase III treatment to obtain esiRNA of the correct length. Agarose gels are presented showing the limited digestion of dsRNA before and after purification. Fragment sizes of the ladder are depicted. dsRNA was digested for 4 h with varying amounts of RNase III and stopped by the addition of EDTA. 1.5 µg RNA was loaded per lane. (A) lane 1: dsRNA not digested; lane 2: 40 mg dsRNA digested with 0.75 mg GST-RNase III; lane 3: 40 µg dsRNA digested with 1.5 µg GST-RNase III; lane 4: 40 µg dsRNA digested with 3 µg GST-RNase III. (B) 1.5 µg esiRNA after column purification.

Protocol 6.3 Generation of PCR Products for *In Vitro* Transcription of Substrate RNAs

The template for *in vitro* transcription can be generated either by amplification of cDNA inserts from clones using primers specific for the vector-backbone or from cDNA preparations using target-specific primers appended with T7 promoter sequences.

When choosing cDNA clones or designing cDNA-specific vectors, it should be taken into account that the amplicon length is critical for the yield of dsRNA. We observed highest yields with amplicon lengths between 400 and 1500 bp. While shorter templates appear to require longer RNase III digestion times, we occasionally observed annealing problems of the two ssRNAs when using templates longer than 1500 bp.

Here we describe the preparation of the template for *in vitro* transcription for the endogenous human gene *Eg5* and the reporter gene *EGFP*. Please note that PCR reaction conditions and cycling parameters depend on the amplicon. Therefore, amplification conditions and buffer composition for other targets may need to be optimized.

Materials

- Taq DNA polymerase and buffers (Bioline)
- Forward and reverse primers with T7 promoter sequence 5'-CGTAATACGACTCACTATAGGG added 5' to a gene-specific or vector-backbone-specific primer sequence; internal T7 promoter sites that are often present in vector backbones can be utilized by choosing a primer to these sites
- Thermal cycler (MJ Research)
- Gene-specific template plasmid

Experimental Procedures

1. Mix the following reagents in a PCR tube:
   - 5 µl 10 × PCR buffer
   - 1.5 µl 50 mM $MgCl_2$
   - 4 µl 10 mM dNTP mix
   - 1 µl 10 µM Forward primer
     5'-CGTAATACGACTCACTATAGGGTGAGCAAGGGCGAGGA (for *EGFP*)
     or
     5'-CGTAATACGACTCACTATAGGGAGAATGTGCTGCAAGGCGATTAAGT (for *Eg5*)
   - 1 µl 10 µM Reverse primer
     5'-CGTAATACGACTCACTATAGGGTACAGCTCGTCCATGCCGA (for *EGFP*) or 5'-TGTGTGGAATTGTGAGCGGATA (for *Eg5*)
   - 2 µl Taq DNA polymerase (1 U/µl)
   - 1–10 ng plasmid template (IMAGE clone 825606 for *Eg5* and for *EGFP*, pEGFPLuc; Clontech or any other plasmid containing the *EGFP* gene)
   - PCR grade water to 50 µl.
2. Amplify as follows: initial denaturation step of 94°C for 3 min; then 40 cycles of 94°C/30 sec, 60°C/30 sec and 72°C/90 sec (*Eg5*) or 72°C/30 sec (*EGFP*); final elongation step of 72°C for 10 min.
3. Analyze products by standard agarose gel electrophoresis.

Protocol 6.4 *In Vitro* Transcription and the Production of dsRNA

PCR products can be used directly for *in vitro* transcription reactions without purification. We routinely use 2.5 µl of the PCR product directly in *in vitro* transcription reactions. The production of dsRNA, including the synthesis of the two RNA strands and their annealing, is performed in a single tube. We have further streamlined this step by skipping the conventional purification steps such as removal of DNA, nucleotides, and precipitation after *in vitro* transcription.

Materials

- MEGAscript™ *in vitro* transcription kit (Ambion)

Experimental Procedures

1. Prepare a 10-µl reaction mix at room temperature as follows:

10 × transcription buffer	1.0 µl
75 mM ATP, GTP, CTP, and UTP	1.0 µl each
PCR product	2.5 µl (~ 0.5 µg DNA template)
Nuclease-free water	1.5 µl
T7 RNA polymerase enzyme mix	1.0 µl

2. Incubate the reaction at 37°C for 4 to 12 h.
3. Perform annealing in a thermal cycler as follows: 90°C for 3 min, ramp to 70°C with 0.1°C/sec, 70°C for 3 min, ramp to 50°C with 0.1°C/sec, 50°C for 3 min, ramp to 25°C with 0.1°C/sec.

Protocol 6.5 Digestion and Purification of dsRNA

Long dsRNA is partially digested to esiRNAs with a length of about 18 to 25 bp. The subsequent purification effectively removes remaining DNA template, unincorporated nucleotides, and dsRNAs longer than 40 bp.

Materials

- dsRNA digestion buffer, pH 7.9 (20 mM Tris-HCl, 0.5 mM EDTA, 5 mM $MgCl^2$, 1 mM DTT, 140 mM NaCl, 2.7 mM KCl, 5% (v/v) glycerol)
- 0.5 M EDTA, pH 8.0
- 4% (v/v) agarose in 1 × TBE electrophoresis buffer
- 25 bp DNA ladder (e.g., HyperLadder V from Bioline)
- TE buffer, pH 7.9 (10 mM Tris-HCl, 1 mM EDTA)
- Isopropanol
- 70% (v/v) ethanol
- Micro Bio-Spin Chromatography Columns (BioRad Laboratories)
- Q Sepharose FastFlow (Amersham Biosciences)
- Equilibration buffer (20 mM Tris, 1 mM EDTA, 300 mM NaCl, pH 8.0)
- Wash buffer (20 mM Tris, 1 mM EDTA, 400 mM NaCl, pH 8.0)
- Elution buffer (20 mM Tris, 1 mM EDTA, 520 mM NaCl, pH 8.0)

- Ethidium bromide, 10 mg/ml (Sigma-Aldrich)
- Orange G (Sigma-Aldrich); marker dye for esiRNAs

Experimental Procedures

1. Mix 10 μl of the *in vitro* transcription reaction (~25-50 μg dsRNA) with 3 μg of GST-RNase III protein in 90 μl digestion buffer (see also Figure 6.3).
2. Incubate at 23°C for 4 h.
3. Run a 2–4 μl aliquot in 4% (w/v) agarose gel along with a 25-bp DNA ladder to check the size range of digestion products. If the size range is not appropriate (i.e., digestion products are too long), incubate for additional 2 h at 37°C and check again.
4. Purify the digestion products immediately as follows, or terminate the reaction by adding 4 μl 0.5 M EDTA.
5. Prepare spin columns for purification. Add 200 μl Q-Sepharose and 500 μl equilibration buffer to the column. Spin at 1000 g for 1 min and discard the flow-through.
6. Again, add 500 μl equilibration buffer, spin at 1000 g for 1 min, and discard the flow-through.
7. Load all of the digested dsRNA onto the column and incubate for 5 min at room temperature.
8. Spin at 1000 g for 1 min and discard the flow-through.
9. Add 500 μl wash buffer, spin at 1000 g for 1 min, and discard the flow-through.
10. Add 300 μl elution buffer, spin at 1000 g for 1 min, and **collect** the flow-through.
11. Repeat step 10.
12. Add 500 μl isopropanol to the 600 μl eluted esiRNA and vortex. Store on ice for at least for 30 min.
13. Spin at 16,000 g for 15 min at 4°C. Discard the supernatant and wash the pellet twice with cold 70% (v/v) ethanol.
14. Air-dry the esiRNA pellet for 10 to 15 min at room temperature and dissolve in 50 μl of 1 × TE buffer.
15. Run a 2–4 μl aliquot on a 4% (w/v) agarose gel, and measure OD_{260} in order to quantify the esiRNA concentration.

Protocol 6.6 Transfection of esiRNA and Analysis of Silencing in Cultured Mammalian Cells

Optimal transfection conditions for esiRNA are highly dependent on the cell line used. Here we describe the silencing of the endogenous gene *Eg5* in HeLa cells (Figure 6.4). The knockdown of *Eg5* results in a mitotic arrest phenotype within 24 hours after transfection and is, therefore, suitable for optimizing transfection conditions in human cell lines.[22] We also describe inhibition of the reporter gene *EGFP*. These two genes can serve as the template for designing other esiRNAs directed against specific genes of interest. However, also consider the following when designing esiRNAs against specific genes. First, each esiRNA directed against a specific gene may have a different silencing capacity. This silencing capacity can be evaluated by quantitative RT-PCR,

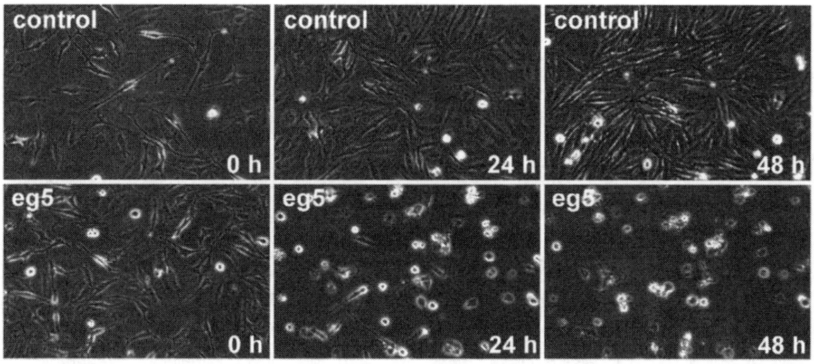

FIGURE 6.4 Transfection of esiRNA into mammalian cells and phenotypic analysis. The upper panel (control) shows HeLa cells transfected with 0.5 µg of luciferase esiRNA photographed at indicated time points. The lower panel (eg5) shows HeLa cells transfected with 0.5 µg esiRNA directed against the kinesin eg5 at the same time points.

northern blot or western blot analyses. Second, each gene product has a different half-life within the cell. Therefore, some phenotypes may be observed earlier then others. Hence, cells should be examined at different time points post-esiRNA transfection. Third, RNAi leads to gene knockdown and not knockout. Therefore, some gene product may still be present in the cell at sufficient concentrations to fulfill its function.

The experiments described here are for transfections in 6-well dishes. Adjust volumes and amount of reagents for tissue culture dishes of a different size.

Materials

- Oligofectamine and Lipofectamine 2000 (Invitrogen); other transfection reagents and protocols should work accordingly
- Dulbecco's Mod Eagle Medium (D-MEM) (InVitrogen)
- Fetal bovine serum (FBS) (InVitrogen)
- Other tissue culture ingredients; L-Glutamine, penicillin/streptomycin (InVitrogen, Sigma-Aldrich)

Experimental Procedures

*Knockdown of an endogenous gene (*Eg5*)*

1. Plate cells ~16 h before transfection in a 6-well culture dish to obtain 30 to 50% confluency on the day of transfection.
2. Change the medium (10% (v/v) FBS, 2 mM glutamine, 100 U/ml penicillin, 100 µg/ml streptomycin, DMEM) before transfection. The volume of the medium for transfection should be 800 µl.
3. Dilute 0.5 µg of esiRNA directed against *Eg5* with 185 µl of serum-free DMEM.
4. Dilute 4 µl of Oligofectamine with 11 µl of serum-free medium and incubate at RT for 5 min.
5. Add diluted esiRNA to diluted Oligofectamine. Incubate at RT for 20 min.
6. Add esiRNA/Oligofectamine mix to the cells. Gently mix and incubate at 37°C for 4 h.

7. Add 1000 μl serum-containing medium.
8. Observe knockdown of *Eg5* by light microscopy within 24 h posttransfection. At this time, many of the cells should be arrested in mitosis, indicated by the rounded-up appearance. Most of the cells should die within 48 hours posttransfection (see also Figure 6.4).

Knockdown of EGFP expressed from a plasmid

1. Plate cells ~16 h before transfection into a 6-well culture dish to achieve 80 to 90% confluency on the day of transfection.
2. Change the medium (10% (v/v) FBS, 2 mM glutamine, 100 U/ml penicillin, 100 μg/ml streptomycin, DMEM) before transfection. The volume of the medium for transfection should be 2000 μl.
3. Dilute 1 μg of esiRNA directed against *EGFP* or a control esiRNA (e.g., firefly luciferase) and 3 μg of EGFP expression plasmid with 250 μl of serum-free DMEM.
4. Dilute 10 μl of Lipofectamine 2000 in 250 μl of serum-free medium, and incubate at RT for 5 min.
5. Add diluted esiRNA-plasmid to diluted Lipofectamine 2000. Incubate at RT for 20 min.
6. Add esiRNA-plasmid/Lipofectamine 2000 mix to the cells. Gently mix and incubate at 37°C for 4 hours.
7. Remove old medium and add 2500 μl serum-containing medium.
8. Observe the cultures using appropriate filter sets for GFP visualization 24 h post-transfection. One should see a clear fluorescence intensity difference between cells transfected with control esiRNA and EGFP esiRNA.

ACKNOWLEDGMENTS

We would like to thank D. Yang and J.M. Bishop for helpful discussions.

REFERENCES

1. Hannon, G.J., RNA interference, *Nature,* 418, 244, 2002.
2. Dykxhoorn, D.M., Novina, C.D., and Sharp, P.A., Killing the messenger: short RNAs that silence gene expression, *Nat. Rev. Mol. Cell Biol.,* 4, 457, 2003.
3. Gonczy, P. et al., Functional genomic analysis of cell division in *C. elegans* using RNAi of genes on chromosome III, *Nature,* 408, 331, 2000.
4. Kamath, R.S. et al., Systematic functional analysis of the *Caenorhabditis elegans* genome using RNAi, *Nature,* 421, 231, 2003.
5. Lum, L. et al., Identification of Hedgehog pathway components by RNAi in *Drosophila* cultured cells, *Science,* 299, 2039, 2003.
6. Stark, G.R. et al., How cells respond to interferons, *Annu. Rev. Biochem.,* 67, 227, 1998.
7. Zamore, P.D. et al., RNAi: double-stranded RNA directs the ATP-dependent cleavage of mRNA at 21 to 23 nucleotide intervals, *Cell* 101, 25, 2000.
8. Hammond, S.M. et al., An RNA-directed nuclease mediates post-transcriptional gene silencing in *Drosophila* cells, *Nature,* 404, 293, 2000.

9. Yang, D., Lu, H., and Erickson, J.W., Evidence that processed small dsRNAs may mediate sequence-specific mRNA degradation during RNAi in *Drosophila* embryos, *Curr. Biol.,* 10, 1191, 2000.
10. Bernstein, E. et al., Role for a bidentate ribonuclease in the initiation step of RNA interference, *Nature,* 409, 363, 2001.
11. Conrad, C. and Rauhut, R., Ribonuclease III: new sense from nuisance, *Int. J. Biochem. Cell Biol.,* 34, 116, 2002.
12. Elbashir, S.M., Lendeckel, W., and Tuschl, T., RNA interference is mediated by 21- and 22-nucleotide RNAs, *Genes. Dev.,* 15, 188, 2001.
13. Kittler, R. and Buchholz, F., RNA interference: gene silencing in the fast lane, *Seminars in Cancer Biology,* 13, 259, 2003.
14. Harborth, J. et al., Sequence, chemical, and structural variation of small interfering RNAs and short hairpin RNAs and the effect on mammalian gene silencing, *Antisense Nucleic Acid Drug Dev.,* 13, 83, 2003.
15. Kawasaki, H. et al., SiRNAs generated by recombinant human Dicer induce specific and significant but target site-independent gene silencing in human cells, *Nucleic Acids Res.,* 31, 981, 2003.
16. Myers, J.W. et al., Recombinant Dicer efficiently converts large dsRNAs into siRNAs suitable for gene silencing, *Nat. Biotechnol.,* 21, 324, 2003.
17. Yang, D. et al., Short RNA duplexes produced by hydrolysis with *Escherichia coli* RNase III mediate effective RNA interference in mammalian cells, *Proc. Natl. Acad. Sci. U. S. A.,* 99, 9942, 2002.
18. Calegari, F. et al., Tissue-specific RNA interference in postimplantation mouse embryos with endoribonuclease-prepared short interfering RNA, *Proc. Natl. Acad. Sci. U. S. A.,* 99, 14236, 2002.
19. Ketting, R.F. et al., Dicer functions in RNA interference and in synthesis of small RNA involved in developmental timing in *C. elegans*, *Genes. Dev.,* 15, 2654, 2001.
20. Zhang, H. et al., Human Dicer preferentially cleaves dsRNAs at their termini without a requirement for ATP, *EMBO J.,* 21, 5875, 2002.
21. Apirion, D. and Miczak, A., RNA processing in prokaryotic cells, *Bioessays,* 15, 113, 1993.
22. Weil, D. et al., Targeting the kinesin Eg5 to monitor siRNA transfection in mammalian cells. *Biotechniques,* 33, 1244, 2002.

7 Plasmid-Mediated Intracellular Expression of siRNAs

Renu Wadhwa, Masayuki Sano, Makoto Miyagishi, and Kazunari Taira

CONTENTS

7.1 INTRODUCTION

RNA interference (RNAi) refers to obstruction of the flow of genetic information at the RNA level, resulting in the silencing of gene expression. It is triggered by the presence of double-stranded RNA. Synthetic short interfering RNAs (siRNAs) of ~21 nt were recently shown to be effective for specific gene silencing in mammalian cells without evoking the interferon response.[1] Following this discovery, many reports describing successful expression and effect of vector-based siRNA in mammalian cells allowing delivery of siRNAs into mammalian cells by a DNA vector (low-cost alternative to the chemically synthesized siRNAs) have appeared. This has initiated a new way of studying gene function in higher eukaryotes. The effect of RNAi strongly depends on the target sequence and vector systems. Various groups, including our own, have developed plasmid based expression systems for siRNA.[2–8] Viral vectors including retero-, adeno- and lentivirus are useful in facilitating the entry of siRNA in cells and tissues that are difficult to transfect. Retroviral vectors have been favored for

0–8493–2141–7/05/$0.00 + $.50

their ability to integrate exogenous genes into the genome of host cells. Replication-defective lentiviral vectors can be used to introduce exogenous sequences into non-proliferating ones such as haematopoietic stem cells. Adenoviral vectors that are able to transduce exogenous sequences into quiescent cells and sustain their long-term expression can also be used.[9–11] Such rapidly developing virus-based siRNA vectors may become potentially useful tools for therapy of dominant human diseases. This chapter describes the strategies and construction of plasmid-based expression systems for siRNA in mammalian cells.

7.2 SYNTHETIC AND VECTOR-BASED EXPRESSION OF siRNA

7.2.1 Synthetic siRNA

Chemically synthesized siRNAs are probably the easiest and most direct approach to induce artificial RNA interference in mammalian cells. Synthetic siRNA duplexes with 3'-dinucleotide overhangs are incubated with cationic lipids formulations to generate liposomes and are introduced into cells by liposome-mediated transfections. Specific siRNA can also be transcribed *in vitro* using RNA polymerase that offers a cheaper alternative to obtain a large quantity of siRNA.[2–3] Efficiency of uptake of siRNA depends not only on the combination of siRNA and liposome mixtures but also on the cell type. Alternatives to the liposome-mediated transfections include electroporation, covalent binding of siRNA to carrier molecules that mediate their cellular uptake by enhanced hydrophobic properties, covalent linkage of oligonucleotides with cell penetrating peptides, and protein transduction domains that help in traversing the plasma membrane.[4] In a modification to enhance the activity of antisense oligonucleotides, a helicase-binding constitutive transport element (CTE) motif was linked and was seen to be more effective than the unlinked controls, possibly due to improved target accessibility of the siRNA owing to the coupling of helicase.[5] Synthetic or *in vitro* transcribed siRNAs have been successfully used in many cases to knockdown gene expression in mammalian cells [iGENE: http://www.igene-therapeutics.co.jp]. However, the cost for chemical synthesis of siRNAs for each experiment and their transient effect urged many groups, including our own, to invent vector-based expression systems that are time- and cost-effective and stable in terms of the suppression of gene function. The vector systems can be used for transient and stable expression; thus, the effect of gene suppression can be studied in short-term and long-term loss of function scenarios.

7.2.2 Vector-Based Expression of siRNAs

7.2.2.1 Choice of Promoter

Many groups have engaged in developing effective vector-based expression of siRNAs and have preferred RNA polymerase III (pol III) promoters (U6 snRNA and the H1 promoter often employed for the expression of siRNA in mammalian cells) because of the following reasons:

(i) Pol III promoters generate relatively small RNAs such as tRNA and snRNA (small nuclear RNA).
(ii) Pol III promoter-driven transcripts are neither capped at the 5'end nor polyadenylated at the 3' end.
(iii) RNA pol III starts transcription at G or A; thus, the start site of the siRNA can be predetermined and offer a system in which any sequences can be inserted downstream of the transcription start site.
(iv) RNA pol III recognizes poly (T) sequence as a termination codon. Thus, it is possible to design small RNAs synthesized by pol III that carry 3' overhangs of one or two uridines (a structural feature for siRNA to be effective *in vitro*).
(v) Controlled RNA pol III promoters can be generated for the analysis of temporally expressed genes or lethal genes. Some examples of these include tetracycline and Cre-loxP controlled expressions.

Brummelkamp and Bernards[6] developed an expression system called pSUPER (suppression of endogenous RNA) to generate endogenous siRNAs for transient and stable suppression in mammalian cells. These were targeted to *CDH1* gene (involved in the activation of the anaphase-promoting complex). They used pol III H1-RNA promoter and generated a transcript that folds back into a strategically designed stem-loop structure. The dsRNA initiates Dicer-mediated events leading to the expression of siRNAs. They also generated expression constructs with variable loop length and reported that the construct with nine-nucleotide loop was most effective and was least toxic to cells. Others, our own laboratory included, have made successful plasmid expression systems using U6 promoter, driving the transcription of stem loop DNA or of the tandemly linked sense and antisense sequences driven by two independent promoters.[6–8, 12–13] The two U6 promoters can be placed either in tandem in one vector or in two separate vectors to direct the transcription of a sense and antisense strand of a small RNA with a 19-nt specific gene-targeting sequence containing 3'UUU overhangs (Figure 7.1). Sense and antisense strands form duplexes *in vivo* and function similarly to the *in vitro* siRNA.

In most of the studies, both tandem and hairpin loop vector–driven siRNA suppressed the target gene expression. The hairpin-type siRNA-expression vector had significantly higher suppressive activity than the tandem-type siRNA-expression vector at a low concentration.[7] Furthermore, modifications of the hairpin-type target sequences that do not affect their activity but make them more stable during vector amplification in bacteria were suggested. In similar experiments, variations of the loop sequence (4 to 11 nt) were tried and an 11-nucleotides-loop sequence (5'GUGUGCUGUCC) was seen to be the most effective.[7,14] Cre-loxP controlled pol III promoter–driven (U6 promoter) expression of siRNA has also been generated. It was tested for firefly luciferase reporter and was effective only in the presence of Cre recombinase.[15]

7.2.2.2 Selection and Modification of the Target Sites

SiRNA generated by Dicer in flies has 3' dinucleotide overhangs. Based on this, the *in vitro* synthesized siRNA with 3' dinucleotide overhangs were designed, and 3' UU was

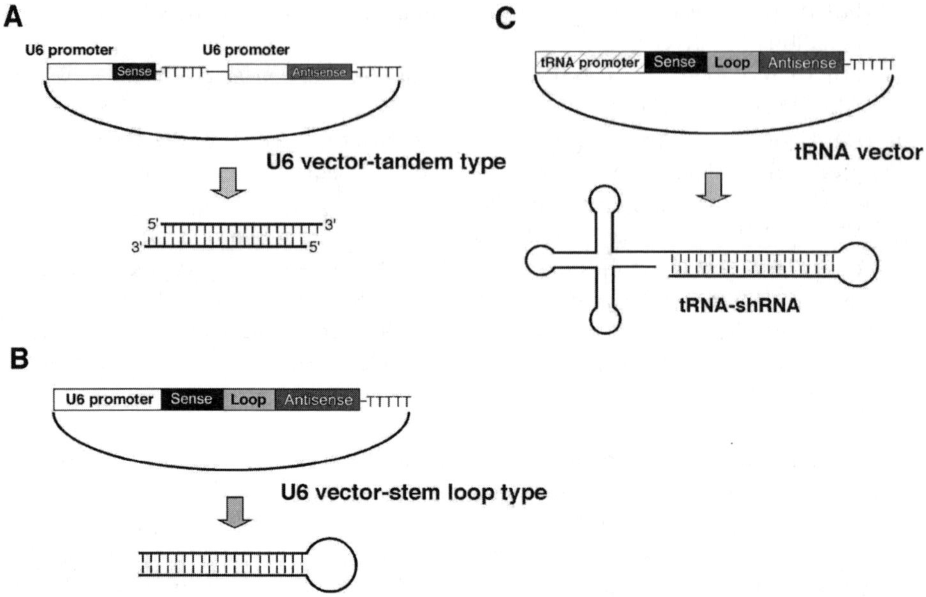

FIGURE 7.1 Diagrammatic presentations of tandem (**A**) and stem loop (**B** and **C**) type siRNA expression vectors driven by U6 (**A** and **B**) or tRNA (**C**) promoter.

seen to be more efficient than AA, CC, or GG. An empirical target site (AAX19UU) is often used as a start point for vector-driven siRNA expression. The efficacy of any siRNA target site is examined by the level of protein that is left after RNAi and is determined by factors such as the half-life of the protein. Furthermore, the functional efficacy of siRNA also depends on the level of protein required for its function. For example, proteins such as enzymes that might function even at a very low level may not yield a null phenotype, while in other cases "knockdown" may generate a phenotype similar to knockout. Although a chosen siRNA works with high probability, it is not always the best. Shifting of siRNA by only one or a few nucleotides may significantly affect its silencing function [iGENE: http://www.igene-therapeutics.co.jp].[14] Thus, the selection of target sites is the most critical issue in application of siRNA. Since the pol III recognizes TTTT as the termination signal, the TTTT-containing regions are avoided in the target sites. The success rate of effective siRNA has been high enough to encourage the researchers to try a few sites and select the most effective one, resulting in a collection of data that can be used as guidelines. Although earlier studies preferred target sites near the initiation codon, the sites at the 3' end or even at the UTR have been shown to be effective.

Approaches similar to the ones used in antisense and ribozyme efficacy such as estimation of the accessibility to the target site for base pairing by a computer-assisted RNA folding program (MFOLD), are often used as a start point. However, these predictions are not always correct. Assays for estimation of the target accessibility include *in vitro* binding of the oligonucleotides to *in vitro*–transcribed RNA and are followed by its incubation with RNase H from *E. coli.* RNase H cleaves RNA-DNA hybrids that can be analyzed on a polyacryamide gel, thus offering a comparative test to access the efficacy of binding of oligonucleotides to the target.[16] In a modification of this method, cellular extracts that provide a native state of RNA-protein complexes are incubated with the oligonucleotides and subjected to cleavage by endogenous RNase H.[4]

Prediction modelling of the RNAi effect by partial least squares (PLS) is also in practice. This method is based on the insight obtained from target inhibition by siRNA constructs against several genes as well as single nucleotide walking constructs of a single gene and takes into account the GC content of the target sites [iGENE: http://www.igene-therapeutics.co.jp]. Based on the cross validation of the predictions, an algorithm by nonlinear regression methods was established[7,14] and was tested to predict the activity of the target site well within the reliable limits. Since the stable stem structures in DNA are unfavorable for construction of siRNA vector, mutations of C to U or A to G on the sense strand were recommended. It was shown that as many as five C to T point mutations (avoiding T clustering to more than 3) that generated G:U base-paring in the sense strand did not affect the silencing effect of a hairpin siRNA. In contrast, insertion of a single C to T mutation in the antisense strand resulted in reduced suppression. Such mismatch studies have also suggested the two strands of siRNA have different functions. Whereas 3' chemical modification of siRNA's sense strand did not affect its activity, a similar modification of antisense strand abolished its RNAi effect, suggesting that a free 3' end of the antisense might be required for recognition/interaction with RISC.[17] Thermodynamic properties of siRNA were reported to play a critical role in determining its function, stability, and strand bias in duplex unwinding and retention by RISC.[18,19]

7.2.2.3 Cytoplasmic Expression of siRNA

Since mRNA targets are transported to the cytoplasm where protein synthesis takes place, it was anticipated that siRNAs that are expressed from the vectors could be more effective if they were transported to the cytoplasm quickly. In a study, five kinds of dsRNA expression vectors that were controlled by $tRNA^{Val}$ (Figure 7.1) or widely used U6 promoter were constructed in order to examine the relationship between the localization and a target cleavage efficacy. Transcripts of tRNA-dsRNA were localized in the cytoplasm and were processed more efficiently by Dicer than the U6-siRNA.[20] Similarly, U6-driven siRNA with a microRNA loop was seen to localize in the cytoplasm and was effectively processed by Dicer.[20] Recently, RNA polymerase II (CMV) promoter-based expression of siRNA was successfully employed,[21,22] suggesting the possibility of organ-, tissue- and cell-specific suppression of genes.

The high efficacy of siRNA for suppression of gene expression has encouraged us, and others, to initiate a genome-wide RNAi library approach by an siRNA oligonucleotide and expression library in mammalian cells. Expression libraries of siRNA driven by U6 and tRNA promoter have been generated.[23] These can be powerful tools to isolate novel genes and regulating factors involved in various biological phenotypes.

7.3 CONCLUDING REMARKS

Sequence-specific gene suppression is a valuable tool in the basic understanding of the gene functions and abrogation of aberrant gene expression. Initiated only recently, this field has achieved enough momentum and vigor to be at the forefront of choices in therapeutics and novel drug designs.

PROTOCOLS

PROTOCOL 7.1 CONSTRUCTION OF U6 PROMOTER-DRIVEN siRNA EXPRESSION VECTOR

Materials

- Custom-synthesized oligonucleotides corresponding to the mRNA target site (see experimental procedures below)
- piGENE™ hU6 plasmid (iGENE Therapeutics, Inc.)
- MinElute Gel extraction (Qiagen)
- MinElute PCR purification kit (Qiagen)
- Pyrobest (Takara Shuzo Co.)
- Ligation high kit (e.g., TOYOBO) or an alternative
- Bacterial alkaline phosphatase (BAP) or Calf intestine alkaline phosphatase (CIAP) and reaction buffers (TOYOBO)

Experimental Procedure

1. Select target sites for tandem-type and stem-loop-type vector as follows. For tandem type, find a sequence of 5'AAG (N17-19) C in the target RNA; in case of absence of this sequence, alternatives such as (i) 5'AA (G/A) (N19-23) (C/T) or (ii) 5'CA (G/A) (N19-23), or of 5'A (G/A) (N19-23), can be chosen. For stem loop type, search for 5'AAG (N21) sequence on the target RNA. In case of absence of this sequence, 5'AAA (N21) or 5'CA (G/A) (N21) can be considered. *N.B.* The U6 promoter starts transcription from purine residue (G or A); thus, G or A should be the start of siRNA. TTTT is added to the 3' of the sense and strand to act as terminal codon.

Preparation of the vector

2. Digest 3–5 μg of piGENE™ hU6 plasmid with *Bsp*M I at 37°C overnight. The vector has two *Bsp*M I recognition sites downstream of a human U6 promoter. *Bsp*M I is a class IIS restriction enzyme (has different recognition and cleavage sites). Owing to this feature of *Bsp*M I, siRNA-encoding sequence can be inserted immediately after the U6 promoter. In case of the tandem-type vector, PCR product containing sense, antisense sequences, and a portion of U6 promoter is inserted into the plasmid. In case of stem-loop-type vector annealed DNA oligonucleotide that contains sense, antisense and loop sequences are inserted.

3. Resolve the cut vector on a 1% (w/v) agarose gel, excise the gel, and extract DNA using the MinElute Gel extraction kit.
4. Dephosphorylate eluted DNA with BAP/CIAP according to the manufacturer's instructions. The BAP dephosphorylation reaction mixture (100 μl) contains *Bsp*M1-digested vector DNA (3 μg), 1 × BAP buffer, and BAP enzyme (3–10 U); incubate at 65°C for 1 h. In case of stem-loop-type BAP, treatment should be skipped.
5. Purify DNA from step 4 using MinElute PCR purification kit only in case of tandem-type siRNA-expression vector.
6. Dissolve the *Bsp*M I cut and purified DNA in 30 μl of TE, and use 1 μl (about 100 ng) for each ligation.

Preparation of the insert DNA

For tandem type

7. Prepare synthetic oligonucleotides primers *A* (sense) and *B* (antisense):
   A: 5'ggctctagaACCTGCcggccacc(N_{19})tttttcaattcaaggtcgggcag
   B: 5'ggctctagaACCTGCtagcgcataaaaa(N_{19})ggtgtttcgtcctttccacaag
   (*N.B.* N_{19} is the selected siRNA sequence from the target mRNA. The first N should be G or A to serve as the transcription start site. Directions of sense and antisense sequences should be taken care of as demonstrated in the following example. For the target sequence GTGCGCTGCTGGTGCCAAC, *primer A* should be 5'ggctctagaACCTGCcggccaccGTGCGCTGCTGGTGCCAACtttttcaattcaaggtcgggcag3' and *primer B* should be 5'ggctctagaACCTGCtagcgcataaaaaGTGCGCTGCTGGTGCCAACggtgtttcgtcctttccacaag3'.)
8. Perform PCR (95°C for 1 min, 30 cycles of 98°C for 10 sec, 55°C for 1 min, and 72°C for 1 min each) using each primer and piGENE™ hU6 as a template.
9. Purify the PCR products by MinElute PCR purification kit and digest them with *Bsp*M I for at least 2 to 4 h at 37°C in a 100-μl volume.
10. Separate *Bsp*M 1 cut PCR product on 2% (w/v) agarose gel, excise gel, and elute DNA using the MinElute Gel extraction kit.
11. Dissolve the DNA in 10 μl of TE (ready to ligate to the vector).

Stem loop type

12. Custom-synthesize two oligonucleotides as demonstrated, for example, target sequence 5'GTGCGCTGCTGGTGCCCAACCC below. Dissolve the oligonuceotides in 100 to 150 mM NaCl to obtain 100 μM solution.
    Oligo-1:
    5'caccGTGCGGCGtTGtTGGTGttAAtCCgtgtgctgtccGGGTTGGCACCAGCAGCGCACttttt3'
    (*N.B.* the lower-case letters in the target sequence denote C → T point mutations.)
    Oligo-2:
    5'gcataaaaaGTGCGCTGCTGGTGCCAACCCggacagcacacGGaTTaaCACCAaCAaCGCAC3'
13. Mix 5 μl of 100 μM stock of each oligonucleotide.
14. Incubate the oligonucleotide mix at 99°C for 2 min, followed by rapid cooling to 72°C and gradual (ramp 2 h) cooling from 72°C to 4°C. Use about 1 μl of the 1:200 diluted annealed oligonucleotides for ligation.

Ligation and transformation

15. Incubate 1 µl of the plasmid and 1 µl of the insert DNA with 1 µl of Ligation high in a total volume of 10 µl at 16°C for 30 min.
16. Transform *E. coli* host cells (e.g., DH5α) with the ligation mixture and select on ampicillin (100 µg/ml) plates.
17. Extract plasmid DNA from bacterial colonies with standard miniprep method.
18. Sequence the plasmid DNA using vector primer 5'CCAGTGAATTCAAGGTCGGGCAGGAAG.

PROTOCOL 7.2 CONSTRUCTION OF THE tRNA PROMOTER-DRIVEN siRNA-EXPRESSION VECTOR

Materials

- piGENE™ tRNA plasmid (iGENE Therapeutics, Inc.)
- Custom-synthesize oligonucleotides as below:
  - i. *A template of stem loop*, 5'X (30 nt) AAAACTCGAGAAAAY (30 nt) GGTACCCCGGATATCTTTTTTT3' (*N.B.* X (30 nt) corresponds to 30 nucleotides of the sense sequence of target RNA, and Y (30 nt) corresponds to 30 nucleotides of the antisense sequence for target RNA; bold letters correspond to the sequence of loop region including a site for *Xho* I.)
  - ii. *Forward primer*, 5'CCAACTTTGAGCTCZ (20 nt)3' (*N.B.* Z (20 nt) corresponds to 20 nucleotides of sense sequence of target RNA; the underlined sequence correspond to *Sac* I site.)
  - iii. *Reverse primer*, 5'AAAAAAAGATATCCGGGGTACC3' (common primer including a site for *Kpn* I)
- Pyrobest (Takara Shuzo Co.)
- Ligation Kit Ver.2 (Takara)
- *E. coli* (JM109, DH5α, or other competent cells)

Experimental Procedure

Preparation of the vector and insert DNA

1. Perform PCR (25 cycles of 95°C for 1 min, 58°C for 1 min, and 72°C for 1 min 30 sec) in a 100-µl volume using forward and reverse primers and a temple DNA.
2. Resolve the PCR product on a 2.0% (w/v) agarose gel, excise the gel, and extract DNA.
3. Digest PCR fragment with *Sac* I and *Kpn* I. In parallel, digest the piGENE™ tRNA plasmid with the same restriction enzymes.
4. Treat the digested plasmid with BAP or CIAP followed by purification.
5. Mix the digested plasmid and the digested insert DNA at a molar ratio of approximately 1:10 and ligate at 16°C for 30 min using the Ligation Kit Ver. 2.
6. Transform *E. coli* host cells with the ligation mixture and plate the cells on LB agar plates containing ampicillin at 100 µg/ml.
7. Isolate plasmid DNA by the standard miniprep method and determine the nucleotide sequence. Use for transfections of the cells using an appropriate transfection reagent.

Technical notes. Analysis of the siRNA effect. Expression of the gene to which siRNA is targeted against is analyzed by western blotting using protein-specific antibodies, northern blotting with gene specific probes, or RT PCR assays with gene-specific primers. Standard protocols for each of these assays can be used. Protein stability and turnover should be kept in mind to see the effect of siRNA on the level of expressed protein and its loss of function phenotype.

ACKNOWLEDGMENTS

Authors are thankful to Hiroyuki Kawasaki for useful information and Sunil Kaul for critical reading of the manuscript.

REFERENCES

1. Tuschl T., RNA interference and small interfering RNAs, *Chembiochem*, 2, 239–245, 2001.
2. Paul, C.P. et al., Effective expression of small interfering RNA in human cells, *Nat. Biotechnol.,* 20, 505, 2002.
3. Yu, J.Y., DeRuiter, S. L., and Turner, D.L., RNA interference by expression of short-interfering RNAs and hairpin RNAs in mammalian cells, *Proc. Natl. Acad. Sci. U.S.A*, 99, 6047, 2002.
4. Scherr, M., Morgan, M.A., and Eder, M., Gene silencing mediated by small interfering RNAs in mammalian cells, *Curr. Med. Chem.*, 10, 245, 2003.
5. Futami, T. et al., Stimulatory effect of an indirectly attached RNA helicase-recruiting sequence on the suppression of gene expression by antisense oligonucleotides, *Antisense Nucleic Acid Drug Dev.,* 13, 9, 2003.
6. Brummelkamp, T.R. and Bernards, R., New tools for functional mammalian cancer genetics, *Nat. Rev. Cancer*, 3, 781, 2003.
7. Miyagishi, M. and Taira, K., Development and application of siRNA expression vector, *Nucleic Acids Res. Suppl.*, 113, 2002.
8. Miyagishi, M. and Taira, K., U6 promoter-driven siRNAs with four uridine 3' overhangs efficiently suppress targeted gene expression in mammalian cells, *Nat. Biotechnol.*, 20, 497, 2002.
9. Qin, X.F. et al., Inhibiting HIV-1 infection in human T cells by lentiviral-mediated delivery of small interfering RNA against CCR5, *Pro. Natl. Acad. Sci. U.S.A.*, 100, 183, 2003.
10. Wall, N.R. and Shi, Y., Small RNA, can RNA interference be exploited for therapy?, *Lancet*, 362, 1401, 2003.
11. Tiscornia, G., Singer, O., Ikawa, M., and Verma, I.M., A general method for gene knockdown in mice by using lentiviral vectors expressing small interfering RNA, *Proc. Natl. Acad. Sci. U.S.A.,* 100, 1844, 2003.
12. Shi, Y., Mammalian RNAi for the masses, *Trends Genet.*, 19, 9, 2003.
13. Arenz, C. and Schepers, U., RNA interference, from an ancient mechanism to a state of the art therapeutic application?, *Naturwissenschaften*, 90, 345, 2003.
14. Miyagishi, M., Hayashi, M., and Taira, K., Comparison of the suppressive effects of antisense oligonucleotides and siRNAs directed against the same targets in mammalian cells, *Antisense Nucleic Acid Drug Dev.,* 13, 1, 2003.
15. Kasim, V., Miyagishi, M., and Taira, K., Control of siRNA expression utilizing Cre-loxP recombination system, *Nucleic Acids Res. Suppl.*, 255, 2003.
16. Birikh, K.R. et al., Probing accessible sites for ribozymes on human acetylcholinesterase RNA, *RNA*, 3, 429, 1997.

17. Hamada, M. et al., Effects on RNA interference in gene expression (RNAi) in cultured mammalian cells of mismatches and the introduction of chemical modifications at the 3'-ends of siRNAs, *Antisense Nucleic Acid Drug Dev.*, 12, 301, 2002.
18. Khvorova, A., Reynolds A., and Jayasena, S.D., Functional siRNAs and miRNAs exhibit strand bias, *Cell*, 115, 209, 2003.
19. Schwarz, D. S. et al., Asymmetry in the assembly of the RNAi enzyme complex, *Cell*, 115, 199, 2003.
20. Kawasaki, H. and Taira, K., Short hairpin type of dsRNAs that are controlled by tRNA(Val) promoter significantly induce RNAi-mediated gene silencing in the cytoplasm of human cells, *Nucleic Acids Res.*, 31, 700, 2003.
21. Xia, H. et al., siRNA-mediated gene silencing *in vitro* and *in vivo*, *Nat. Biotechnol.*, 20,1006, 2002.
22. Kato, Y. and Taira, K., Expression of siRNA from a single transcript that includes multiple ribozymes in mammalian cells, *Oligonucleotides*, in press, 2003.
23. Miyagishi, M., Matsumoto, S., and Taira, K., Generation of an shRNA: expression library against the whole human transcripts, *Virus Res.*, 102, 117, 2004.

8 Lentiviral Vector-Mediated Delivery of si/shRNA

Dong Sung An, Xiao-Feng Qin, Vincent C. Auyeung, David Baltimore, and Irvin S.Y. Chen

CONTENTS

8.1 INTRODUCTION

RNA interference (RNAi) is a highly conserved mechanism in most eukaryotic cells thought to serve as a regulator of gene expression and an antiviral defense mechanism.[1–4] RNAi is a multi-step process that begins with the cleavage of double-stranded (dsRNA) by an RNase III-like enzyme, Dicer,[5] into small interfering RNAs (siRNAs).[6] The resulting 21 to 23 nt siRNAs are then incorporated into the RNA-induced silencing complex (RISC),[7] which directs siRNAs to their cognate mRNAs for destruction.[3]

0–8493–2141–7/05/$0.00 + $.50

In 2001, Elbashir et al. found that gene-specific suppression in mammalian cells could be achieved by transient transfection of chemically synthesized siRNAs 21 nt in length.[8] These siRNAs are long enough to induce gene-specific suppression but too short to activate the host interferon response. Alternatively, siRNAs can be derived from dsRNAs containing a stem-loop hairpin structure. These short hairpin RNAs (shRNAs) can be expressed *in vivo* using plasmid DNA vector-based strategies driven by RNA polymerase III promoters (Pol III).[9–14] The shRNAs are subsequently processed into siRNAs by Dicer and incorporated into RISC complex in the cytoplasm where mRNA degradation occurs. These studies indicated that it was feasible to produce si/shRNA in mammalian cells. However, silencing was transient and transfection of primary cells was relatively inefficient. To achieve higher levels of transduction in primary cells and stable long-term expression of siRNAs, viral vectors have been adapted for use as delivery vehicles for siRNAs.

8.2 VIRAL VECTORS FOR SHRNA DELIVERY

Long-term gene suppression is essential to determining gene function and requires an efficacious method for creating continuous cell lines, transgenic animals, and therapeutic strategies in which a target gene is stably silenced. Several groups have successfully utilized viral vector systems that can mediate long-term gene-specific silencing in a wide variety of cells and animals, making RNAi more feasible as a therapeutic agent and a reverse genetic tool in animal models. We will outline the pros and cons of the viral systems designed for siRNA-based gene therapy.

Adenoviral vectors expressing shRNAs have been used successfully to reduce levels of their target proteins *in vitro* and *in vivo*.[15,16] This vector system promotes highly efficient delivery of shRNAs into dividing and nondividing cells in a wide variety of tissues.[17] However, expression of shRNAs is transient since adenovirus-based vectors are maintained as episomes and are diluted out as cells divide. Low-level expression of adenoviral proteins and preexisting immunity to adenoviral proteins also result in a diminished population of cells expressing shRNA and limit their use *in vivo*.

The adeno-associated viral (AAV) vector system is an attractive choice,[18,19] as the vectors are capable of infecting both dividing and nondividing cells, do not cause disease in humans, and do not trigger an immune response. The vector can persist for a long time in infected cells in episomal form and occasionally may integrate into the host genome, enabling long-term gene expression.[17] Although AAV vectors can transduce muscle and brain cells efficiently, transduction of hematopoietic stem cells is poor or controversial, limiting the usefulness of the vector.[20]

Since oncoretroviral vectors integrate into the host genome and establish stable gene transduction, they have been utilized for gene transfer in numerous experimental settings and in human gene therapy clinical trials.[21] Several groups have developed oncoretroviral vectors expressing an shRNA.[9,22–24] The main drawback of oncoretroviruses is their inability to transduce nondividing cells, which severely restricts their potential targets and limits their usefulness in transduction of primary cells *in vivo*.[25]

Furthermore, transcriptional inactivation of integrated transgene expression *in vivo* is frequently observed.[26] For therapeutic applications, oncoretroviral vectors may have a serious drawback due to the potential activation of cellular oncogenes by insertional mutagenesis, resulting in malignancies.

Over the last several years, lentiviral vector–mediated gene transfer has been applied with great success in many cell types, including primary cells and tissues.[27] Lentiviral vectors are capable of transducing nondividing cells because of nuclear localization signals within the vectors.[28,29] Lentiviral vectors have been generated from many different lentiviruses; among these, HIV-1 has been extensively utilized for the development of efficient and safe vectors.[30] Transduction efficiency has been improved by the incorporation of the HIV-1 central polypurine tract (cppt) element, resulting in more efficient nuclear translocation of the vector preintegration.[31] The woodchuck hepatitis virus posttranscriptional regulatory element (WPRE) also serves to enhance posttranscriptional transgene expression.[32] To ensure the safety of the vector, the gag/pol and *rev* genes, which are essential for HIV-1 replication, are expressed from two different plasmid DNAs.[30] Deletion of viral enhancers and promoter elements present in the LTRs of self-inactivating (SIN) vectors further improved the safety of this lentiviral vector system.[33] To enable the transduction of a variety of cell types, lentiviral vectors can be pseudotyped with many different envelopes and cell surface molecules. Among them, vesicular stomatitis virus protein G (VSV-G) has the highest and broadest host tropism.[34] Direct injection of lentiviral vectors in rodents resulted in sustained transgene expression in multiple organs, including the brain, liver, muscle, kidney, and retina.[29,35] Lentiviral vectors are capable of transducing hematopoietic stem/progenitor cells without cytokine stimulation and are effective in expressing transgenes in multiple differentiated lineages of hematopoietic cells in a number of animal models.[36–38] Lentiviral vectors are also capable of infecting macrophages and dendritic cells.[38,39] Production of transgenic mice utilizing lentiviral vectors is far more efficient when compared to the conventional pronuclear-injection/homologous recombination technique.[40,41] Unlike oncoretroviral vectors, lentiviral vectors are highly resistant to transgene inactivation and permit stable nonvariegated expression in transgenic mice.[40] Thus, lentiviral vectors should have a broad range of applications in delivering genes for gene therapeutic regimes. We, and others, have developed lentiviral vectors for delivering si/shRNAs, and have successfully silenced various genes, including a CCR5 chemokine receptor and HIV-1 regulatory genes as potential antiviral targets.[38,42–51] The ability to target a critical gene implicated in cancer development, p53, also demonstrates its potential as an anticancer therapeutic agent.

8.3 LENTIVIRAL VECTORS FOR SHRNA DELIVERY

8.3.1 DESIGN AND CONSTRUCTION OF LENTIVIRAL VECTORS FOR SHRNA DELIVERY AND EXPRESSION

Expression of si/shRNAs from lentiviral vectors requires the identification of an effective siRNA target sequence and an efficient RNA polymerase III promoter. The requirements of an effective shRNA target sequence are currently under investigation[52] and discussed in Chapter 2; we have successfully utilized one published by Elbashir's group.[53] Our experience, and that of others, indicates that an siRNA targeting a particular mRNA sequence is successful at reducing expression of the target about 30% of the time. Therefore, we suggest testing at least a minimum of five siRNAs to identify a highly effective siRNA. We perform BLAST searches to exclude siRNA target sequences that overlap significantly with other mRNA sequences.

To construct a pol III promoter shRNA expression unit, annealed dsDNA oligonucleotides containing a 19-nt siRNA target sense stem, a 9-nt loop, a 19-nt antisense stem, and a termination signal of 5 thymidine residues are inserted downstream of the pol III promoter, described in detail in our published studies[42] and in Chapter 7. At present, U6 or H1 pol III promoters are used for shRNA lentiviral vectors. In addition to the U6 and H1 promoters, a CMV pol II and an tRNA pol III promoter have been utilized for driving shRNA expression in other systems.[15,43,54] It would be interesting to compare these promoters for relative strength in the expression of shRNA from a lentiviral vector and the subcellular localization of their transcripts.

To express an si/shRNA from a lentiviral vector, we, and others, have inserted a pol III shRNA expression unit between the 5' and 3' long terminal repeats (LTRs) in 5' to 3' orientation.[38,42,44–47] Li et al. compared the efficiency of such an shRNA expression unit relative to an siRNA expression unit in which the sense and antisense strands of an siRNA are expressed from separate U6 pol III promoters in a lentiviral vector.[46] Although an siRNA effect was observed in both instances, they found that expression of an shRNA from a U6 pol III promoter was more effective in HIV-1 inhibition.

The other place for the insertion of a pol III promoter shRNA expression unit is in the U3 region of the 3' LTR.[48–50,60] Following reverse transcription and integration of the shRNA vector, the expression unit is duplicated into both the 5'and 3'LTRs. In both vector designs, generation of shRNA was successful and resulted in >90% reduction in target gene expression.

Inserting a pol III-driven shRNA expression unit in reverse orientation was also effective for specific gene reduction.[49,61] We have examined whether the orientation of the siRNA sequence has an effect on the level of shRNA expression within our vector. Our results indicate that the forward orientation have a stronger RNAi effect (Qin et al., unpublished observations). Thus, our current lentiviral vector has a forward orientation.

We have generated a dual luciferase reporter shRNA lentiviral vector system in which firefly luciferase is targeted by the shRNA, while *Renilla* luciferase serves as an internal control.[56] The assay provides rapid, sensitive, and quantitative controls for the development of new shRNA vectors.

8.3.1.1 Effects on Viral Titer

Several groups have reported a 3 to 8-fold reduction in titer from their lentiviral vector expressing si/shRNA, raising the possibility that shRNAs expressed from transfected lentiviral plasmid DNA during viral production target their own vector RNA genome.[46,61]To inhibit conversion of shRNAs into siRNAs in vector-producing cells, we co-transfected a plasmid expressing an shRNA against human Dicer during virus production. We found the co-transfection of shRNA against Dicer returned vector titers close to the original levels (An et al., unpublished observations).

We have found that a higher multiplicity of infection would increase the RNAi effect in transduced cells, which appears to correlate with the elevated level of siRNA production (An et al., unpublished observations).

8.3.1.2 Production of High-Titer Lentiviral Vectors

Our current protocol to produce high-titer lentiviral vectors utilizes an efficient and cost-effective calcium phosphate–mediated (CaP) transfection into 293T cells. We found that high-titer vector production can be greatly influenced by the passage number of 293T cells and differences between lots of FCS and 2 × HBS reagents. Therefore, it is important to optimize the transient transfection condition for vector production (see Protocols).

Pseudotyping lentiviral vectors with VSV-G protein enables us to concentrate vectors by ultracentrifugation to 109/ml. We titrate our concentrated viruses using 293T cells based on GFP expression in serially diluted transduced 293T cells.

In our experience, we can achieve close to 100% transduction efficiency of VSV-G pseudotyped lentiviral vector in most cell lines and 50 to 80% efficiency in PHA-activated human primary lymphocytes (PBLs) (see Protocols).[42] Based on previous reports, lentiviral vector delivery of shRNA can inhibit gene expression by over 90% in a number of different target genes, as measured by northern, RT-PCR, western, or flow cytometric analysis of target gene expression.

8.3.2 Applications

8.3.2.1 Conditional Suppression of Cellular Genes

In many situations, for example, identifying the function of genes during development or in therapeutic applications, the controlled suppression of a cellular gene is desirable. Wiznerowicz et al. developed a lentiviral vector in which shRNA expression can be induced from a modified H1 promoter following doxycycline treatment.[60] p53 and lamin gene expression were successfully reduced in a controlled fashion and, furthermore, the degree of reduction could be regulated by the dose of the drug.

8.3.2.2 Generation of Transgenic Knockout Mouse Models

The conventional technique for generating knockout mice requires microinjection of DNA into ES cells and screening for homologous recombination.[63] This method is labor intensive, time consuming, and, in general, limited to the mouse model. The application of lentiviral vectors to ES cells or single fertilized egg transduction provided a rapid and efficient protocol for generating transgenic mice and rats.[40,41]

By utilizing a lentiviral vector expressing shRNA, it is now feasible to generate a gene knockout phenotype via lentiviral vector infection of embryonic stem cells and embryos. We, and others, have generated several knockouts from transduced fertilized eggs of reporter lines, including lacZ knockouts (Qin et al., manuscript in preparation) and EGFP knockouts.[49] Rubinson et al. utilized a lentiviral vector expressing shRNA against CD8 for the transduction of mouse ES cells or single fertilized eggs and showed reduction of CD8 expression in thymocytes and in splenocytes.[38] In a different experiment in which the p53 gene was targeted, reduction of p53 expression was observed in the liver and brain.[38] Processed 21 to 23 nucleotide siRNAs were detected in multiple organs, including the brain, heart, kidney, and testes. The gene reduction was maintained in nine-week-old adult mice. Unlike conventional knockout mice, shRNA-mediated gene suppression is not a complete knockout; nevertheless, this technology has great potential for generating disease models in larger animals.

8.3.2.3 Antiviral Applications and Gene Therapy

Gene therapy is an attractive strategy for treating genetic disorders, cancer, and infectious diseases. Since the idea of intracellular immunization was proposed,[64] numerous antiviral genes have been created and have exhibited some degree of potency *in vitro*; however, the effects were modest in human clinical trials.[65] One of the reasons for the failure is the low effectiveness of antiviral genes *in vivo*. Another limitation is the lack of an effective gene delivery system. Because most target cells for human gene therapy are nondividing cells, lentiviral vectors that infect nondividing cells are the vectors of choice. Thus, the lentiviral vector–mediated transduction and expression of shRNAs are an attractive therapeutic strategy to genetically suppress viral replication and treat other diseases. In fact, we, and other investigators, have recently developed successful model therapeutic intervention strategies for the suppression of an HIV-1 co-receptor and HIV-1 regulatory genes.

For the treatment of HIV-1-related disease, we, and others, have chosen to inhibit a chemokine receptor, CCR5, one of the major co-receptors for primary HIV-1 infection in primary T cells and macrophages.[42,49] Individuals who are homozygous for the CCR5Δ32 mutation lack CCR5 expression on cell surfaces and are resistant to CCR5-tropic HIV-1 infection; importantly, the CCR5Δ32 mutation does not appear to cause any adverse effects, even in homozygous carriers of the mutated gene. To test our hypothesis, we transduced PHA/IL2-activated PBLs with VSV-G-pseudotyped lentiviral vector bearing an shRNA against CCR5. CCR5 expression, but not CXCR4 or CD4 expression, was reduced 8- to 10-fold in transduced PBLs during 8 to 10 days in culture.

The transduced PBLs were then challenged with a CCR5-tropic HIV-1 virus expressing murine heat stable antigen (HSA). Using HSA positivity as an indicator of productive HIV-1 infection, we found that CCR5-tropic HIV-1 infection was reduced to 3- to 7-fold. We also observed 3-fold reduction of p24 production in the culture supernatant. In contrast, CXCR4-tropic HIV-1 infection was not significantly affected, again confirming the specificity of shRNA against CCR5. Lee et al. also demonstrated lentiviral vector–mediated transduction of CCR5 shRNA into primary macrophages and inhibition of CCR5 tropic HIV-1 reporter virus.[49]

Another strategy to inhibit HIV-1 replication is to target viral genes essential for replication. Li et al. achieved HIV-1 inhibition by targeting HIV-1 *tat* and *rev* genes using lentiviral vectors expressing an shRNA in stimulated PBMC *in vitro*.[46] Transduction of PBMC with their si/shRNA *tat/rev* lentiviral vector resulted in 15 to 20% GFP positive cells. After enrichment to achieve 100% transduction, a potent 3-log HIV-1 inhibition was observed upon HIV-1 challenge in their system.

Hematopoietic stem cells are ideal targets for introducing therapeutic genes for AIDS and other hematopoietic diseases. Transduction of such cells with anti-HIV shRNAs would result in the reconstitution of a hematopoietic system that, in theory, would be protected from the effects of HIV-1. Banerjea et al. demonstrated the successful transduction of cytokine-stimulated fetal liver cells derived from CD34+ hematopoietic stem/progenitor cells with a lentiviral vector expressing siRNA against HIV-1 *tat* and *rev* and protection from HIV-1 infection in differentiated transduced cells.[47] Differentiation of vector-transduced CD34+ cells into macrophages indicated that 70% of the macrophages expressed GFP and were resistant to HIV-1 infection for 14 days *in vitro*. Transduced CD34+ cells were also injected into irradiated SCID-hu thy/liv implants to examine stem cell differentiation into normal thymocytes *in vivo*. After 60 days, in one thy/liv implant 53% of differentiated thymocytes maintained GFP expression, and thymocyte differentiation seemed to occur normally. GFP+ thymocytes were sorted for GFP+ and challenged by HIV-1 *in vitro*; these cells were highly resistant to HIV-1 infection (3-log HIV-1 inhibition). These results demonstrate that lentiviral vector–mediated siRNA expression could remain stable during differentiation of human thymocytes and macrophages from hematopoietic progenitor cells.

8.3.2.4 Anticipated Problems and Possible Adverse Effects

8.3.2.4.1 HIV-1 Escape Mutants from siRNA Strategies

Since RNAi does not result in the complete silencing of target gene expression, the emergence of a mutant resistant to siRNA is a serious concern when shRNAs are used to target HIV-1 genes. Boden et al. recently reported that HIV-1 acquired a mutation in the *tat* gene and became resistant to their shRNA in an *in vitro* culture system.[19] In H9 cells, AAV vector delivery of shRNA against *tat* inhibited HIV-1 replication by 95% until day 14 of culture. However, HIV-1 replication gradually increased from day 21, and at day 25 they found a single nucleotide mutation of A to T within the *tat* target sequence. These results highlight the importance of selecting target sequences in HIV-1 genome that cannot be varied without loss of viral potency. Delivering multiple siRNAs against several targets or combining siRNA therapy with existing antiviral agents may help alleviate this type of resistance.

8.3.2.4.2 Interferon Response in shRNA-Transduced Cells

Early reports suggested that dsRNAs shorter than 29 bp were not capable of inducing an interferon response in mammalian cells.[66] However, more recent reports have indicated that even transfected 21-nt synthetic siRNAs, as well as shRNAs delivered by lentiviral transduction, could result in induction of interferon responses.[67,68] Using microarray analysis, Bridge et al. showed induction of many interferon target genes in human lung fibroblasts after transduction of one of their lentiviral vectors expressing an shRNA against MORFAL1.[68] Induction of 2'5'-oligoadenylate synthetase (OAS1) was specific to the shRNA target sequence, as one of the four vectors expressing an shRNA against MORFAL1 induced OAS1 50-fold while another vector expressing an shRNA against MORFAL2 did not. They also compared OAS1 induction by lentiviral vectors bearing either an H1 or a U6 promoter-driven shRNA and found that the U6 promoter caused a higher frequency of OAS1 induction than H1, presumably due to higher expression of shRNA by the U6 promoter. The OAS1 induction was dose dependent; increasing the lentiviral moi also increased OAS1 induction. The induction of IFN seems to be only a partial response, as some IFN-responsive genes were not affected. It will be important to examine whether the partial induction of the INF response is deleterious to normal cell function.

8.3.2.4.3 Insertional Mutagenesis and Safety Concerns

Serious concerns over the safety of retro/lentiviral vectors recently emerged after two patients developed a leukemia-like disease during an X-linked SCID gene therapy trial in France, apparently due to the upregulation of LMO-2 genes by oncoretroviral vector insertional mutagenesis.[69] Lentiviral vectors tend to integrate more frequently into active genes,[70] while oncoretroviral vectors favor integration near transcriptional start sites.[71] Although no report has directly linked lentiviral insertional mutagenesis to a tumor malignancy, the safety of lentiviral vector-mediated gene transduction needs to be rigorously examined in appropriate animal models. We have used lentiviral vectors to transduce hematopoietic stem/progenitor cells in a rhesus macaque reconstitution experiment; five years after transduction, no adverse side effects have been observed.[37] Further tests for the efficacy and longevity of RNAi using the lentiviral vector system in many different animal models will ensure the usefulness and safety of lentiviral vectors.

8.4 CONCLUDING REMARKS

Since the discovery of RNA interference, siRNA delivery methods have rapidly progressed beyond the transfection of synthetic siRNA. Endogenous expression of si/shRNA, driven by RNA polymerase III, can now be achieved through transient transfection of plasmid DNA and viral vector systems. Lentiviral vector–mediated RNAi has been successfully applied to the efficient and stable transduction of shRNA in nondividing cells and primary cells, generation of transgenic mice with knockout phenotypes, and therapeutic applications. Also, inducible shRNA vectors are now available.

Beyond the current vector design, lentiviral vectors hold great potential for further development of shRNA delivery techniques. For example, multiple shRNAs may be stably expressed from a single vector. For genome-wide screening, an shRNA library could be inserted into the vector. Some dominant or codominant gene could be reduced by suppressing mutant gene expression through RNAi while expressing the wild type gene from an internal pol II promoter. This method would be useful for diseases like sickle cell anemia and Marfan syndrome.

Because RNAi potently suppresses specific gene expression, there is strong hope for applying the technology to human gene therapy. The steps to a successful gene therapeutic approach are: (1) identification of an effective therapeutic gene, (2) identification of an effective delivery vehicle, and (3) use of effective animal-model systems to test the efficacy and the safety of the gene therapeutic agent *in vivo* before its use in humans. siRNA provides an attractive alternative to other gene therapeutic agents due to its small size, higher success ratio, and ease of manipulation. However, current gene therapy vectors still need to overcome transduction efficiency difficulties in animal models. For example, p53 and BCR/Abl are excellent targets for siRNA therapy, but the delivery of shRNA into tumor cells *in vivo* is a major hurdle to successful cancer gene therapy.

Although the combination of RNAi and lentiviral vectors has opened new windows of opportunity in many areas, global induction of interferon-induced genes by certain siRNAs expressed from lentiviral vectors may hinder the application of the technology for biomedical research and human gene therapy. The induction seems to be partial, but further studies need to examine whether this induction is problematic for normal cell function. Likewise, it is uncertain whether insertional mutations caused by lentiviral integration will cause malignancies. Despite these issues, lentiviral vectors will serve as a potent delivery system for RNAi in mammalian cells for genetic studies, developmental biology, the generation of animal models, and therapeutic applications.

PROTOCOLS

Protocol 8.1 Lentiviral Vector Production by Calcium Phosphate Precipitation for DNA Transfection into 293T Cells

Materials

- 2 × HEPES buffered saline (HBS) (100 ml ddH_2O, 1 g HEPES (acid, Sigma H3375), 1.6 g NaCl, 0.72 ml Na_2HPO_4 (0.25 M), 1 ml KCl (1 M)); roughly adjust pH to near 7.12 (crucial) by adding 5 M NaOH; once pH is near 7.12, use 1 M NaOH for fine pH adjustment (crucial to adjust at pH 7.12); bring volume up to 100 ml; filter solution through a 0.2-µm pore size filter and split into 50-ml aliquots; store at 4°C; check efficiency of new 2 × HBS by transfecting a CMV promoter–driven EGFP expression plasmid; if the transfection efficiency is over 90% in 293T cells, the 2 × HBS is good for making vectors
- 2 M $CaCl_2$ tissue culture grade (Sigma C7902); filter and store in the dark at room temperature
- 10 mM chloroquine (filter and store in the dark at –20°C); prepare by mixing 51.6 mg chloroquine (Sigma C6628) in 10 ml ddH_2O
- Appropriate cell line, e.g., 293T cells

- Iscov's modified Dulbecco's medium with 10% (v/v) fetal calf serum, 100 U/ml of penicillin and 100 μg/ml of streptomycin (Iscov10FCS)

Experimental Procedure

We start a culture of 293T cells from frozen stocks (1×10^7 per vial) to avoid poor transfection efficiency caused by prolonged culture of the cells.

1. Inoculate the cell stock in a T175 flask with 25 ml of Iscov10FCS.
2. After 2 days, split the cells 1 to 7 in a new T175 flask. Culture the cells another 2 days.
3. Split 1.5×10^7 293T cells into T175 flasks with 25 ml of Iscov10FCS on the day before transfection.
4. Start early in the morning. When cells reach 70% confluency (day 1), perform transfection as follows. Note that the condition of the cells is very important for optimal transfection. They should be evenly distributed and growing exponentially.
5. Feed the cells with 25 ml fresh warm medium.
6. Add DNA to 980 μl ddH_2O in a 50-ml tube. A total of 42.5 μg (47.5 μg with shRNA Dicer) DNA should be used in one T175 flask. The amounts of each plasmid are: 5 μg VSV-G, 12.5 μg gag/pol packaging plasmid, 5 μg *rev* expression vector, and 20 μg lentiviral vector. In case the vector titer is reduced, compare to the original vector; add 5 μg of plasmid DNA expressing shRNA against Dicer.
7. Add 133 μl of 2 M $CaCl_2$ to the DNA solution, mix well, and incubate on ice for 5 min.
8. Add 1110 μl of 2 × HBS solution drop by drop while simultaneously mixing the solution.
9. Mix again, and incubate on ice for 20 min.
10. Add the mixed precipitates into the culture drop by drop, and mix gently.
11. Add 100 μl of Chloroquine (10 mM) to the culture.
12. Incubate the culture for 6 h at 37°C, 5% CO_2.
13. After the 6-h incubation, change media with 35 ml of fresh medium. A fine precipitate is visible under microscope.
14. On the morning of day 3 (36 h after removing DNA-CaP precipitation), harvest the culture supernatant and filter it through a 0.45-μm pore size filter.
15. Centrifuge vector supernatant at 50,000 × g (17,000 rpm in a Beckman L3-50 ultracentrifuge with an SW28 rotor) at 4°C for 90 min.
16. Remove supernatant by decantation and suck out residual supernatant with a pipetteman. Add 120 μl of PBS.
17. Leave the pellet overnight at 4°C. Seal tubes with Parafilm.
18. On the next day, resuspend the pellet and aliquot 50 μl into tubes. Store aliquots at –70°C. The vector supernatant can be concentrated and stored at –70°C without losing infectivity.

PROTOCOL 8.2 TRANSDUCTION OF PHA/IL2-STIMULATED PRIMARY PBLs

Materials

- Peripheral blood (healthy volunteers)
- Histopaque (Sigma)
- PHA (2.5 μg/ml, Murex Biotech, UK)
- RPMI with 20% (v/v) FCS, 100 U/ml of penicillin, and 100 μg/ml of streptomycin
- RPMI 20% (v/v) FCS with IL2 (20 U/ml, Roche Molecular Biochemicals)
- Polybrene (Sigma)

Experimental Procedures

We have established a protocol for efficient lentiviral vector–mediated shRNA delivery in PHA/IL2-stimulated human peripheral blood–derived lymphocytes (PBLs). By using our optimized protocol, we achieved more than 50 to 80% transduction in PBLs.

1. Isolate PBLs from leukopacks by Histopaque according to manufacturer's instructions.
2. Stimulate PBLs (2×10^6 cells/ml) with PHA (2.5 μg/ml, Murex Biotech, UK) for two days in RPMI 20% (v/v) FCS with 100 U/ml of penicillin and 100 μg/ml of streptomycin.
3. Plate the cells (4×10^5) in 100 μl of RPMI 20% FCS with IL2 in a 96-well plate for 3 h prior to lentiviral vector transduction.
4. After 3 h, add concentrated vector (usually 8 to 10 μl) at an moi of 5 to the PBL culture in the presence of polybrene (8 μg/ml).
5. After 2 h of infection, carefully remove the vector supernatant and add 250 μl fresh RPMI 20% (v/v) FCS with IL2 to the PBL culture.
6. Twelve hours postinfection, transfer the cells to a 24-well plate with 1.5 ml RPMI, 20% (v/v) FCS and IL2.

Technical note. Since the level of transgene expression via lentiviral vector becomes stable after three days posttransduction, we usually analyze silencing of the target gene from three days after transduction.

ACKNOWLEDGMENTS

We thank Elizabeth Withers-Ward, Betty Poon, and Sam K. P. Kung for critically reading the manuscript.

REFERENCES

1. Kim, V.N., RNA interference in functional genomics and medicine, *J. Korean. Med. Sci.*, 18, 309, 2003.
2. Sharp, P.A., RNA interference—2001, *Genes Dev.*, 15, 485, 2001.

3. Hannon, G.J., RNA interference, *Nature*, 418, 244, 2002.
4. Hutvagner, G. and Zamore, P.D., RNAi: nature abhors a double-strand, *Curr. Opin. Genet. Dev.*, 12, 225, 2002.
5. Bernstein, E., Caudy, A.A., Hammond, S.M., and Hannon, G.J., Role for a bidentate ribonuclease in the initiation step of RNA interference, *Nature*, 409, 363, 2001.
6. Zamore, P.D., Tuschl, T., Sharp, P.A., and Bartel, D.P., RNAi: double-stranded RNA directs the ATP-dependent cleavage of mRNA at 21 to 23 nucleotide intervals, *Cell*, 101, 25, 2000.
7. Hammond, S.M., Bernstein, E., Beach, D., and Hannon, G.J., An RNA-directed nuclease mediates post-transcriptional gene silencing in Drosophila cells, *Nature*, 404, 293, 2000.
8. Elbashir, S.M., Harborth, J., Lendeckel, W., Yalcin, A., Weber, K., and Tuschl, T., Duplexes of 21-nucleotide RNAs mediate RNA interference in cultured mammalian cells, *Nature*, 411, 494, 2001.
9. Brummelkamp, T.R., Bernards, R., and Agami, R., Stable suppression of tumorigenicity by virus-mediated RNA interference, *Cancer Cell*, 2, 243, 2002.
10. Paddison, P.J., Caudy, A.A., Bernstein, E., Hannon, G.J., and Conklin, D.S., Short hairpin RNAs (shRNAs) induce sequence-specific silencing in mammalian cells, *Genes Dev.*, 16, 948, 2002.
11. Paul, C.P., Good, P.D., Winer, I., and Engelke, D.R., Effective expression of small interfering RNA in human cells, *Nat. Biotechnol.*, 20, 505, 2002.
12. Sui, G., Soohoo, C., el Affar, B., Gay, F., Shi, Y., and Forrester, W.C., A DNA vector-based RNAi technology to suppress gene expression in mammalian cells, *Proc. Natl. Acad. Sci. U. S. A.*, 99, 5515, 2002.
13. Lee, N.S., Dohjima, T., Bauer, G., Li, H., Li, M.J., Ehsani, A., Salvaterra, P., and Rossi, J., Expression of small interfering RNAs targeted against HIV-1 rev transcripts in human cells, *Nat. Biotechnol.*, 20, 500, 2002.
14. Miyagishi, M. and Taira, K., U6 promoter-driven siRNAs with four uridine 3' overhangs efficiently suppress targeted gene expression in mammalian cells, *Nat. Biotechnol.*, 20, 497, 2002.
15. Xia, H., Mao, Q., Paulson, H.L., and Davidson, B.L., siRNA-mediated gene silencing *in vitro* and *in vivo*, *Nat. Biotechnol.*, 20, 1006, 2002.
16. Shen, C., Buck, A.K., Liu, X., Winkler, M., and Reske, S.N., Gene silencing by adenovirus-delivered siRNA, *FEBS Lett.*, 539, 111, 2003.
17. Lai, C. M., Lai, Y.K., and Rakoczy, P. E., Adenovirus and adeno-associated virus vectors, *DNA Cell Biol*, 21, 895, 2002.
18. Tomar, R.S., Matta, H., and Chaudhary, P.M., Use of adeno-associated viral vector for delivery of small interfering RNA, *Oncogene*, 22, 5712, 2003.
19. Boden, D., Pusch, O., Lee, F., Tucker, L., and Ramratnam, B., Human immunodeficiency virus type 1 escape from RNA interference, *J. Virol.*, 77, 11531, 2003.
20. Srivastava, A., Obstacles to human hematopoietic stem cell transduction by recombinant adeno-associated virus 2 vectors, *J. Cell Biochem. Suppl.*, 38, 39, 2002.
21. McTaggart, S. and Al-Rubeai, M., Retroviral vectors for human gene delivery, *Biotechnol. Adv.*, 20, 1, 2002.
22. Barton, G.M. and Medzhitov, R., Retroviral delivery of small interfering RNA into primary cells, *Proc. Natl. Acad. Sci. U.S.A.*, 99, 14943, 2002.

23. Devroe, E. and Silver, P.A., Retrovirus-delivered siRNA, *BMC Biotechnol.,* 2, 15, 2002.
24. Paddison, P.J., Caudy, A.A., and Hannon, G.J., Stable suppression of gene expression by RNAi in mammalian cells, *Proc. Natl. Acad. Sci. U.S.A.*, 99, 1443, 2002.
25. Miller, D.G., Adam, M.A., and Miller, A.D., Gene transfer by retrovirus vectors occurs only in cells that are actively replicating at the time of infection, *Mol. Cell Biol.*, 10, 4239, 1990.
26. Jahner, D., Stuhlmann, H., Stewart, C.L., Harbers, K., Lohler, J., Simon, I., and Jaenisch, R., De novo methylation and expression of retroviral genomes during mouse embryogenesis, *Nature*, 298, 623, 1982.
27. Hacein-Bey-Abina, S., von Kalle, C., Schmidt, M., Le Deist, F., Wulffraat, N., McIntyre, E., Radford, I., Villeval, J.-L., Fraser, C. C., Cavazzana-Calvo, M., and Fischer, A., A serious adverse event after successful gene therapy for X-linked severe combined immunodeficiency, *N. Engl. J. Med.*, 348, 255, 2003.
28. Naldini, L. and Verma, I.M., Lentiviral vectors, *Adv. Virus Res*, 55, 599, 2000.
29. Naldini, L., Blomer, U., Gallay, P., Ory, D., Mulligan, R., Gage, F.H., Verma, I.M., and Trono, D., In vivo gene delivery and stable transduction of nondividing cells by a lentiviral vector, *Science*, 272, 263, 1996.
30. Dull, T., Zufferey, R., Kelly, M., Mandel, R.J., Nguyen, M., Trono, D., and Naldini, L., A third-generation lentivirus vector with a conditional packaging system, *J. Virol.*, 72, 8463, 1998.
31. Zennou, V., Petit, C., Guetard, D., Nerhbass, U., Montagnier, L., and Charneau, P., HIV-1 genome nuclear import is mediated by a central DNA flap, *Cell*, 101, 173, 2000.
32. Zufferey, R., Donello, J.E., Trono, D., and Hope, T.J., Woodchuck hepatitis virus posttranscriptional regulatory element enhances expression of transgenes delivered by retroviral vectors, *J. Virol.*, 73, 2886, 1999.
33. Miyoshi, H., Blomer, U., Takahashi, M., Gage, F.H., and Verma, I.M., Development of a self-inactivating lentivirus vector, *J. Virol.*, 72, 8150, 1998.
34. Yee, J. K., Friedmann, T., and Burns, J.C., Generation of high-titer pseudotyped retroviral vectors with very broad host range, *Methods Cell Biol.*, 43 Pt. A, 99, 1994.
35. Kafri, T., Blomer, U., Peterson, D.A., Gage, F.H., and Verma, I.M., Sustained expression of genes delivered directly into liver and muscle by lentiviral vectors, *Nat. Genet.,* 17, 314, 1997.
36. Miyoshi, H., Smith, K.A., Mosier, D.E., Verma, I.M., and Torbett, B.E., Transduction of human CD34+ cells that mediate long-term engraftment of NOD/SCID mice by HIV vectors, *Science*, 283, 682, 1999.
37. An, D.S., Wersto, R.P., Agricola, B.A., Metzger, M.E., Lu, S., Amado, R.G., Chen, I.S., and Donahue, R.E., Marking and gene expression by a lentivirus vector in transplanted human and nonhuman primate CD34(+) cells, *J. Virol.*, 74, 1286, 2000.
38. Rubinson, D.A., Dillon, C.P., Kwiatkowski, A.V., Sievers, C., Yang, L., Kopinja, J., Rooney, D.L., Ihrig, M.M., McManus, M.T., Gertler, F.B., Scott, M.L., and Van Parijs, L., A lentivirus-based system to functionally silence genes in primary mammalian cells, stem cells and transgenic mice by RNA interference, *Nat. Genet.,* 33, 401, 2003.

39. Schroers, R., Sinha, I., Segall, H., Schmidt-Wolf, I.G., Rooney, C.M., Brenner, M.K., Sutton, R.E., and Chen, S.Y., Transduction of human PBMC-derived dendritic cells and macrophages by an HIV-1-based lentiviral vector system, *Mol. Ther.*, 1, 171, 2000.
40. Lois, C., Hong, E.J., Pease, S., Brown, E.J., and Baltimore, D., Germline transmission and tissue-specific expression of transgenes delivered by lentiviral vectors, *Science*, 295, 868, 2002.
41. Pfeifer, A., Ikawa, M., Dayn, Y., and Verma, I ., Transgenesis by lentiviral vectors: lack of gene silencing in mammalian embryonic stem cells and preimplantation embryos, *Proc. Natl. Acad. Sci. U.S.A.*, 99, 2140, 2002.
42. Qin, X.-F., An, D.S., Chen, I.S.Y., and Baltimore, D., Inhibiting HIV-1 infection in human T cells by lentiviral-mediated delivery of small interfering RNA against CCR5, *Proc. Nat. Acad. Sci. U.S.A.*, 100, 183, 2002.
43. Kawasaki, H. and Taira, K., Short hairpin type of dsRNAs that are controlled by tRNA(Val) promoter significantly induce RNAi-mediated gene silencing in the cytoplasm of human cells, *Nucleic. Acids. Res.*, 31, 700, 2003.
44. Rubinson, D.A., Dillon, C.P., Kwiatkowski, A.V., Sievers, C., Yang, L., Kopinja, J., Rooney, D.L., Ihrig, M.M., McManus, M.T., Gertler, F.B., Scott, M.L., and Van Parijs, L., A lentivirus-based system to functionally silence genes in primary mammalian cells, stem cells and transgenic mice by RNA interference, *Nat. Genet.*, 33, 401, 2003.
45. Stewart, S.A., Dykxhoorn, D.M., Palliser, D., Mizuno, H., Yu, E.Y., An, D.S., Sabatini, D.M., Chen, I.S., Hahn, W.C., Sharp, P.A., Weinberg, R.A., and Novina, C.D., Lentivirus-delivered stable gene silencing by RNAi in primary cells, *RNA*, 9, 493, 2003.
46. Li, M.J., Bauer, G., Michienzi, A., Yee, J.K., Lee, N.S., Kim, J., Li, S., Castanotto, D., Zaia, J., and Rossi, J.J., Inhibition of HIV-1 infection by lentiviral vectors expressing Pol III-promoted anti-HIV RNAs, *Mol. Ther.*, 8, 196, 2003.
47. Banerjea, A., Li, M.J., Bauer, G., Remling, L., Lee, N.S., Rossi, J., and Akkina, R., Inhibition of HIV-1 by lentiviral vector-transduced siRNAs in T lymphocytes differentiated in SCID-hu mice and CD34+ progenitor cell-derived macrophages, *Mol. Ther.*, 8, 62, 2003.
48. Tiscornia, G., Singer, O., Ikawa, M., and Verma, I.M., A general method for gene knockdown in mice by using lentiviral vectors expressing small interfering RNA, *Proc. Natl. Acad. Sci. U.S.A.*, 100, 1844, 2003.
49. Lee, M.T., Coburn, G.A., McClure, M.O., and Cullen, B.R., Inhibition of human immunodeficiency virus type 1 replication in primary macrophages by using Tat- or CCR5-specific small interfering RNAs expressed from a lentivirus vector, *J. Virol.*, 77, 11964, 2003.
50. Scherr, M., Battmer, K., Ganser, A., and Eder, M., Modulation of gene expression by lentiviral-mediated delivery of small interfering RNA, *Cell Cycle*, 2, 251, 2003.
51. Matta, H., Hozayev, B., Tomar, R., Chugh, P., and Chaudhary, P.M., Use of lentiviral vectors for delivery of small interfering RNA, *Cancer Biol. Ther.*, 2, 206, 2003.
52. Khvorova, A., Reynolds, A., and Jayasena, S.D., Functional siRNAs and miRNAs exhibit strand bias, *Cell*, 115, 209, 2003.

53. Elbashir, S.M., Lendeckel, W., and Tuschl, T., RNA interference is mediated by 21- and 22-nucleotide RNAs, *Genes Dev.*, 15, 188, 2001.
54. Kawasaki, H. and Taira, K., Short hairpin type of dsRNAs that are controlled by tRNA(Val) promoter significantly induce RNAi-mediated gene silencing in the cytoplasm of human cells, *Nucl. Acids Res.*, 31, 700, 2003.
55. Boden, D., Pusch, O., Lee, F., Tucker, L., Shank, P.R., and Ramratnam, B., Promoter choice affects the potency of HIV-1 specific RNA interference, *Nucl. Acids Res.*, 31, 5033, 2003.
56. An, D.S., Xie, Y., Mao, S.H., Morizono, K., Kung, S.K., and Chen, I.S., Efficient lentiviral vectors for short hairpin RNA delivery into human cells, *Hum. Gene Ther.*, 14, 1207, 2003.
57. Li, M.J., Bauer, G., Michienzi, A., Yee, J.K., Lee, N.S., Kim, J., Li, S., Castanotto, D., Zaia, J., and Rossi, J.J., Inhibition of HIV-1 infection by lentiviral vectors expressing Pol III-promoted anti-HIV RNAs, *Mol. Ther.*, 8, 196, 2003.
58. Banerjea, A., Li, M.J., Bauer, G., Remling, L., Lee, N.S., Rossi, J., and Akkina, R., Inhibition of HIV-1 by lentiviral vector-transduced siRNAs in T lymphocytes differentiated in SCID-hu mice and CD34+ progenitor cell-derived macrophages, *Mol. Ther.*, 8, 62, 2003.
59. Tiscornia, G., Singer, O., Ikawa, M., and Verma, I.M., A general method for gene knockdown in mice by using lentiviral vectors expressing small interfering RNA, *Proc. Natl. Acad. Sci. U.S.A.*, 100, 1844, 2003.
60. Wiznerowicz, M. and Trono, D., Conditional suppression of cellular genes: lentivirus vector-mediated drug-inducible RNA interference, *J. Virol.*, 77, 8957, 2003.
61. Abbas-Terki, T., Blanco-Bose, W., Deglon, N., Pralong, W., and Aebischer, P., Lentiviral-mediated RNA interference, *Hum. Gene Ther.*, 13, 2197, 2002.
62. An, D. S., Xie, Y., Mao, S.H., Morizono, K., Kung, S.K., and Chen, I.S., Efficient lentiviral vectors for short hairpin RNA delivery into human cells, *Hum. Gene Ther.*, 14, 1207, 2003.
63. Charreau, B., Tesson, L., Soulillou, J.P., Pourcel, C., and Anegon, I., Transgenesis in rats: technical aspects and models, *Transgenic Res.*, 5, 223, 1996.
64. Baltimore, D., Gene-Therapy–Intracellular Immunization, *Nature*, 335, 395, 1988.
65. Statham, S. and Morgan, R.A., Gene therapy clinical trials for HIV, *Curr. Opin. Mol. Ther.*, 1, 430, 1999.
66. Manche, L., Green, S.R., Schmedt, C., and Mathews, M.B., Interactions between double-stranded RNA regulators and the protein kinase DAI, *Mol. Cell Biol.*, 12, 5238, 1992.
67. Sledz, C.A., Holko, M., Veer, M.J.D., Silverman, R.H., and Williams, B.R.G., Activation of the interferon system by short-interfering RNAs, *Nat. Cell Biol.*, 5, 834, 2003.
68. Bridge, A.J., Pebernard, S., Ducraux, A., Nicoulaz, A.L., and Iggo, R., Induction of an interferon response by RNAi vectors in mammalian cells, *Nat. Genet.*, 34, 263, 2003.
69. Cavazzana-Calvo, M., Hacein-Bey, S., Basile, G.D.S., Gross, F., Yvon, E., Nusbaum, P., Selz, F.C.O., Hue, C., Certain, S.E.P., Casanova, J.-L., Bousso, P.,

Deist, F.C.O.L., and Fischer, A., Gene therapy of human severe combined immunodeficiency (SCID)-X1 disease, *Science*, 288, 669, 2000.

70. Schroder, A.R., Shinn, P., Chen, H., Berry, C., Ecker, J.R., and Bushman, F., HIV-1 integration in the human genome favors active genes and local hotspots, *Cell*, 110, 521, 2002.
71. Wu, X., Li, Y., Crise, B., and Burgess, S.M., Transcription start regions in the human genome are favored targets for MLV integration, *Science*, 300, 1749, 2003.

9 Exogenous siRNA Delivery: Protocols for Optimizing Delivery to Cells

Andrew J. Hollins, Stephen P. Fox, and Saghir Akhtar

CONTENTS

9.1 INTRODUCTION

Posttranscriptional gene silencing by RNA interference (RNAi) appears a promising, novel approach for the targeted inhibition of gene expression in cell culture and *in vivo*.[1–3] RNAi can be mediated by exogenous delivery of small interfering RNA (siRNA), short duplexes RNA about 21mers in size that usually house two nucleotide overhangs at each 3'-end terminus. Inside cells, these duplexes are recognized by the RNA-induced silencing complex (RISC) in the cytosol, whereby the antisense RNA strand is then guided by this complex to the complementary sequence in target mRNA. Provided the complementary sequence in the target mRNA is hybridization accessible, the RISC complex then, by mechanisms that are not fully understood, initiates its hydrolysis and thus prevents or "silences" gene expression. This approach has been investigated for potential use as a biological tool for understanding gene function, as a drug target validation strategy, and as a potential therapeutic agent for the treatment of genetic diseases.[1–3]

Although endogenous expression of double-stranded RNA or short-hairpin RNAs using viral vectors has been successfully used for gene silencing in cells and in animal models,[3] in this chapter we will focus on exogenous delivery systems for effective siRNA uptake into cells. In particular, we provide generic protocols for optimizing the uptake of synthetic siRNA using any delivery system, though we used cationic lipid formulations as a working example.

0-8493-2141-7/05/$0.00 + $.50

9.1.1 Exogenous Cellular Delivery of siRNA

Although no definitive studies on the mechanism of uptake of naked siRNAs have been published to date, it is likely that these duplexed RNA molecules are taken up by cells in a manner similar to that described for RNA enzymes (ribozymes)[4] and other related nucleic acid structures such as antisense oligonucleotides.[5–7] Indeed, like ribozymes and antisense oligonucleotides, naked siRNAs appear to be poorly transported into cells in culture, most likely by some form of endocytosis.[8] What little free nucleic acid that eventually enters cells largely remains sequestered within endosomal/lysosomal vesicles, where it may also undergo degradation by nucleases. Therefore, to achieve siRNA-mediated gene silencing in cells, a delivery system is required to improve cellular uptake and intracellular trafficking of siRNAs in cells. In the case of siRNAs that target mRNAs, delivery needs to be cytosolic rather than nuclear. Some of the obstacles of getting siRNA into cells to exert its intended biological activity are summarized in Figure 9.1.

Biological instability is the first barrier to consider when delivering RNA oligonucleotides to cells (see Figure 9.1). Unmodified deoxynucleotides and ribonucleotides sequences are rapidly degraded in biological fluids[9–11] by a combination of both endo- and exonucleases. Athough duplexed RNAs are expected to be more stable in the biological milieu than single-stranded counterparts, several chemical modifications have been attempted within either the sense or antisense strand of siRNA duplexes to further improve biological stability to levels that allow cellular gene-silencing activity.[12–14] The most tolerated and, thus, widely used chemical modification includes the inclusion of deoxynucleotides within the 3'-terminal nucleotide overhangs,[15] though 2'-O-Methyl[16,17] and other modifications such as 2'-deoxy-2'-fluorouridine, locked nucleic acid (LNA), and phosphorothioate have also been employed recently.[18]

9.1.2 siRNA Delivery Systems

As the cellular uptake of naked nucleic acids is generally inefficient through a combination of fluid-phase (pinocytosis), adsorptive, and receptor-mediated endocytosis,[19,20] improved cellular uptake of siRNA will require a suitable delivery system. Tables 9.1 and 9.2 summarize some of the delivery systems used in recent studies for the effective delivery of siRNAs both *in vitro* and *in vivo*. These include the use of physical transfection methods such as electroporation,[21,22] use of cationic peptides,[23] and polymeric dendrimers (e.g., Superfect).[24] However, the most widely used *in vitro* delivery systems for siRNA appear to be cationic liposome formulations such as Oligofectamine (see Tables 9.1 and 9.2).

Cationic lipids can be classified into several categories based on their structure, size, and charge.[38] Due to their high delivery efficiency, particular interest has been given to the positively charged cationic lipid or liposome formulations for siRNA transfections.[13,26,27,34] These reagents form unilamellar liposomes bearing positive charges that interact spontaneously with polyanionic siRNA to form lipoplexes. The

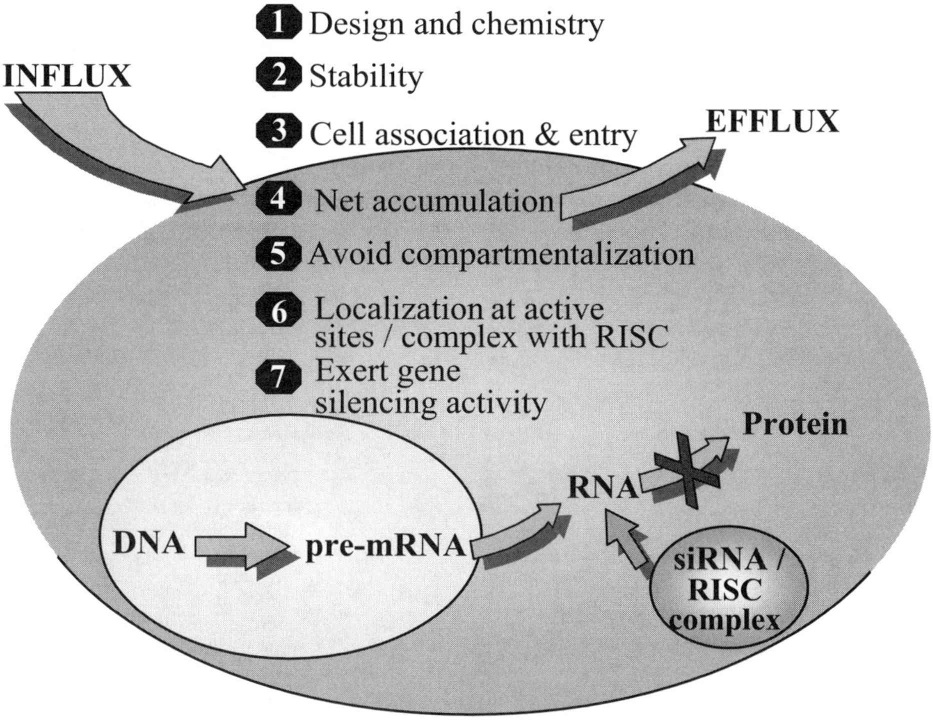

FIGURE 9.1 Issues of concern in the cellular delivery of siRNA. In order for an siRNA to be efficacious it has to: (1) be designed to heteroduplex in hybridization-accessible sites within the targeted mRNA and have the appropriate chemistry for effective hybridization; (2) confer sufficient extra- and intracellular biological stability; and (3) have efficient uptake, usually involving cell association and entry. Once inside the cell, it needs to (4) have net accumulation, where influx is greater than efflux. However, the cellular siRNA must (5) avoid errant compartmentalization in nontarget cellular organelles and remain bioavailable to (6) localize at sites of bioactivity in association with the RNA-induced silencing complex (RISC) within the cell cytoplasm, where it should (7) exert its gene-silencing activity in the intended specific manner. Adapted from Akhtar et al., 2000.[20]

lipid-siRNA lipoplex is thought to enter cells via adsorptive endocytosis and, by mechanisms not fully understood as yet, release nucleic acids out of the endosomal/lysosomal compartments with the net effect of yielding high uptake and intracellular delivery of siRNA (see Figure 9.2; also [27]).

By virtue of their positive charge, these systems readily bind nucleic acids to form complexes termed lipoplexes and have a high affinity for most cell membranes, which are negatively charged under physiological conditions. They usually gain entry into cells by markedly enhanced adsorptive endocytosis. To help facilitate the subsequent release from endosomal/lysosomal compartments, many commercially available lipoplex transfection agents (aka transfectins or cytofectins) contain a helper lipid such as dioelylphosphatidylethanolamine (DOPE). This is an inverted-cone-shaped

TABLE 9.1
Selected Example Studies Delivering siRNA to Cells in Cultures

Delivery System	Method of delivery	Target	Model	Reference
Cationic Lipids	Lipofectamine, Lipofectamine 2000	Human tissue factor (HTF), human protein serine kinase (PSKH1)	HeLa, Cos-1, HEK 293, human keratinocytes (HaCaT)	13
	Lipofectamine and propriety plus reagent	Green fluorescent protein (GFP), neomycin phospho-transferase (neo), chloramphenicol acetyl transferase (CAT)	HeLa, human embryonic kidney (HEK 293), primary mouse embryonic fibroblasts (MEF)	25
	Oligofectamine	Type 1 insulin-like growth factor (IGF1R)	Human ER negative breast cancer (MDA-MB-231), human myeloma (ME), human nonsmall cell lung carcinoma (A549), human androgen-resistant prostate cancer (DU145), human ovarian cancer (UC101)	26
	Lipofectin, Oligofectamine	Epidermal growth factor (EGFR)	Human epidemoid squamous cell carcinoma (A431)	27
Physical Method	Electroporation	Akt 1& 2	Adipocytes (3T3-L1) differentiated from fibroblasts	22
Cationic peptide	Peptide-mediated delivery (MPG)	GAPDH	Human fibroblasts (HS-68), Cos-7, HeLa	23

lipid thought to facilitate cytosolic release through the fusion to and/or the disruption of the endosomal membrane,[39] possibly by the formation of hexagonal lipid structures.[40] Studies have also shown that bound oligonacelotides can readily dissociate from the complexes and are in a bioavailable form within the cell.[41-43]

Although the exact mechanism of how these delivery systems function is unknown, it is quite clear that a significant optimization, in terms of charge ratio between cationic lipid and nucleic acid at given dose, is essential for effective delivery and activity. The effectiveness of lipoplexes is also dependent on the type and nature of the cationic lipid, the cell type, nucleic acid chemistry, length, and even the method of formation of the complexes. Thus, optimization of delivery with lipids should be performed each time a variable is altered.

TABLE 9.2
Selected Example Studies Delivering siRNA to Animal Models. Where *in vitro* studies were performed together with the *in vivo,* these have also been included for completeness ([a]*in vitro* studies included with *in vivo* siRNA articles; [b]*in vivo* study details).

Method of delivery	Target	Model	Reference
[a] Cytofectin GSV [b] Subcutaneous injection into tumor	Green fluorescent protein (GFP)	[a] HeLa-GFP cells [b] Xenografted mice	24
[a] Transit TKO transfection reagent [b] Sub-retinal injection with TKO reagent	[a] hVEGF5 [b] Human vascular epidermal growth factor (hVEGF5) and murine (mVegf)EFGP	[a] Human Embryonic Kidney 293 & HeLa cells [b]C57Bl/6 Jackson mice	28
i.v. tail-vein (hydrodynamic) and i.m. injections	Reporter plasmid *Photinus pyralis* luciferase (pGL3)	Female ddY Shizuoka mice, 4 weeks, ~20g	29
High-pressure tail-injection	[a] Luciferase reporter plasmid and human placental alkaline phosphatase (SEAP) [b] Enhanced green fluorescent protein (EGFP)	[a] Postnatal mice, ICR strain, Harlan 4 to 6 weeks [b] Mice C57BI/6-TgN (ACTcEGFP)10 sb 10 weeks	30
Hydrodynamic liver injection of naked siRNA	Luciferin (luc), and non-structural protein 5B, viral-polymerase-encoding (NS5B) region of hepatitis C virus (HCV)	Adult Jackson mice	31
Intra-arcuate nucleus bilateral injection	Agouti-related peptide (AGRP)	Jackson mice	32
Intravenous injections	Malaria parasite cysteine protease genes and *Berghepain* 1& 2	Female BALB/c mice, 4 to 6 weeks, 20 to 25g	33
Tail vein injection (hydrodynamic transfection method)	Apoptosis-regulating gene, Fas (Tnfrsf6)	Male Jackson BALB/c mice, 8 to 10 weeks, 20 to 25 g	34
[a] DOTAP transfection [b] Intravenous (EGFP) and intraperitoneal (TNFα) injections withDOTAP	[a] TNFα [b] Tumor necrosis factor-alpha (TNFα), EGFP	[a] Murine peritoneal macrophages (10^4 cells) [b] BALB/c mice	35
[a] Oligofectamine [b] Oligofectamine i.p. injections to HCT116 Xenograft	[a] β-Catenin 1 & 2, APC, NF-kB p65, HTLV Tax [b] β-Catenin	[a] Colon adenocarcinoma SW480 (APC mutant) HCT116 (β-Catenin mutant) [b]Female Charles River nude$^{nu/nu}$ mice, 4 to 6 weeks	36
Tail and hepatic portal vein injections	Caspase 8	Male BALB/c and C57BL/6JTgN (MtnlacZ) mice, 6 to 8 weeks	37

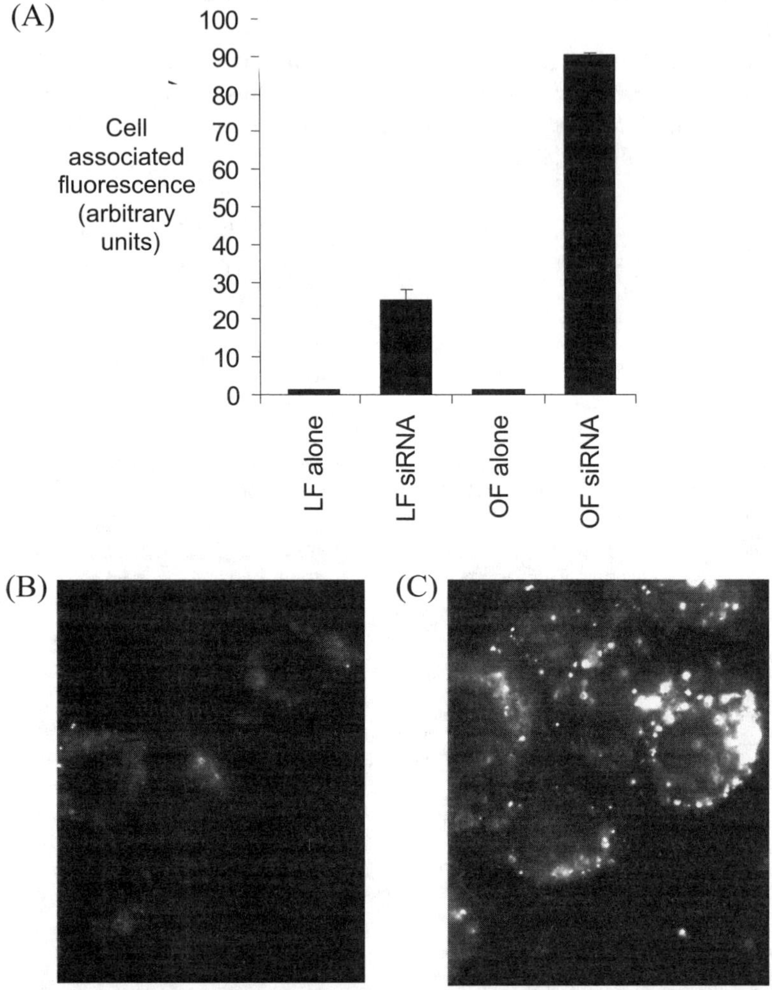

FIGURE 9.2 Typical results of fluorescent siRNA (f-siRNA) uptake complexed with Oligofectamine (OF) and Lipofectin (LF), assessed by both flow cytometry (A) and live cell fluorescent microscopy (B and C). **(A)** Flow cytometry histogram showing the fluorescence associated with A431 cells incubated with and without f-siRNA complexes. **(B)** Live cell fluorescent micrograph of A431 cells treated with Oligofectamine: f-siRNA complexes. **(C)** Live cell fluorescent micrograph of A431 cells treated with Lipofectin: f-siRNA complexes.

Recent studies have shown that only cationic lipoplex formulations optimized for these variables will significantly improve the intracellular bioavailability and activity of nucleic acids.[44,45] Although Table 9.2 shows that Oligofectamine complexes with siRNA are also effective when delivered via intraperitoneal injection, past experience with antisense oligonucleotides and ribozymes nucleic acids suggest that most cationic lipid delivery systems are either ineffective or too toxic when used *in vivo*. Indeed, despite recent efforts to identify serum stable cytofectins, the toxicity and serum sensitivity of nucleic–acid-lipoplex systems is unlikely to lead to their widespread use *in vivo*.[46] Indeed, cationic lipid-mediated cellular toxicity has been observed,[38,47-50] and more recently some cationic lipid formulations including Oligofectamine have been shown to elicit inadvertent gene expression in cells, which may impact on gene inhibition studies.[51] Although further studies are needed to fully understand the impact of gene expression changes to siRNA efficacy experiments, it is conceivable that induced nontarget gene expression changes may potentially mask or interfere with the desired or intended gene inhibition by siRNA.

For *in vivo* delivery, the most commonly used approach has been the use of high-volume (hydrodynamic) tail vein injections,[29,30,34,37] but this strategy is only suitable for animals and cannot be extended to clinical use. Success has also been reported with siRNA delivered locally to the liver,[31] brain,[32] and eye,[28] and local delivery of siRNA may represent a viable option for future clinical applications.

9.2 PROTOCOLS FOR OPTIMIZING siRNA DELIVERY TO CELLS IN CULTURE

In this chapter we will limit our discussions to a generic stepwise approach to optimizing siRNA delivery to cells in culture. This approach will be pertinent to optimizing siRNA activity with any delivery system including cationic lipids, polymers, and peptides.

9.2.1 The Optimization Strategy

The following steps should be followed for optimization of siRNA delivery to cells in culture.

Step 1: Determine a maximal nontoxic (MNT) or acceptable toxic (AT) concentration of the delivery system in the target cell culture system. This can be done by MTT assay or trypan blue exclusion assay (Protocols 9.1 and 9.2). Figure 9.3 shows a hypothetical toxicity curve for a given delivery system.

Step 2: At concentration up to the MNT or AT, optimize the ratios of delivery system to siRNA that gives maximal uptake. Ratios can be either based on relative number of positive to negative charges (charge ratio) or weight (weight ratio). In the case of some proprietary delivery preparations (e.g., Oligofectamine), only volume information is provided and, therefore, increasing volume per fixed siRNA concentration or charge can be used. Uptake in the target cell type is usually determined using fluorescently labeled siRNA or radio-labeled siRNA in the nucleic acid delivery system complex (Protocols 9.3 to 9.8). Figure 9.4 shows a hypothetical cellular uptake profile for increasing rations of an siRNA delivery system.

Step 3: Determine the cellular distribution of the optimal ratio of the siRNA delivery system in target cells by confocal or fluorescent microscopy (Protocol 9.9). Figure 9.5 shows the typical images of cellular distributions obtained by naked nucleic acid (siRNA) and complexes with the cationic lipid (Oligofectamine). It is also

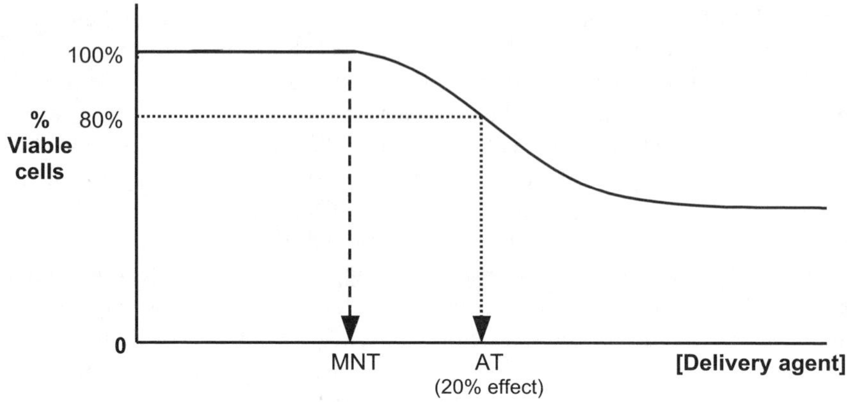

FIGURE 9.3 A hypothetical toxicity curve determination of maximal nontoxic (MNT) and acceptable toxic (AT) concentration of a delivery system in the target cell system. This can be achieved using methods set out in Protocol 9.1, MTT assay, or trypan blue exclusion assay.

recommended to include, as a control, a marker for fluid phase endocystosis such as FITC-dextran.

Step 4. Determine cellular activity in terms of protein and mRNA knockdown (Protocols 9.10 and 9.11). Figure 9.6 shows typical western blot and RT-PCR analyses of gene-silencing nucleic acids delivered by a cationic lipid.

N.B. The above optimization steps should ideally be repeated whenever a new variable is introduced into the siRNA (e.g., when the sequence, chemistry, or length is altered), the delivery system (e.g., when lipid type, supplier, or buffers are altered), or the cell culture conditions (e.g., when cell type, culture media, or pH are altered).

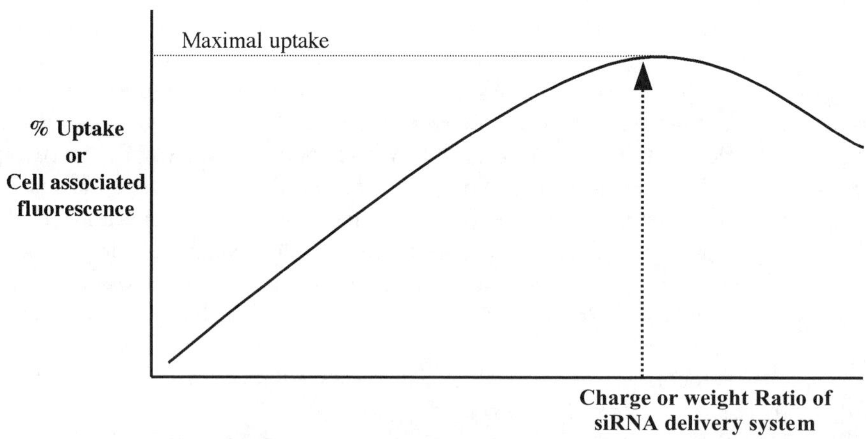

FIGURE 9.4 A hypothetical uptake profile for nucleic acids shown as percent uptake/cell associated fluorescence against concentration of delivery agent (nucleic acid concentration constant). *N.B.* It is normal to take the lowest concentration of the delivery agent that yields the maximal uptake.

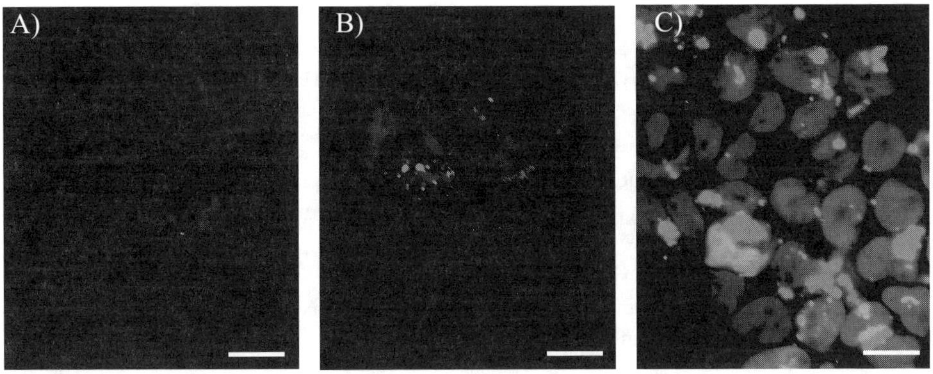

FIGURE 9.5 Typical effect of lipoplex delivery upon the uptake/cellular distribution of nucleic acids (also see Beale et al. 2003).[27] (A) control, (B) nucleic acid alone, (C) delivery system: nucleic acid complexes. White bars represent 10 μm.

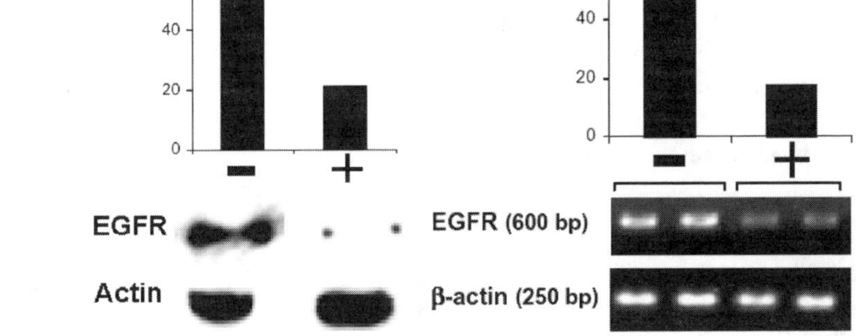

FIGURE 9.6 Typical effect of siRNA targeting the epidermal growth factor receptor (EGFR) on the expression of receptor protein (Western blot shown in (A), or mRNA levels shown in (B) when delivered as cationic lipoplexes.) (–) control for delivery system alone, (+) with lipoplex delivery system complexed with siRNA. Graphs plot the densitometry, showing intensity of each band expressed with respect to the housekeeping gene (in this example, β-actin).

PROTOCOLS

PROTOCOL 9.1. MTT CYTOTOXICITY ASSAY

Materials

- Anthos HTII plate reader (570 nm)
- MTT, (3-[4,5-dimethylthiazol-2-yl]-2,5-diphenyltetrazolium bromide thiazolyl-blue (Sigma-Aldrich)
- Dimethylsulphoxide (DMSO) (Sigma-Aldrich)
- Sorensen's glycine buffer (0.1 M Glycine, 0.1 M NaCl, pH 10.5) (components from Fisher Life Sciences)

Experimental Procedures

1. Seed 96-well microtiter plates with 10,000 cells per well and culture for 24 h prior to undertaking transfection protocols/treatments.
2. Following treatments, replace with fresh normal culture media and incubate for 20 h.
3. After this recovery period, add 50 µl of the MTT indicator dye (5 mg/ml) to each well and incubate with the cells for 4 h in the dark.
4. Carefully remove the medium and indicator dye, then lyse the cells with 200 µl DMSO.
5. After addition of (25 µl) Sorensen's glycine buffer, measure the absorbance at 570 nm using the plate reader.
6. Express the values as a percentage of the appropriate control.

PROTOCOL 9.2 TRYPAN BLUE CELL COUNTS

Materials

- Phosphate-buffered saline (PBS)
- Trypsin/EDTA (Invitrogen)
- Culture media
- Trypan blue (Sigma-Aldrich)
- Haemocytometer (Sigma-Aldrich)

Experimental Procedures

1. Seed 24-well plates with 50,000 cells per well and incubate for 24 h prior to undertaking transfection protocols/treatments.
2. Following treatments, replace with fresh normal culture media and incubate for 20 h.
3. Wash 3 times with prewarmed PBS before dissociation with trypsin/EDTA at 37°C.
4. Add a volume of culture media to stop the trypsin and collect the detached cells in microfuge tubes.

5. Stain with an equal volume of Trypan blue and perform a differential count (blue cells nonviable).
6. Express the values as a percentage of the appropriate control.

Protocol 9.3 Fluorescence-Activated Cell Sorting (FACS or Flow Cytometry) Analysis of Fluorescent-Labeled siRNA Uptake

Materials

- FACScalibar flow cytometer (BD Biosciences)
- FACS flow, FACS rinse, FACS safe fluids (BD Biosciences)
- Fluorescently labeled probe siRNA (e.g., FITC-labeled siRNA, MWG-Biotech)
- Phosphate-buffered saline (PBS)
- FACS tubes (Fahrenheit)
- WinMDI 2.8 (Internet freeware, http://facs.scripps.edu/software.html)

Experimental Procedures

1. Seed cells in a 24-well format, e.g., A431 cell line, and culture overnight prior to study.
2. Treat cells with siRNA, chosen transfection protocol, and conditions.
3. After incubation with siRNA complexes, wash the cells 3 times with ice-cold PBS before dissociation with trypsin/EDTA at 4°C.
4. Once cells are dissociated, add ice-cold culture media.
5. Transfer the cells to FACS tubes and place on ice for immediate analysis.
6. Determine the cell-associated fluorescence distributions using the flow cytometer, obtaining a minimum of 10,000 events per cell sample through the appropriate bandpass filter (for FITC, the FL1 filter).
7. Analyze using a validated analysis software package, e.g., WinMDI 2.8, taking the fluorescence of gated cell populations with statistical analysis performed on the raw (median fluorescence intensity) data.

Technical note. Alternatively, uptake of radio-labeled siRNA (5'-end or 3'-end; Protocols 9.4 to 9.7) can be determined (Protocol 9.8).

Protocol 9.4 T4 Kinase Nucleic Acid Labelling Reaction, ^{32}P 5′-End Label

Materials

- Nucleic acids, either siRNA (Qiagen), RNA (MWG-Dharmacon), or Oligonucleotide (MWG-Biotech)
- γ-^{32}P [ATP] (Amersham)
- T4 Kinase (ABgene)
- Forward kinase buffer (ABgene)
- Double distilled water (Millipore MilliQ DNase/RNase free or equivalent)

Experimental Procedures

1. Assemble the mix of reaction components.

Nucleic acid (0.2 mM)	2 μl
γ-^{32}P [ATP]	2 μl
T4 polynucleotide kinase	1 μl
Forward kinase buffer	5 μl
Double-distilled water	10 μl

2. Incubate at 37°C for 1 h.
3. Purify using either column or gel purification (see Protocols 9.6 and 9.7).

Technical note. To minimize contamination with free label, gel purification is our preferred method.

Protocol 9.5 Terminal Deoxynucleotidyl Transferase Labelling Reaction, ^{35}S Internal Label

Materials

- Nucleic acids
- α-^{35}S [dATP] (10 pmol/μl, Amersham)
- Terminal deoxynucleotidyl transferase (13 U/μl) (TdT, Amersham)
- 5 × Terminal transferase buffer
- Double-distilled water (Millipore MilliQ DNase/RNase free or equivalent)

Experimental Procedures

1. Assemble the mix of reaction components.

Nucleic acid (0.2 μM)	10 μl
α-^{35}S [dATP] (10 pmol/μl)	1 μl
Terminal transferase (13 U/μl)	1 μl
5 × Terminal transferase buffer	4 μl
Double-distilled water	4 μl

2. Incubate the reaction mixture for up to 1.5 h.
3. Purify using either column or gel purification (see Protocols 9.6 and 9.7).

Technical note. To minimize contamination with free label, gel purification is our preferred method.

PROTOCOL 9.6 COLUMN PURIFICATION

Materials

- Micro-centrifuge (Eppendorf)
- Labeled nucleic acids
- G-25 sephedex column (Amersham)
- Double-distilled water (Millipore MilliQ DNase/RNase free or equivalent)

Experimental Procedures

1. Briefly vortex a G-25 Sephedex column and then prespin at 2500 rpm for 1 min.
2. Pool labeled nucleotides (2 × 20 ml) and then dilute by the addition of double-distilled water (40 μl).
3. Add the diluted sample to the column and spin at 2500 rpm for 2 min (removing free label).
4. Pass the samples through two further columns to ensure removal of free label.

Protocol 9.7 Gel Purification

Materials

- Vertical gel electrophoresis apparatus (Fisher Life Sciences)
- Autoradiography film (Kodak)
- Sealed cassette (Amersham)
- Orbital shaker (Techne)
- SpeedVac (ThermoLife Science)
- Double-distilled water (Millipore MilliQ DNase/RNase free or equivalent)

Experimental Procedure

1. Make a 20% (w/v) polyacrylamide (PAGE) gels composed of:

Acrylamide	200 g
Bis-acrylamide	6.6 g
5 × TBE	200 ml
Double-distilled water	to 1000 ml

2. Add polymerizing reagents (300 μl of 10% (w/v) ammonium persulfate and 60 μl TEMED) to 10 μl of gel solution.
3. Quickly add this solution to both sides of the assembled electrophoresis glass plates, thus forming a bung to prevent the gel leakage.
4. Mix 60 ml of acrylamide solution with double the amounts of polymerizing reagents; place on top of the polymerized bung.
5. Position on top of the gels and leave for 30 min to fully polymerize.
6. Prerun the gel at 300 V for a least 1 h at 4°C.
7. Load the labeled nucleic acid samples and run at 300 V for 3 h.
8. Expose autoradiography film to the gel in a sealed cassette.
9. Replace the developed film on top of the gel and excise the bands corresponding to labeled nucleic acids (place a fresh film over the gel to confirm the complete removal of the bands).
10. Crush the excised gel and place in three 1.5-ml centrifuge tubes, each containing 1 ml of ddH_2O.
11. Place the tubes in liquid nitrogen to crack the gel; allow the nucleic acids to elute for 1 h on an orbital shaker at maximum speed.
12. Briefly centrifuge and then remove the eluted labeled nucleic acids.

13. Place another 1 ml of ddH_2O and repeat the elution process twice to recover ample nucleic acid.
14. Dry overnight in a SpeedVac at a low temperature setting.
15. Pool the samples and make up to a known concentration with ddH_2O.

Technical note. The acrylamide solution volumes quoted are relevant to the 16.5-cm x 17.5-cm × 0.01-cm gel system used in our laboratory.

Protocol 9.8 Uptake of siRNA Labeled with Radionucleotide

Materials

- Phosphate-buffered saline (PBS)
- PBS/sodium azide (0.05%, w/v, NaN_3/PBS) (sodium azide; Sigma Aldrich)
- Scintillation vials (Fisher Scientific)
- Triton X-100 (Sigma Aldrich)
- Optiphase Hi-safe 3 (Pharmacia-Wallace)
- Scintillation counter

Experimental Procedure

1. Culture cells in 24-well format until cells are near confluence.
2. Wash × 3 with prewarmed PBS (37°C).
3. Aspirate and replace with serum-free media containing radio-labeled nucleic acids.
4. Incubate at 37°C for an appropriate duration.
5. Upon completion of incubation, wash × 3 with ice-cold PBS/sodium azide, collecting washes in scintillation vials.
6. Harvest cells by solubilizing 0.5 ml of 3% (v/v) Triton X-100 in ddH_2O for 1 h at room temperature.
7. Wash the wells × 2 with Triton X-100 and pool all of the cellular fractions in scintillation vials.
8. Add 5 ml of Optiphase to each vial.
9. Perform radioactivity count for 5 min, using the relevant radionuclide program on a scintillation counter.

Technical note. In our work, we determined the appropriate number of washes through light microscopic observations of the cell monlayer and scintillation counting of an increased number of washes. Included in our experimental design were fluid phase markers such as [^{14}C] Mannitol over the same time courses.

Protocol 9.9 Cellular Distribution (Fluorescently Labeled siRNA Uptake)

Materials

- 30-mm glass-bottom culture dishes (MatTek Corporation)
- Phosphate-buffered saline (PBS)

- Leica DM IRB inverted epifluorescent microscope fitted with a 12-bit cooled monochrome QImaging camera
- ImproVision, Openlab software (Version 3.0.9)

Experimental Procedures

1. Conduct live-cell microscopic analysis of fluorescent siRNA intracellular distribution in A431 cells in 30-mm glass-bottom culture dishes at a seeding density of 25,000 cells/cm^2.
2. Carry out transfection with appropriate siRNA complexes (and controls).
3. Wash cells three times with sterile phosphate buffered saline (PBS).
4. Capture all images using an inverted epifluorescent microscope of live cells maintained in PBS.
5. Collect and handle images using ImproVision.

Protocol 9.10 Measurement of siRNA Activity by Western Blotting — EGFR Model

Materials

- Lysis buffer (50 mM Tris-base, 5 mM EGTA, 150 mM NaCl, 1% Triton 100, 2 mM Na_3VO_4, 50 mM NaF, 1 mM PMSF, 20 mM phenylarsine, 10 mM sodium molybdate, 10 mg/ml leupeptin, and 8 mg/ml aprotinin; pH 7.6 (components from Sigma Aldrich)
- Standard BCA assay (BioRad)
- Sample loading buffer (60 mM Tris, pH 6.8, containing 2% (w/v) SDS, 10% (v/v) glycerol, 0.005% (w/v) bromophenol, and 250 mM DTT)
- Protran nitrocellulose membrane (Schleicher & Schuell)
- SuperSignal chemiluminescent substrate (Pierce)
- BM chemiluminescence blotting buffer POD (Roche)
- Rabbit polyclonal anti-EGFR antibody (Upstate Cell Signaling)
- Sheep antimouse IgG-HRP-conjugated secondary antibody (Amersham Biosciences)
- Kodak autoradiography film (G.R.I.)
- GS-700 densitometer with Molecular Analyst software (BioRad)

Experimental Procedures

1. Seed cells in a 24-well plate and perform desired transfections/treatments.
2. Lyse cells by the addition of lysis buffer (50 to 100 µl/well).
3. Remove cell debris by centrifugation at 13,000 rpm for 15 min at 4°C.
4. Assay the protein content of the supernatants using the standard BCA assay.
5. Mix total protein lysate samples with loading buffer.
6. Boil samples for 10 min at 100°C.
7. Load equal amounts of protein (for example, 50 mg/lane) onto an SDS-PAGE (5% stacking gel and 8% resolving gel).

8. Resolve at 100V until the dye front reaches the bottom of the gel.
9. Blot transfer the proteins onto nitrocellulose membrane.
10. Block membranes with 5% chemiluminescence blotting buffer.
11. To detect total EGFR, probe with rabbit polyclonal anti-EGFR antibody, followed by the antirabbit IgG-HRP-conjugated secondary antibody.
12. Detect immunoreactive bands with chemiluminescent substrate using autoradiography film.
13. Scan and quantify images using the densitometer.

Technical note. To ensure equal loading of proteins, β-actin can be detected using primary mouse antihuman β-actin monoclonal antibody, followed by the sheep antimouse IgG-HRP-conjugated secondary antibody.

Protocol 9.11 Determination of Messenger RNA Levels by Reverse Transcriptase Polymerase Chain Reaction (RT-PCR)

Materials

- Double-distilled water (Millipore MilliQ DNase/RNase free or equivalent)
- 5 × first strand buffer (Invitrogen)
- DTT (0.1 M) (Invitrogen)
- pdN6 (Amersham)
- Total RNA (harvested from cell cultures using Trireagent (Sigma))
- RNasin (Promega)
- MMLV-rt (Invitrogen)
- dNTPs (Amersham)
- Primer3 (http://www.genome.wi.mit.edu/genome_software/other/primer3.html)
- Forward and reverse primers (MWG-Biotech)
- Taq polymerase enzyme (5 units/ml) (Qiagen)
- Taq 10 × buffer containing $MgCl_2$ (15 mM) (Qiagen)
- $MgCl_2$ (25 mM) (Qiagen)
- GS-700 densitometer with Molecular Analyst software (BioRad)

Experimental Procedures

1. Heat-denature the samples for 4 min before adding the reverse transcriptase mix consisting of:

Total RNA	1 μg
5 × first strand buffer	8 μl
dNTPs (10 mM)	4 μl
DTT (0.1 M)	4 μl
pdN6 (100 ng)	4 μl
RNasin (20 U)	1 μl

   RNase/DNase free water, in a total volume of 40 μl
   MMLV-rt (200 U)
2. Incubate at 25°C for 10 min, 42°C for 50 min, 99°C for 2 min, and 4°C for 2 min.

3. Design specific PCR primers using, for example, Internet-based primer design program Primer3.
4. Set up a PCR reaction mix as follows (for 25-ml reaction):

RNase/DNase free ddH_2O	8.5 μl
Taq 10 × buffer containing $MgCl_2$ (15 mM)	2.5 μl
$MgCl_2$ (25 mM)	1 μl
dNTPs (10 mM)	2 μl
cDNA (100 ng/ml)	2 μl
Forward and reverse primers (10 pmol/ml)	0.02–0.5 μl
Taq polymerase enzyme (5 units/ml)	0.1 μl

5. Cycle in a thermal cycler (denaturation at 94°C for 30 sec, annealing at 57 to 64°C for 45 sec, and extension at 72°C for 45 sec through a total of 30 cycles).
6. Resolve the PCR products in a 1.2% (w/v) agarose gel in the presence of 1 mg ethidium bromide/1 ml gel and visualized under UV light. Include appropriate positive control total RNA and suitable negative controls (such as RT-PCR reactions performed without the desired cDNA and/or primers).
7. Measure the density of expressed bands with a densitometer.
8. Normalize the density of each spot by subtracting local background density and measuring its ratio over the density of a housekeeping gene such as actin.

Technical notes. Northern blotting or real time PCR may also be used to determine gene expression levels (for methods, see [52]). siRNA activity may also be measured using cell number, depending upon the target and its effect on cell number as per Protocol 9.2.

REFERENCES

1. Dykxhoorn, D.M., Novina, C.D., and Sharp, P.A., Killing the messenger: short RNAs that silence gene expression, *Nat. Rev. Mol. Cell Biol.*, 4, 457, 2003.
2. McManus, M.T., and Sharp, P.A., Gene silencing in mammals by small interfering RNAs, *Nat. Rev. Genet.*, 3, 737, 2002.
3. Lieberman, J. et al., Interfering with disease: Opportunities and roadblocks to harnessing RNA interference, *Trends Mol. Med.*, 9, 397, 2003.
4. Fell, P.L. et al., Cellular uptake properties of a 2'-amino/2'-*O*-methyl-modified chimeric hammerhead ribozyme targeted to the epidermal growth factor receptor mRNA, *Antisense Nucl. Acid Drug Dev.*, 7, 319, 1997.
5. Shoji, Y. et al., Mechanism of cellular uptake of modified oligodeoxynucleotides containing methylphosphonate linkages, *Nucleic Acids Res.*, 19, 5543, 1991.
6. Hughes, J. et al., In vitro transport and delivery of antisense oligonucleotides, *Methods Enzymol.*, 313, 342, 2000.
7. Loke, S.L. et al., Delivery of C-Myc antisense phosphorothioate oligodeoxynucleotides to hematopoietic-cells in culture by liposome fusion–specific reduction in C-Myc protein expression correlates with inhibition of cell-growth and DNA-synthesis, *Curr. Top. Microbiol. Immunol.*, 141, 282, 1988.
8. Hollins, A.J. and Akhtar, S., unpublished data, 2003.
9. Wickstrom, E., Oligodeoxynucleotide stability in subcellular extracts and culture media, *J. Biochem. Biophys. Methods,* 13, 97, 1986.

10. Akhtar, S., Kole, R., and Juliano, R.L., Stability of antisense DNA oligodeoxynucleotide. Analogs in cellular extracts and sera, *Life Sci.*, 49, 1793, 1991.
11. Hudson, A.J. et al., Cellular delivery of hammerhead ribozymes conjugated to a transferrin receptor antibody, *Int. J. Pharm.*, 182, 49, 1999.
12. Parrish, S. et al., Functional anatomy of a dsRNA trigger: differential requirement for the two trigger strands in RNA interference, *Mol. Cell*, 6, 1077, 2000.
13. Holen, T. et al., Positional effects of short interfering RNAs targeting the human coagulation trigger Tissue Factor, *Nucleic Acids Res.*, 30, 1757, 2002.
14. Chiu, Y.L., and Rana, T.M., RNAi in human cells: basic structural and functional features of small interfering RNA, *Mol. Cell*, 10, 549, 2002.
15. Elbashir, S.M. et al., Functional anatomy of siRNAs for mediating efficient RNAi in Drosophila melanogaster embryo lysate, *EMBO J.*, 20, 6877, 2001.
16. Amarzguioui, M. et al., Tolerance for mutations and chemical modifications in a siRNA, *Nucleic Acids Res.*, 31, 589, 2003.
17. Chiu, Y.L., Rana, T.M., siRNA function in RNAi: a chemical modification analysis, *RNA*, 9, 1034, 2003.
18. Braasch, D.A. et al., RNA interference in mammalian cells by chemically-modified RNA, *Biochemistry*, 42, 7967–7975, 2003.
19. Akhtar, S. and Juliano, R.L., Cellular uptake and intracellular fate of antisense oligonucleotides, *Trends Cell Biol.*, 2, 139, 1992.
20. Akhtar, S. et al., The delivery of antisense therapeutics, *Adv. Drug Deliv. Rev.*, 44, 3, 2000.
21. Hemmings-Mieszczak, M. et al., Independent combinatorial effect of antisense oligonucleotides and RNAi-mediated specific inhibition of the recombinant rat P2X3 receptor, *Nucleic Acids Res.*, 31, 2117, 2003.
22. Jiang, Z.Y. et al., Insulin signaling through Akt/protein kinase B analyzed by small interfering RNA-mediated gene silencing, *Proc. Natl. Acad. Sci. U.S.A.*, 100, 7569, 2003.
23. Simeoni, F. et al., Insight into the mechanism of the peptide-based gene delivery system MPG: implications for delivery of siRNA into mammalian cells, *Nucleic Acids Res.*, 31, 2717, 2003.
24. Bertrand, J.R. et al., Comparison of antisense oligonucleotides and siRNAs in cell culture and *in vivo*, *Biochem. Biophys. Res. Commun.*, 296, 1000, 2002.
25. Caplen, N.J. et al., Specific inhibition of gene expression by small double-stranded RNAs in invertebrate and vertebrate systems, *Proc. Natl. Acad. Sci. U.S.A.*, 98, 9742, 2001.
26. Bohula, E.A. et al., The efficacy of small interfering RNAs targeted to the type 1 insulin-like growth factor receptor (IGF1R) is influenced by secondary structure in the IGF1R transcript, *J. Biol. Chem.*, 278, 15991, 2003.
27. Beale, G. et al., Gene silencing nucleic acids designed by scanning arrays: Anti-EGFr efficacy of siRNA, Ribozyme and DNA enzymes targeting the same hybridization-accessible site using the same delivery system, *J. Drug Target.*, 11, 449, 2003.
28. Reich, S.J. et al., Small interfering RNA (siRNA) targeting VEGF effectively inhibits ocular neovascularization in a mouse model, *Mol. Vis.*, 9, 210, 2003.
29. Kobayashi, N. et al., Vector-based *in vivo* RNA interference: dose- and time-dependent suppression of transgene expression, *J. Pharmacol. Exp. Ther.*, 2003.
30. Lewis, D.L. et al., Efficient delivery of siRNA for inhibition of gene expression in postnatal mice, *Nat. Genet.*, 32, 107, 2002.
31. McCaffrey, A.P. et al., RNA interference in adult mice, *Nature*, 418, 38, 2002.
32. Makimura, H. et al., Reducing hypothalamic AGRP by RNA interference increases metabolic rate and decreases body weight without influencing food intake, *BMC Neurosci.*, 3, 18, 2002.

33. Mohmmed, A. et al., *In vivo* gene silencing in *Plasmodium berghei*–a mouse malaria model, *Biochem. Biophys. Res. Commun.*, 309, 506, 2003.
34. Song, E. et al., RNA interference targeting Fas protects mice from fulminant hepatitis, *Nat. Med.*, 9, 347, 2003.
35. Sorensen, D.R., Leirdal, M., and Sioud, M., Gene silencing by systemic delivery of synthetic siRNAs in adult mice, *J. Mol. Biol.*, 327, 761, 2003.
36. Verma, U.N. et al., Small interfering RNAs directed against beta-catenin inhibit the *in vitro* and *in vivo* growth of colon cancer cells, *Clin. Cancer Res.*, 9, 1291, 2003.
37. Zender, L. et al., Caspase 8 small interfering RNA prevents acute liver failure in mice, *Proc. Natl. Acad. Sci. U.S.A.*, 100, 7797, 2003.
38. Pedroso de Lima, M.C. et al., Cationic lipid-DNA complexes in gene delivery: from biophysics to biological applications, *Adv. Drug Deliv. Rev.*, 47, 277, 2001.
39. Farhood, H. et al., Effect of cationic cholesterol derivatives on gene-transfer and protein-kinase-C activity, *Biochim. Biophys. Acta*, 1111, 239, 1992.
40. Zelphati, O. and Szoka, F.C., Intracellular distribution and mechanism of delivery of oligonucleotides mediated by cationic lipids, *Pharm. Res.*, 13, 1367, 1996.
41. Roh, H., Pippin, J., and Drebin, J.A., Down-regulation of HER2/neu expression induces apoptosis in human cancer cells that overexpress HER2/neu, *Cancer Res.*, 60, 560, 2000.
42. Abe, T. et al., Specific inhibition of influenza virus RNA polymerase and nucleoprotein gene expression by liposomally encapsulated antisense phosphorothioate oligonucleotides in MDCK cells, *Antiviral Chem. Chemother.*, 9, 253, 1998.
43. Alahari, S.K. et al., Inhibition of expression of the multidrug resistance-associated P-glycoprotein by phosphorothioate and 5' cholesterol-conjugated phosphorothioate antisense oligonucleotides, *Mol. Pharmacol.,* 50, 808, 1996.
44. Islam, A. et al., Studies on uptake, sub-cellular trafficking and efflux of antisense oligodeoxynucleotides in glioma cells using self-assembling cationic lipoplexes as delivery systems, *J. Drug Target.*, 7, 373, 2000.
45. Williams, S.A. and Buzby, J.S., Cell-specific optimization of phosphorothioate antisense oligodeoxynucleotide delivery by cationic lipids, *Methods Enzymol.*, 313, 388, 2000.
46. Akhtar, S., Antisense technology: Selection and delivery of optimally acting antisense oligonucleotides, *J. Drug Target.*, 5, 225, 1998.
47. Campbell, P.I., Toxicity of some charged lipids used in liposome preparations, *Cytobios*, 37, 21, 1983.
48. Nagahiro, I. et al., Toxicity of cationic liposome-DNA complex in lung isografts, *Transplantation*, 69, 1802, 2000.
49. Scheule, R.K. et al., Basis of pulmonary toxicity associated with cationic lipid-mediated gene transfer to the mammalian lung, *Hum. Gene. Ther.*, 8, 689, 1997.
50. Filion, M.C. and Phillips, N.C., Toxicity and immunomodulatory activity of liposomal vectors formulated with cationic lipids toward immune effector cells, *Biochim. Biophys. Acta*, 1329, 345, 1997.
51. Omidi, Y. et al., Toxicogenomics of non-viral vectors for gene therapy: a microarray study of Lipofectin- and Oligofectamine-induced gene expression changes in human epithelial cells, *J. Drug Target.*, 11, 449, 2003.
52. Sambrook, J. and Russell, D.W., *Molecular cloning: a laboratory manual.* 3rd ed. Cold Spring Harbor Laboratory Press, New York, 2001, Vols 1–3.

10 RNAi in *Drosophila* Cell Cultures

Mónica Bettencourt-Dias, Rita Sinka, Lisa Frenz, and David M. Glover

CONTENTS

10.1 INTRODUCTION TO THE USE OF RNA INTERFERENCE IN *DROSOPHILA*-CULTURED CELLS

RNA interference (RNAi) is a recent technological advance that allows the reduction of the expression of a gene of interest at the posttranscriptional level.[1–3] Initially

0–8493–2141–7/05/$0.00 + $.50

discovered in *Caenorhabditis elegans*, its use has been extended to *Drosophila* embryos and cell lines, mammalian embryos and cell lines, plants, fungi, and eukaryotic pathogens such as trypanosomes.[1,4] Recent findings indicate that RNA silencing is an evolutionarily conserved pathway that participates in the regulation of gene expression and that protects genomes from genomic parasites such as viruses and transposons.[5]

In several species the effects of the introduction of dsRNA may be sustained for long periods of time. Studies in *C. elegans* have proposed that there is a mechanism of producing additional siRNA molecules, through amplification of the target, in a reaction catalyzed by a putative RNA-directed RNA polymerase (RdRP).[2,6] However, in both *Drosophila* and humans there is evidence against an obligatory role of an RdRP in RNAi.[3,5] Expression of dsRNA-producing transgenes specific for a determined region of an RNA species in *Drosophila* does not generate siRNAs for regions upstream and downstream of the targeted region. This allows the design of isoform-specific targeted dsRNAs.[7]

Invertebrate cell culture systems, particularly those from *Drosophila melanogaster*, that are well established and grow easily and economically in the laboratory have become a valuable tool in analyzing biological function. The fact that RNAi can be done in cultures of *Drosophila* cell lines, together with its specificity, dose dependence, and ability to use more than one species of dsRNA at the same time, allows a range of applications, including analysis of a phenotype due to the reduction in the expression of a particular gene product;[8,9] analysis of protein complexes in the absence of a subunit; study of downstream targets of enzymes by proteomics and epistasis analysis;[10,11] pooling of different dsRNAs in screens;[10] and targeting of specific alternatively spliced mRNA isoforms.[7] All of these properties and the availability of fully sequenced and annotated genomes make RNAi in *Drosophila*-cultured cells a powerful tool for rapid screens of several genes for a common function (e.g., cell cycle genes) or a particular biological or biochemical property. In fact, screens have been published for a wide range of processes from cell cycle to morphogen signaling.[10–13]

In this chapter we will discuss the basics of performing RNAi in *Drosophila*-cultured cells. Whichever cell type (Section 10.2) and assay is chosen, we recommend using robust negative and positive controls. We suggest dsRNA for GFP or for an intronic sequence as the negative control. In the absence of a good positive control for RNA interference suitable for the aspect of cell biology in which you are interested (a gene that shows a phenotype in the selected assay), we suggest using dsRNA for Aurora B, a protein kinase that is involved in several steps in mitosis. The absence of Aurora B leads to a very striking and easily scored phenotype, with formation of cells with several nuclei or giant nuclei.[8]

10.2 CHOOSING A CELL LINE AND CULTURING *DROSOPHILA* CELLS

Drosophila has been traditionally chosen as a model system because of the versatile genetic approaches with the organism. Work to establish *in vitro* culture systems for insect cells only began in the late 1960s, more than 10 years after mammalian cell lines. Specific media had to be developed for *Drosophila* cells, and that was only possible after the biochemical features of insect body fluids were established. Several *Drosophila* cell lines were derived from specific primary cultures and have been kept in culture, frozen, for more then 30 years.[14] Most available *Drosophila* cell lines, including haploid cell lines, have their origin in cultures generated through the dissociation of embryos.[15,16] However, other cell lines have been produced with other origins, such as wing and leg discs and the larval central nervous system.[17,18] A complete list of cell lines, including cells carrying particular mutations, can be found in Echalier[14] and Ashburner.[19] *Drosophila* cell lines are easy to grow (Table 10.1): all of the cells can be kept in a 25 to 27°C degree incubator; they tolerate wide changes of pH and most *Drosophila* tissue culture media do not use bicarbonate buffering, so there is no need for a controlled CO_2 atmosphere; and most of the cell lines grow as monolayers, do not require trypsinization, and detach by simple agitation. Even though *Drosophila* cells grow at a lower temperature then mammalian cell lines, their doubling time is similar (about 24 h or less).[14] RNAi has been shown to be effective in different *Drosophila* cell lines, including S2, S2R+, Dmel-2, Kc, and Cl-8 lines.[10–13,20] Table 10.1 contains information about these different cell lines. Once it is decided which cells to use, the plastic ware and media needed for culturing the cells should be obtained. There should be access to a clean incubator at 25°C and a tissue culture flow hood where sterile manipulation of the cells can be performed. Cells may change phenotype considerably with time in culture, and they may be lost by contamination. As soon as cells are received it is very important that they are expanded by subculturing so that aliquots can be frozen and retrieved later (Protocol 10.1). It is advisable to freeze sufficient cells of a low passage number for the entire experiment/screen.

10.3 MAKING THE DSRNA

DNA template for dsRNA production from any *Drosophila* gene can be obtained from Cyclacel Ltd. through the MRC geneservice (www.geneservice.mrc.ac.uk) or from Open Biosystems (http://www.openbiosystems.com). If it is decided to design the dsRNA, the optimal length for maximum interference activity in *Drosophila* is 300–1000 bp.[21]

10.3.1 PREPARATION OF THE TEMPLATE FOR TRANSCRIPTION

In general, only two PCR primers and a single PCR reaction are necessary to make the template for the dsRNA (Figure 10.1). Genomic DNA template can be used to produce the PCR product if there is an exon of more than 400 bp at the beginning of the gene (see Protocol 10.2 for genomic DNA isolation). Using the Apollo program helps to visualize intron/exon boundaries and ORFs (http://www.fruitfly.org/annot/apollo). Alternatively, cDNA template can be used, which can be an expressed sequence tag (EST) from the *Drosophila* Genome Project (http://www.fruitfly.org/sequence) or any other cDNA (see Protocol 10.3 for cDNA isolation). Amplification of the selected

TABLE 10.1
***Drosophila* Commonly Used Cell Lines, Their Specifications, and Requirements**

	S2 (also known as SL2 or Dm2)	S2R+	Kc	DMEL-2	Clone 8
Origin	Oregon R-20 to 24-hour embryos	Clone derived from S2	Disaggregated young embryos (8 to 12hr)	Derived from S2 but adapted to serum-free media	Dissociated larval wing discs
Particularities	Do not respond to Wg	Respond to Wg. More adherent than Kc or S2	Sensitive to ecdysteroids Respond to Wg Less adherent	Grow in serum-free media; used for protein expression	Respond to Wg
Ease of culturing	Very easy	Easy	Very easy	Very easy and inexpensive (do not require serum)	Difficult
Karyotype	Female, 60–80% tetraploid		Fundamentally diploid (up to 90%), female chromosomal set, one single IV chromosome	As S2, but polyploidy can arise after long passaging	Largely diploid
Medium	Schneider medium (Sigma) 10% heat-inactivated FBS (Sigma) optional 1/100 (0.01 ×) volume penicillin, streptomycin (PS; Gibco)	Schneider medium (Sigma) 10% heat-inactivated FBS (Sigma) optional 1/100 (0.01 ×) volume PS	Schneider's medium (Gibco) 5% FCS optional 1/100 (0.01 ×) volume PS	Serum-free medium (Gibco) 1/10 volume 200-mM L-glutamine optional 1/100 (0.01 ×) volume PS	Shields and Sang Medium 3 (SIGMA) 2% heat-inactivated FCS 2.5% FE2 (see reference) 12.5IU insulin/100ml

Growing conditions	monolayer–can grow in suspension 1:7 dilution every 3 to 4 days 25°C 20 to 24 hours doubling time	monolayer– needs to be trypsinised for subculturing 1:6 dilution every 3 to 4 days 20 to 24 hours doubling time	monolayer–can grow in suspension 1:9 dilution every 3 to 4 days 18 hrs doubling time	monolayer–can grow in suspension 27°C 1:9 dilution every 3 to 4 days 14 to 24 hours doubling time	monolayer–very adherent Needs to be trypsinized for subculturing 24 hrs with a lag of about a day after passaging
Transfection of dsRNA	Slightly higher efficiency with transfection reagent (Transfast-Promega) but works OK without	Slightly higher efficiency with transfection reagent (Transfast-Promega) but works OK without	Works OK without transfection reagent	Same efficiency with/without transfection reagent	Low efficiency Needs transfection reagent Effectene (Qiagen)
Available at	ATCC:CRL-1963		ECACC :90070550	Invitrogen: 10831-014 http://www.invitrogen.com/	Contact: Deborah Cottam (dc6@st-andrews.ac.uk) or Martin Milner (mjm5@st-and.ac.uk)
References	29	30	31		17 see also http://biology.st-and.ac.uk/sites/ flycell/

ATCC: American Tissue Culture Collection; ECACC: European Collection of Animal Cell Cultures; Wg: wingless

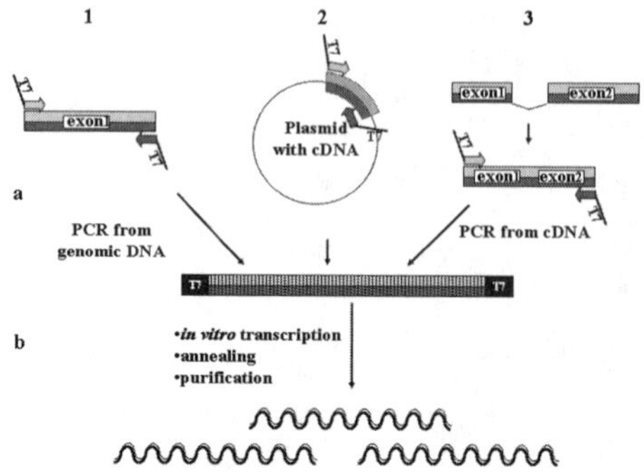

FIGURE 10.1 Amplification strategies and general process of dsRNA production. (a) PCR templates can be generated from genomic DNA (1) and cDNA (3) using primers with gene-specific sequences and T7 polymerase-binding sites; alternatively, the PCR template can be generated with T7 and/or SP6 primers from plasmids having the cDNA flanked by those sequences (2) (for example, in the case of constructs from the *Drosophila* Gene Collection). (b) After the PCR amplification, there are three steps to produce the final dsRNA: *in vitro* transcription, annealing, and purification.

cDNA from the appropriate cell line is a good indication that the candidate gene is expressed in that cell line. We have found that the success rate of the PCR amplification is high where genomic DNA is used as the template (97 to 100%) but lower where cDNA is used (75 to 90%), with the level and quality of the resulting PCR products also being poorer. In these instances, it is sometimes possible to improve yields by a second round of amplification using the purified PCR product as the template together with the original primers. The PCR product may include the 5' or 3' untranslated sequence, depending on the cDNA length. PCR primers can be designed with a primer design program (e.g., DNASTAR). The length of the sequence of the primers that is gene specific is about 16 to 25 bp, with an additional 23 bp corresponding to the T7 RNA polymerase-binding site added to the 5' end (Table 10.2). The melting temperature of the gene-specific part of the PCR primers should be between 50 to 60°C for an optimal PCR reaction. If possible, it is best to avoid any conserved region of the gene to guarantee the specificity of the dsRNA. In Table 10.2, examples of primers for a possible negative control, GFP, and a potential positive control, aurora B, are shown. Protocol 10.4 provides details of the PCR reaction.

10.3.2 *In Vitro* Transcription and Determination of dsRNA Concentration

There are several ways of producing a dsRNA and all involve an intermediary step of having a DNA sequence of the gene of interest flanked by RNA polymerase (T7 or SP6) binding sites (Figure 10.1). This is then used as a template for the *in vitro* transcription reaction. We have successfully used T7 RiboMAX™ Express RNAi System (Promega) or Megascript RNAi Kit (Ambion) to produce dsRNA, following the manufacturer's recommendations (see Protocol 10.5).

TABLE 10.2
Primer Sequence Used for the Preparation of Aurora B and Control GFP Templates

AurBF	5'-TAATACGACTCACTATAGGGAGA**TTCTCCTTGCCGCGATCCCGTTTCTG**-3'
AurBR	5'-TAATACGACTCACTATAGGGAGA**ACATCCTGCGCCTGCTCACTTGGTTCC** -3'
GFPF	5'-TAATACGACTCACTATAGGGAGA**CTTCAGCCGCTACCCC**-3'
GFPR	5'-TAATACGACTCACTATAGGGAGA**TGTCGGGCAGCACG**-3'
eIF4aF	5'-TAATACGACTCACTATAGGGAGA**GAAATGAGATACCTCAGGATGGCCC**-3'
eIF4aR	5'-TAATACGACTCACTATAGGGAGA**ACGTTAGTGCCGCCAATGCA**-3'

The T7 RNA polymerase-binding site sequence is underlined. Specific primers for the eukaryotic initiation factor-4a (eIF4a) can be used in the RT-PCR reaction as an internal control for gene expression in *Drosophila* cells.

10.4 PERFORMING RNAi ON *DROSOPHILA* CELLS

It is not clear how the dsRNA enters the cells. In *C. elegans* a transmembrane protein, SID-1, was found that is essential for systemic but not cell-autonomous RNAi.[22] Members of the same family of proteins are present in humans but not in the fly genome. Most *Drosophila* cell lines will incorporate the dsRNA after soaking in serum-free media containing the dsRNA.[11,13] This is a good strategy for carrying out large-scale screens. We found that addition of transfection reagents to the media, although more expensive, can potentiate RNAi in serum-growing cells. There are several types of transfection reagents; the ones we have used are synthetic, cationic lipids that facilitate delivery of the dsRNA to the cells. In the case of clone 8 cells, transfection reagents[10] or expression of *C. elegans* SID-1[22] are essential for efficient RNAi. Because of the low efficiency of transfection of clone 8 cells, a reporter can be used to select which cells have been transfected.[10] The amount of dsRNA and duration of the assay depends largely on the gene that is being studied. In general, we and others have found that a range of 5 to 25 μg dsRNA/well (6-well plate; see Table 10.3) and an incubation of 3 to 4 days are sufficient for a large reduction in protein levels.[8,9,20,23–25]

10.4.1 Transfection of S2 Cells

We present here the transfection protocol for S2 cells in 6-well plates (Protocol 10.6), but the conditions for different plates can be extrapolated from Table 10.3.

10.4.2 Transfection of Other Cell Lines

The same method (Protocol 10.6) can be used to transfect S2R+ cells. The protocol for transfecting DMEL and Kc cells is very similar to that described for S2 cells but with no need to use transfection reagent.[20] As with S2 cells, DMEL and Kc cells have to be incubated with the dsRNA in serum-free media. If the Kc cells are not very

TABLE 10.3
Experimental Details for Transfection of Cells in Different Volumes

Size of plate	Growth area (cm²)	Relative area	Number of cells to plate a day before transfection/ volume of media to use per well	Amount of the dsRNA in transfection medium/ total volume per well	Volume of Transfast
96 well	0.32	0.2 ×	10 ,000/150 μl	680 ng/50 μl	0.68 μl
24 well	1.88	1	50, 000/750 μl	4 μg/250 μl	4 μl
6 well	9.4	5	1 million/3000 μl	20 μg/1000 μl	20 μl

Note that after incubation of the cells in the transfection media for 1 hour, fresh serum containing media should be added (see protocol 10.6) for a total volume of 150, 750, and 3 ml in wells of 96-, 24-, and 6-well plates, respectively.

adherent, pellet them by centrifugation and resuspend them in 5 ml serum-free media. Calculate the cell density and adjust it to 2×10^6 cells/ml. Add 1 ml of cells per well (6-well plate). Add the dsRNA and mix well. After 1 h, add 2 ml of serum containing media (15%). Clone 8 transfection has been performed using Effectene (Qiagen) transfection reagent according to manufacturer's instructions.[10]

10.4.3 Analysis of the Transfection Efficiency and Cells after Transfection

There are several ways of checking the efficiency of transfection. First, fluorescently labeled dsRNA can be used to establish the proportion of cells incorporating dsRNA. One way to produce labeled dsRNA uses the Silencer™ siRNA Labelling Kit adapted to long dsRNA. Use 10 μg dsRNA (GFP) with 15 μl labeling reagent in a 100-μl reaction volume and follow the instructions provided in the kit. Use this dsRNA to transfect the cells in the same way as the dsRNA specific to the gene of interest. After the transfection, wait 4 h and wash the cells with PBS to remove the labeled dsRNA from the surface of the cells and from the solution. Fix the cells and check them under the fluorescent microscope. It is helpful to stain cellular DNA to identify the fluorescent dsRNA within the cells more easily. Second, the strength of the phenotype observed in cells transfected with the positive control dsRNA should give an idea of the success of RNAi (see Figure 10.2 for our example of positive control).

The incorporation of the dsRNA and the phenotype of the positive control may not reflect how effective any sample of dsRNA is in reducing the levels of mRNA. To this end, the protein level can be checked by western blot analysis or, in case there is no available antibody able to recognize the protein under study, the RNA level can be monitored by semiquantitative RT-PCR or northern hybridization. If the decision is made to perform semiquantitative RT-PCR, Protocol 10.3 can be used to isolate the RNA and to synthesize the first strand of cDNA. Additionally, an internal control that is constitutively expressed in the cells should be used, such as eukaryotic initiation factor-4a (eIF4a; Table 10.2).

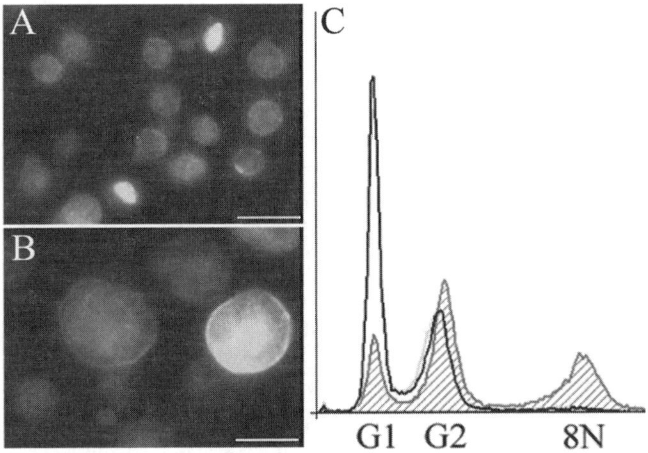

FIGURE 10.2 Using dsRNA of Aurora B kinase as a positive control for dsRNA production and transfection of the cells. Cells in parallel wells were transfected with dsRNA for GFP or Aurora B. After three days, cells were fixed for immunostaining or for FACS analysis. (A, B) Cells were fixed using –20°C methanol for 3 minutes and stained with an antibody for α-tubulin. Note that the cells treated with dsRNA for Aurora B (B) are much larger than the control cells (A). This strong effect of the dsRNA can also be observed using bright field microscopy of live cells or by staining the DNA. (C) The effect of dsRNA for Aurora B is also easily seen by FACS analysis after DNA staining with propidium iodide. Cells treated with dsRNA for Aurora B become polyploid (profile with cross-hatch; note peak with 8N DNA content), while control cells (solid line) show a normal cell cycle profile. Scale bar is 10 µm.

10.4.4 Silencing Multiple Genes

More than one gene can be targeted by RNAi simultaneously. This is particularly convenient for large screens, where several genes can be pooled in one batch,[10] and studies of gene interaction.[11,12] Because RNAi depends on the cleavage of the dsRNA into siRNAs by DICER, there is a limit of dsRNA that can be added to the cells, after which the effect observed for each individual dsRNA may decrease.[2,26–28] In S2 and clone 8 cells, 2–3 dsRNA species have been successfully used (25 µg dsRNA/well (6-well plate) for each species).[10,11]

10.5 Large-Scale RNAi

The features of an RNAi screen are essentially the same as for the analysis of individual gene RNAi knockdown, although the steps are scaled up to accommodate the larger number of reactions and performed largely in a 96-well plate format for convenience. There are also available facilities where conditions and reagents for whole genome RNAi screen are present (http://www.flyrnai.org). We will follow with an analysis of each of the steps of performing RNAi and how these require modifying for a genome-wide screen.

10.5.1 Choosing a Cell Line and Culturing Cells for Large-Scale Screens

Several criteria should be taken into account when choosing a cell line for a screen (see Table 10.1), relating to the cost and ease of culturing and transfecting the cells and the assay used to analyze the resulting phenotype. For example, DMEL-2 cells are well suited for large screens due to a number of properties that make transfection of these cells with dsRNA relatively cheap, quick, and easy; the cells do not require serum for their growth, they can be transfected with dsRNA without inclusion of a transfection reagent, and they grow as a monolayer (large-scale transfections of cells growing as a suspension could prove difficult and time consuming). Additionally, adherent cells are better for microscopy-based assays. There is a list of equipment that may need to be bought or organized for a large-scale screen, such as multichannel pipettes, a DNA electrophoresis system compatible with multichannel pipettes, freezer space at –80°C for storing a large number of 96-well plates, a fluorimeter with 96-well plate capacity, and a PCR machine with 96-well plate capacity.

10.5.2 Preparation of dsRNA

10.5.2.1 Primer Design

While it is possible to use cDNA libraries to prepare gene-specific dsRNA for large-scale screening, those currently available often contain inserts of a size not ideally suited for dsRNA preparation. You can now obtain a library of dsDNAs for production of dsRNAs or, alternatively, you can design the primers as discussed in section 10.3 or purchase a purpose-built program for their design (e.g., SysArris software). Irrespective of how the primers are designed, they should be ordered arrayed in a 96-well plate, with each plate coded to enable identification of the location of each primer. For a genome wide screen, the large number of 96-well plates generated means that it is advisable to set up a database to log the details and storage location of each plate. In the case of the primer plates, details of whether genomic DNA or cDNA should be used as the template in the PCR reaction should also be assigned to each of the primer pairs.

10.5.2.2 PCR Amplification of Gene-Specific Sequence

The preparation of gene-specific PCR products for a large-scale screen uses genomic DNA or cDNA prepared as detailed in Protocols 10.2 and 10.3, using the same reaction mixtures and PCR program, with the only exception being the scale of the preparation, which must take into account the number of PCR reactions to be performed. For consistency purposes, it is preferable to use a single preparation of genomic DNA and cDNA for the entire screen. While it is possible to perform much of the liquid handling using a multichannel pipette, for a genome wide screen a robot designed to handle plates, aliquot reagents, and log the details into a database is a better option. The details of performing the various steps using a robot will be unique to each robot and so will not be covered here, particularly since the cost of such equipment may make it an unaffordable luxury.

There are some general considerations to be taken into account when using 96-well plates in these large-scale approaches. Transfer of liquid from one plate to another can be carried out on a robot or using a multichannel pipette; at each step, the solutions in each well of a plate should be transferred to the equivalent well of the next plate in order to maintain the gene identity of each well from primer plates all the way through to dsRNA plates. Plates should always be sealed (we found a heat sealer to be the most efficient in this respect) before being placed on a thermal cycler. To purify the resulting PCR products, we have used a filter-based purification step, such as the Millipore Montage PCR_{96} Cleanup Kit. This is compatible with some robot systems, or a separate vacuum manifold can be purchased. The PCR products are purified following the manufacturer's recommendations and finally eluted in a volume of 50 µl. PCR products can be analyzed by running 2 µl on a 1% (w/v) 1 × TBE-agarose gel and staining with 0.5 µg/ml ethidium bromide. Loading each sample individually is possible but time consuming, so we advise purchasing a BioRad Sub-Cell GT DNA Electrophoresis Cell (cat no. 170-4403) together with Multichannel Pipette Compatible Combs (cat no. 170-4456), which enable 12 samples to be loaded simultaneously using a 12-channel pipette. As discussed in a previous section, RNase- and DNase-free plastic ware, tips, and reagents should be used throughout.

10.5.2.3 dsRNA Preparation

The Ambion T7 MegaScript reagent can be used for dsRNA preparation, but equivalent high-yield RNA transcription kits from other manufacturers would be equally suitable. The transcription reaction and DNase treatment can be performed as discussed previously but in 96-well plates instead of Eppendorf tubes. Plates can then be sealed and the RNA annealed. This is done most easily in a thermal cycler programmed accordingly. The dsRNA can be analyzed by running 2 µl on a 1% (w/v 1 × TBE-agarose gel and staining with 0.5 µg/ml ethidium bromide, preferably using a multichannel Pipette Compatible DNA electophoresis system. We routinely analyze 2 µl of the final 50 µl dsRNA preparation, aliquoting this into equivalent wells of a fresh plate containing 2 × DNA loading buffer, ready for loading onto the gel.

10.5.2.4 Quantification of dsRNA

To analyze the quantities of dsRNA prepared in 96-well plates, a fluorescent 96-well plate reader is essential. With some such plate readers it is possible to assess concentrations and quality by A_{260}/A_{280} readings, but this requires relatively high concentrations of purified dsRNA and expensive 96-well plates engineered with special optical properties to permit these readings. As we have found no effect of using unpurified dsRNA in our RNAi experiments in terms of transfection efficiency or phenotypic effects, we routinely use the annealed *in vitro* transcription reaction preparations without further purification.

To evaluate the concentrations of these dsRNA samples, we use Pico-green according to the manufacturer's instructions (Cambridge Bioscience. Cat No. P-7581). This dye is an ultrasensitive fluorescent nucleic acid stain, and the method is tolerant of impurities within the sample and requires very little of the valuable dsRNA preparations. For that method we use as standards samples of dsRNA that have been purified and the concentration assessed by A_{260}/A_{280} values.

A comparison of the assigned dsRNA concentrations with the dsRNA yield analyzed by agarose gel electrophoresis can be used to verify the concentrations, with those in which there is an apparent disparity being adjusted accordingly prior to their transfection into *Drosophila* cells. Having assessed the concentration of dsRNA in each well of the dsRNA plate, this must next be diluted to a standard concentration (67 ng/μl) in a separate 96-well plate for use in transfections. Where no robot is available, this is an arduous task, as the concentration in each well is likely to be unique, resulting in a large number of separate dilutions (usually 96) to be performed. The calculations of quantities of dsRNA product and RNase-free water required for this purpose can be calculated from an Excel spreadsheet, but there are no shortcuts for performing the dilutions, which, by necessity, must be performed individually.

10.5.3 Transfection of Cell Line with dsRNA in Large-Scale Screens

Transfection of *Drosophila* tissue culture cells with dsRNA in a large-scale screen is similar to the process described in Protocol 10.6, with all of the same considerations applying. We have concentrated on immunofluorescent microscopy-based assays of RNAi-induced phenotypes, so we have performed transfections in triplicate in flat, clear- bottomed 96-well plates suited to this purpose (Perkin Elmer Life Sciences, Black Viewplate-96; Greiner 96-well micro-clear bottom imaging plates). This could be adapted for other assays with different requirements. Brief details of the steps involved in transfection of DMEL-2, which does not require a transfection reagent, in a 96-well plate format were given in Table 10.3 and Protocol 10.6. For cells requiring transfection reagent, additional steps to remove FCS from the cells and relating to manufacturers' recommendations for the particular reagent will be required.

10.5.4. Analysis of Phenotype in Large-Scale Screens

While there are any number of possible means of analyzing phenotypes resulting from RNAi knockout of specific genes, with many of those used to study RNAi-induced knockdown phenotypes for 1 or 2 genes adaptable to a large-scale screen, we have concentrated on those based on immunofluorescent microscopy. These adapt well to large-scale analysis and can be designed to assess a large number of functions using specific antibodies directed against proteins involved in the cellular process of interest. For example, we have used an antiphosphohistone antibody that identifies cells in mitosis to screen for genes involved in this stage of the cell cycle (available from Upstate Biotechnologies). RNAi knockdown of such a gene will result in a change in the percentage of mitotic cells in the population. While it is possible to assess this manually, this is a labor-intensive task and there are a number of automated systems on the market that can help to speed up the process. One example is Cellomics® ArrayScan HCS System, which is able to carry out automated assessment of a number of cell parameters, including mitotic index, activity of various cell-signalling pathways, apoptosis, cell motility, and viability.

While it is possible to analyze RNAi-induced phenotypes using flow cytometric techniques, as many of these machines have the capacity to process small quantities of cells directly from 96-well plates, such assays will require that cells grown as a suspension are used in the initial transfection and may prove limited and practically difficult. In addition, it should prove relatively straightforward to develop biochemical-type assays based upon the activity of a particular enzyme involved in the process of interest, provided a suitable substrate and some form of readout, e.g., fluorescent or radioactive emission or microscopy-based assays, is available.

PROTOCOLS

PROTOCOL 10.1 SUBCULTURING AND FREEZING OF CELLS

Materials

- 75 cm^2 tissue culture flasks
- 70% (v/v) ethanol
- Media and PBS (see Table 10.1; varies for each cell line)
- DMSO (tissue culture grade) (Sigma)
- Typsin-EDTA (0.1% (w/v) Trypsin in 2 mM EDTA; Gibco)
- Cryovials (Nunc)
- Tissue culture hood
- Tissue culture incubator at 25°C
- Inverted microscope
- Access to –140°C freezer or liquid nitrogen container
- Centrifuge (with adaptors for 50-ml and 15-ml Falcon tubes) and microcentrifuge

Experimental Procedures

Cells should be routinely subcultured once to twice a week (Table 10.1). Use an aseptic technique (wipe everything with 70% (v/v) ethanol and always work inside the tissue culture hood).

1. Check the cells on the microscope to make sure they look healthy and that there is no contamination. Resuspend cells from the bottom of the flask in 10 ml of medium.
2. Add 1:4 to 1:10 of the cells (see Table 10.1) to a fresh tissue culture flask (15 ml of warmed media (25°C) in a 75-cm^2 flask).

Trypsinization

3. If clone 8 or S2R+ cells are used, these may be strongly attached to the flask and they may be detached from the flask/plate by trypsinization. Aspirate the medium and wash the cells with PBS (without calcium or magnesium).

4. Remove PBS and add 2 ml trypsin–EDTA solution. Incubate at RT for 5 min (check for detachment of the cells using a microscope).
5. Add 8 ml of fresh medium and spin cells at 300 × *g* for 5 min.
6. Aspirate supernatant and resuspend the cell pellet in 5 ml of fresh medium. Add 1 ml of this cell suspension to a fresh tissue culture flask containing appropriate media.

Freezing the cells

7. Three aliquots of cells can be frozen from one confluent 75-cm^2 flask. Prepare a box of dry ice and warm the medium to room temperature. Appropriately label the 3 cryo-tubes (e.g., cell type, passage number, name, and date).
8. In a sterile tube, prepare a mix (cryogenic media) of 3 ml of fresh media (in the case of DMEL cells, use the supernatant obtained after pelleting cells by centrifugation) and 260 µl DMSO.
9. Centrifuge cells at 300 x *g* for 5 min and aspirate the medium.
10. Working quickly, carefully resuspend the pellet in 3 ml of cryogenic medium.
11. Place 1 ml of cell suspension into each labelled cryo-tube, place on ice, and immediately transfer to a –80°C freezer.
12. The following day (after ~24 h), transfer the tube to a –140°C freezer (transport on dry ice) or a liquid nitrogen container.

Recovery of cells

13. Recover cultures from frozen storage by rapidly thawing a vial of cells in a 37°C water bath, gently shaking the tube until the medium completely thaws.
14. Immediately transfer into a 15-ml tube with 9 ml of prewarmed (25°C) medium.
15. Centrifuge at 300 × *g* for 5 min, resuspend in 10 ml medium, and inoculate into a small tissue culture flask (25 cm^2).

Protocol 10.2 Preparation of Genomic DNA from *Drosophila* Cells

Materials

- Lysis buffer (100 mM NaCl, 10 mM Tris, pH 8.0, 25 mM EDTA, pH 8.0, 0.5% (w/v) SDS, 0.1 mg/ml Proteinase K)
- Phosphate-buffered saline (PBS), pH 7.4
- Sterile/RNase and DNase-free tips and plastic ware
- Phenol:Chloroform (1:1, v/v)
- 7.5 M Ammonium acetate
- Nuclease-free water
- Ethanol
- Hemocytometer (Sigma)

Experimental Procedures

1. Pellet cells from a 75-cm^2 flask of healthy-growing culture and gently resuspend the cells in 10 ml fresh medium. Transfer the cell suspension to a 50-ml tube.
2. Count cells using a hemocytometer: after dispersing the cells in medium, add 10 µl of the cell suspension to the haemocytometer; each large square of a standard haemocytometer usually contains 10^{-4} ml.

3. Calculate the suspension volume containing 30 × 10^6 cells and centrifuge this volume at 300 × *g* for 5 min at 22°C. Discard the medium and wash the cells twice in 25 ml PBS.
4. Resuspend the cells in 250 µl lysis buffer and transfer into an Eppendorf tube.
5. Incubate the samples at 50°C overnight.
6. Remove proteins by phenol/chloroform extraction. Add an equal volume (250 µl) to cells, mix by inversion, and centrifuge for 5 min at maximum speed in a microcentrifuge.
7. Pipette the upper, aqueous phase to a new tube and repeat the extraction once more. Add an equal volume of chloroform and repeat the purification twice in the same way. Transfer the aqueous phase to a fresh tube.
8. Precipitate the DNA by adding ½ volume of 7.5 M ammonium acetate and 2 volumes of 96% (v/v) ethanol. Invert 6 to 8 times and incubate at –20°C for 15 min.
9. Centrifuge in a microcentrifuge at top speed for 10 min, pour off the supernatant, and rinse the pellet with 1 ml ice-cold 70% (v/v) ethanol.
10. Dry the DNA at room temperature for 10 to 15 min and resuspend in 500 µl nuclease-free water.

Protocol 10.3 Isolation of Total RNA and cDNA Synthesis

There are a variety of methods for the isolation of total RNA from cells. Commercial kits are available from many companies (Qiagen, Gibco, Pharmacia). One simple way that we will describe here is to prepare total RNA from cells using TRIZOL reagent (GIBCO BRL). All plastic ware should be RNase-free and sterile. Filter tips should be used. Gloves must be worn at all times.

Materials

- Trizol (Gibco)
- SuperScript™ First-Strand Synthesis System (Invitrogen)
- Phosphate-buffered saline (PBS), pH 7.4
- Sterile/RNase and DNase-free tips and plastic ware
- Nuclease-free water
- General PCR-dNTPs, Taq polymerase and buffer
- Chloroform
- Ethanol
- Isopropanol
- Thermal cycler (MJ Research)
- Spectrophotometer (Helios)

Experimental procedures

Total RNA isolation

1. Pellet cells from a 75-cm^2 flask of healthy-growing culture and gently resuspend them in 10 ml fresh medium. Transfer the cell suspension to a 50-ml tube.
2. Count the cells in a hemocytometer and calculate the total cell number.

3. Centrifuge the cells at 300 × *g* for 5 min at 22°C. Discard the medium and wash the cells twice in 25 ml PBS.
4. Resuspend 5 to 10 × 10^6 cells in 1 ml TRIZOL reagent, transfer into a sterile Eppendorf tube, and incubate at room temperature for 5 min.
5. Add 0.2 ml chloroform to the sample, shake by hand for 15 sec, and incubate at room temperature for 3 min.
6. Centrifuge at 4°C for 15 min at 12,000 rpm to separate the phases. Pipette the upper aqueous phase to a new tube.
7. Add 0.5 ml isopropanol and mix gently by pipetting. Incubate at room temperature for 10 min.
8. Centrifuge at 4°C for 10 min.
9. Wash the pellet in ice-cold 75% (v/v) ethanol and centrifuge at room temperature for 5 min at 12,000 rpm.
10. Dry the pellet at room temperature for 10 min and dissolve the RNA in 50 μl nuclease-free water.
11. Measure the quality and the quantity of the RNA using a spectrophotometer at 260 nm/280 nm. Ratios of A_{260}/A_{280} of 1.9 to 2.0 indicate highly purified preparations of RNA. The RNA yield can be calculated as follows:
    A_{260} × dilution factor × 40 = RNA concentration in μg/ml

cDNA synthesis

Synthesis of cDNA involves a number of steps, and there is a variety of alternative procedures for each of these. The first step in the synthesis of cDNA is using the mRNA template to produce a complementary single-stranded cDNA. This first strand of cDNA is suitable for PCR amplification. To prepare a first strand of cDNA from total RNA, we have used SuperScript first-strand synthesis system (Invitrogen) according to the manufacturer's instructions.

Protocol 10.4 Production of DNA Template by PCR

It may be necessary to perform at least two or three 50-μl PCR reactions to produce the appropriate amount of template DNA for the *in vitro* transcription.

Materials

- Thermal cycler
- Template DNA (cDNA or genomic DNA)
- Primers containing T7/SP6 polymerase binding site
- Taq polymerase; PCR buffer, dNTPs (10 mM), $MgCl_2$ (25 mM)
- Nuclease-free water

Experimental Procedures

1. Mix the template DNA (50 ng of plasmid cDNA, 500 ng of isolated genomic or cDNA) with 1 μl of Forward primer (20 μM), 1 μl of Reverse primer (20 μM), 5 μl 10 × PCR Buffer, 5 μl 25 mM $MgCl_2$, 1 μl 10 mM dNTP mix, 1 μl Taq polymerase (5U/μl), and water to a final volume of 50 μl.

2. We have successfully used the following PCR program for more than 300 different genes:
   94°C, 2 min
   94°C, 30 sec; 55°C, 30 sec; 72°C, 60 sec; 35 cycles
   72°C, 10 min
3. The PCR product can be purified by ethanol precipitation or with the PCR Clean UP kit (Qiagen), and the quality and the quantity of the PCR product can be tested by gel electrophoresis (1% (w/v) agarose-TAE buffer). Quantity can also be determined by measuring the absorbance at 260 nm (A_{260} × dilution factor × 50 = DNA concentration in μg/ml). The product should only have one band and it should run at the predicted size.

Protocol 10.5 Making dsRNA

Materials

In vitro transcription reagents (e.g., T7 RiboMax™ Express RNAi system)

Experimental Procedure

1. Quantify the DNA template and use 2 μg in 40 μl final volume of the transcription reaction.
2. Thaw the reaction component at room temperature, except the enzyme mix, and add the reaction components for 40 μl reaction volume into an RNase-free 1.5-ml microcentrifuge tube:

RiboMAX™ 2 × buffer	20 μl
Template DNA	1–16 μl ((2μg)
Nuclease-free water	0–15 μl
T7 Express Enzyme Mix	4 μl
Total	40 μl

3. Mix gently and incubate at 37°C for 4 hours.
4. To anneal the RNA strands, heat the reaction mix to 70°C for 20 min and cool down slowly (for example, by incubating for 20 min at 37°C and for 20 min at room temperature).
5. Remove the DNA template and the single-strand RNA by digestion with DNase and RNase. This is performed by diluting the supplied RNase solution 1:200, adding 2 μl freshly diluted RNase solution and 2 μl DNase solution per 40 μl reaction volume, and incubating for 30 min at 37°C.
6. Precipitate the dsRNA with 1/10 volume of 3M Sodium Acetate (pH 5.2) and 2.5 volumes of 95% (v/v) ethanol. Mix and incubate on ice for 10 min.
7. Centrifuge at top speed in a microcentrifuge at 4°C for 10 min.
8. Wash the pellet with 500 μl cold 70% (v/v) ethanol.
9. Remove ethanol completely and dry the dsRNA at room temperature for 15 min.
10. Resuspend dsRNA in 100 μl nuclease-free water.
11. Quantify dsRNA concentration by measuring the absorbance of 1/500th of the volume of dsRNA at 260 nm and calculate the concentration (one A_{260} unit equals ~40 μg/ml of dsRNA).

12. Run 1/100th volume of the dsRNA on an agarose gel; check the quantity and the quality. The dsRNA should be a single band on the gel. Note that dsRNA migrates slightly more slowly than the dsDNA. Compare the amount of the dsRNA to known molecular markers (e.g., Low DNA Mass™ Ladder, Invitrogen).
13. Aliquote and store dsRNA at –20°C. It should be stable for at least 12 months at –20°C without detectable loss of efficiency.

Protocol 10.6 Transfection of S2 Cells

Materials

- 75-cm^2 tissue culture flasks
- Culture media (see Table 10.1)
- Phosphate-buffered saline (PBS), pH 7.4
- 6/24/96-well tissue culture dishes
- Transfast (Promega)
- Effectene (Qiagen)
- Hemocytometer
- Sterile pipette tips and Eppendorf tubes

Experimental Procedure

1. Use exponentially growing cells just about to peak confluence. Remove the old medium from the flask and resuspend the cells in 10 ml fresh medium by pipetting.
2. Count and calculate the cell density using a hemocytometer.
3. Add 1×10^6 cells in 3 ml of medium to each well of a 6-well plate (Table 10.3). The cells should settle overnight.
4. The following day, prepare the transfection medium (Table 10.3). For a 6-well plate, the total volume per well should be 1 ml. For each well, add 20 μl Transfast reagent, 10 μl (20 μg) dsRNA, and 970 μl serum-free medium to a sterile tube and vortex.
5. Incubate for 15 min at room temperature.
6. Briefly remove the serum-containing medium from the cells; gently wash with 2 ml serum-free medium.
7. Remove the medium and add the transfection medium to the cells and swirl the plate to produce a uniform distribution of the liquid.
8. Incubate the plate at 25°C for 1 h.
9. Add 2 ml of medium, supplemented with 15% (v/v) FBS.
10. Place the cells in the incubator at 25°C until analysis, which is generally 3 to 5 days postincubation.

REFERENCES

1. Cottrell, T.R. and Doering, T.L., Silence of the strands: RNA interference in eukaryotic pathogens, *Trends Microbiol.,* 11, 37, 2003.
2. Hannon, G.J., RNA interference, *Nature,* 418, 244, 2002.

3. Cerutti, H., RNA interference: traveling in the cell and gaining functions? *Trends Genet.,* 19, 39, 2003.
4. Fire, A. et al., Potent and specific genetic interference by double-stranded RNA in *Caenorhabditis elegans*, *Nature,* 391, 806, 1998.
5. Zamore, P.D., Ancient pathways programmed by small RNAs, *Science,* 296, 1265, 2002.
6. Sijen, T. et al., On the role of RNA amplification in dsRNA-triggered gene silencing, *Cell,* 107, 465, 2001.
7. Roignant, J.Y. et al., Absence of transitive and systemic pathways allows cell-specific and isoform-specific RNAi in *Drosophila*, *RNA,* 9, 299, 2003.
8. Giet, R. and Glover, D.M., *Drosophila* aurora B kinase is required for histone H3 phosphorylation and condensin recruitment during chromosome condensation and to organize the central spindle during cytokinesis, *J. Cell Biol.,* 152, 669, 2001.
9. Giet, R. et al., *Drosophila* Aurora A kinase is required to localize D-TACC to centrosomes and to regulate astral microtubules, *J. Cell Biol.,* 156, 437, 2002.
10. Lum, L. et al., Identification of Hedgehog pathway components by RNAi in *Drosophila* cultured cells, *Science,* 299, 2039, 2003.
11. Goshima, G. and Vale, R.D., The roles of microtubule-based motor proteins in mitosis: comprehensive RNAi analysis in the Drosophila S2 cell line, *J. Cell Biol.,* 162, 1003, 2003.
12. Rogers, S.L. et al., Molecular requirements for actin-based lamella formation in *Drosophila* S2 cells, *J. Cell Biol.,* 162, 1079, 2003.
13. Somma, M.P. et al., Molecular dissection of cytokinesis by RNA interference in *Drosophila* cultured cells, *Mol. Biol. Cell,* 13, 2448, 2002.
14. Echalier, G., *Drosophila Cells in Culture*. San Diego: Academic Press, 1997.
15. Debec, A. and Abbadie, C., The acentriolar state of the *Drosophila* cell lines 1182, *Biol. Cell,* 67, 307, 1989.
16. Debec, A., Haploid cell cultures of *Drosophila melanogaster*, *Nature,* 274, 255, 1978.
17. Peel, D.J. et al., The ultrastructure of imaginal disc cells in primary cultures and during cell aggregation in continuous cell lines, *Tissue Cell,* 22, 749, 1990.
18. Ui, K. et al., Newly established cell lines from *Drosophila* larval CNS express neural specific characteristics, *In Vitro Cell Dev. Biol. Anim.,* 30A, 209, 1994.
19. Ashburner, M., *Drosophila. A Laboratory Handbook*, Cold Spring Harbor, NY: Laboratory Press, 1989.
20. Kiger, A. et al., A functional genomic analysis of cell morphology using RNA interference, *J. Biol.,* 2, 27, 2003.
21. Yang, D. et al., Evidence that processed small dsRNAs may mediate sequence-specific mRNA degradation during RNAi in *Drosophila* embryos, *Curr. Biol.,* 10, 1191, 2000.
22. Feinberg, E.H. and Hunter, C.P., Transport of dsRNA into cells by the transmembrane protein SID-1, *Science,* 301, 1545, 2003.
23. Worby, C.A. et al., RNA interference of gene expression (RNAi) in cultured *Drosophila* cells, *Sci. STKE,* 2001, PL1, 2001.
24. Muda, M. et al., Use of double-stranded RNA-mediated interference to determine the substrates of protein tyrosine kinases and phosphatases, *Biochem. J.,* 366, 73, 2002.
25. Clemens, J.C. et al., Use of double-stranded RNA interference in Drosophila cell lines to dissect signal transduction pathways, *Proc. Natl. Acad. Sci. U.S.A.,* 97, 6499, 2000.
26. Dykxhoorn, D.M. et al., Killing the messenger: short RNAs that silence gene expression, *Nat. Rev. Mol. Cell Biol.,* 4, 457, 2003.
27. Hutvagner, G. and Zamore, P.D., RNAi: nature abhors a double-strand, *Curr. Opin. Genet. Dev.,* 12, 225, 2002.

28. Gonczy, P. et al., Functional genomic analysis of cell division in *C. elegans* using RNAi of genes on chromosome III, *Nature,* 408, 331, 2000.
29. Schneider, I., Cell lines derived from late embryonic stages of *Drosophila melanogaster*, *J. Embryol. Exp. Morphol.,* 27, 353, 1972.
30. Yanagawa, S. et al., Identification and characterization of a novel line of *Drosophila* Schneider S2 cells that respond to wingless signaling, *J. Biol. Chem.,* 273, 32353, 1998.
31. Echalier, G. and Ohanessian, A., *In vitro* culture of *Drosophila melanogaster* embryonic cells, *In Vitro,* 6, 162, 1970.

11 RNAi in *Caenorhabditis Elegans*

Femke Simmer and Ronald H.A. Plasterk

CONTENTS

11.1 INTRODUCTION

The nematode *Caenorhabditis elegans* is a useful animal model organism to study complex biological phenomena (Figure 11.1).[1] Most *C. elegans* genes have homologues in other organisms, including humans, and many biological mechanisms are well conserved. Importantly, *C. elegans* is easily maintained in the laboratory and is amenable to a large range of experimental methods. Experiments in *C. elegans* have, for instance,

0–8493–2141–7/05/$0.00 + $.50

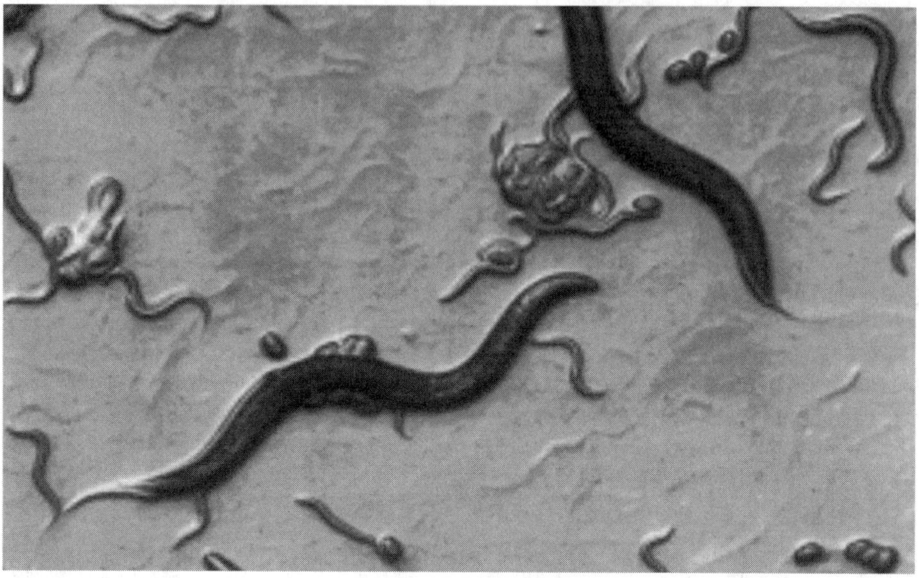

FIGURE 11.1 *C. elegans* is a small transparent roundworm (adults are 1 mm) with a short life cycle (~3 days). This picture shows a *C. elegans* culture growing on the surface of an NGM agar plate. Populations of *C. elegans* consist almost entirely of diploid self-fertilizing hermaphrodites. A low percentage of males do exist, which can cross-fertilize hermaphrodites. Development and physiology of *C. elegans* are understood at the single-cell resolution. Every adult hermaphrodite worm contains 959 somatic cells that have arisen in an essentially invariant manner.

played an important role in elucidating the RAS-signaling pathway and aspects of apoptosis and in the discovery of RNA interference (RNAi).

Traditionally, gene function in *C. elegans* has been analyzed by classical forward genetics, in which mutant animals with a specific phenotype are identified following treatment with a mutagen. The mutated gene is subsequently isolated by positional cloning. Identification of one gene is then the starting point for further genetic analysis, such as suppressor or enhancer screens to identify other factors involved in the process causing the phenotype. The complete *C. elegans* genome has now been sequenced (approximately 19,000 genes),[2] which makes it possible to start experiments from a different angle. In reverse genetics a gene is identified first through its sequence and then mutations are recovered, for instance, by targeted gene disruption[3] that inactivate or alter its function. Rather than examining a single gene, it has also become possible to do large-scale experiments with subsets of genes with certain sequence motifs, such as gene families.[4] In addition, it is now possible to perform even larger experiments, such as DNA microarray analysis, to assess the transcriptional response of the entire genome under different conditions.[5] Furthermore, several experiments become easier and faster with the knowledge of an entire genome sequence. For instance, the genome sequence is now used to identify the mutated gene in mutants obtained in forward genetic screens; once the mutated gene has been mapped to a region of the genome, sequencing of candidate genes in that region makes the identification of the mutated gene easier. RNAi is an important discovery in *C. elegans* that also allows new approaches for studying gene function.[6] It is a useful tool for the transient inactivation of a specific gene. RNAi can provide information on gene function for genes that have interesting sequence motifs. Alternatively, RNAi can be used in combination with genetic or biochemical experiments. For example, RNAi can be used as a method to investigate the function of genes identified by yeast two-hybrid experiments. Furthermore, genome-wide RNAi screens can be carried out to identify genes involved in a specific biological process.[7–11] Finally, RNAi can be useful to find genes that suppress or enhance specific mutant phenotypes or induce a synthetic phenotype.

11.2 DOUBLE-STRANDED RNA

RNAi is induced by double-stranded RNA (dsRNA). Introduction of dsRNA causes degradation of mRNA with complementary sequence to the dsRNA. When the mRNA is lost, the production of the encoded protein stops. This results in a transient loss-of-function effect.[12] dsRNA targeting the mRNA of each gene of *C. elegans* can be made. The sequence to be targeted for gene-specific RNAi should be carefully selected to avoid cross-interference between highly homologous sequences.[13] In addition, it can be worthwhile to choose more than one sequence, because differences in efficiency between different dsRNAs that target the same gene are observed.[13] The length of dsRNA can also have an effect on the potency of RNAi.[14] Most studies utilize dsRNA that is between 0.5 and 1 kb in length, although smaller fragments can be successful.

There are four different ways to introduce the dsRNA into *C. elegans*: (1) injection into different body parts;[6] (2) soaking in a solution of dsRNA;[13] (3) feeding of *Escherichia coli*-producing dsRNA;[15] and (4) expression of dsRNA from a transgene in the worm.[16,17] In most labs, RNAi induced by feeding is primarily used. A protocol for making dsRNA-expressing bacteria and for RNAi induced by feeding is described in detail here. Other techniques are discussed briefly, and references to the literature are included in Section 11.6 (also see Chapter 12).

11.2.1 *In Vitro* Synthesis of dsRNA

dsRNA for injection or soaking can be synthesized *in vitro*.[6] First, a PCR product is made from cDNA or genomic DNA using specific primers that also contain T7, T3, or SP6 promoter sequence. The PCR product is subsequently used for *in vitro* transcription with T7, T3, or SP6 RNA polymerase. dsRNA is ideally injected at a concentration between 0.5 and 1.0 mg/ml. A drawback of the induction of RNAi by soaking is that relatively large amounts of dsRNA are needed (between 1 and 5 mg/ml).

11.2.2 Bacteria Expressing dsRNA

For the induction of RNAi by feeding, bacteria should be made that express dsRNA (Figure 11.2).[15,18] A DNA fragment of *C. elegans* coding sequence has to be cloned

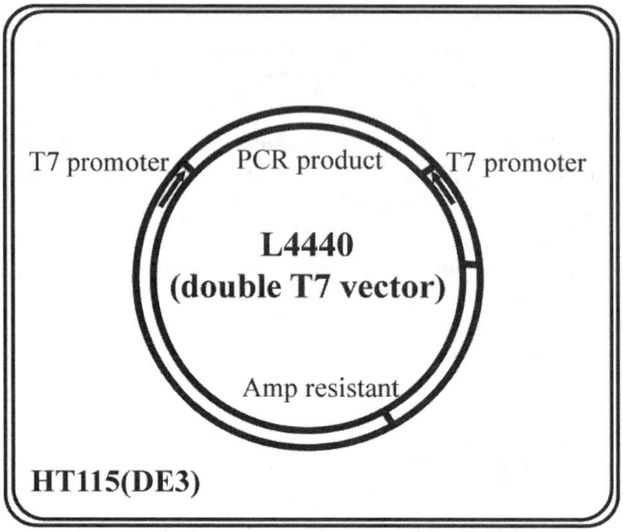

FIGURE 11.2 Scheme for dsRNA production by bacteria. A fragment from the gene of interest is amplified by PCR and cloned into the L4440 vector. The vector with insert is subsequently transformed into HT115(DE3), an RNase III deficient *E. coli* strain with IPTG inducible expression of T7 polymerase. The L4440 vector has two T7 promoters in inverted orientation flanking the multiple cloning site. Transcription will result in dsRNA of the insert.

into an expression vector with a multiple cloning site flanked by two copies of the T7 promoter in opposite orientation. The constructed vector is preferably transformed into the *E. coli* strain HT115 that contains a T7 RNA polymerase under the control of an inducible promoter. This *E. coli* strain is also deficient in RNase III, since RNase III normally degrades most dsRNA in the bacteria; this causes an increase the concentration of dsRNA. Protocol 11.1 describes the construction of dsRNA-expressing bacteria in more detail.

Once the gene of interest is cloned into a vector for dsRNA expression and the vector is transformed into bacteria, the dsRNA-expressing bacteria can be frozen and reused. Therefore, this method of dsRNA production is relatively inexpensive compared to *in vitro* synthesized dsRNA for injection or soaking. In addition, it is not necessary to work in an RNase-free manner. Even more convenient is that bacterial strains for 86% of the *C. elegans* genes are already available from a library of dsRNA-expressing bacteria.[8,19] This library was made by the Ahringer laboratory (Cambridge, UK), and is available by individual chromosome sets (I, II, III, IV, V, and X) or as single bacteria strains from MRC geneservice (http://www.hgmp.mrc.ac.uk/geneservice/index.shtml). Thus, a great majority of *C. elegans* genes can be studied using RNAi induced by feeding these bacterial strains.

11.2.3 Transgene-Derived dsRNA

C. elegans is typically transformed by injecting DNA fragments into the gonads of the animals.[20] Transgenic arrays that produce dsRNA can be obtained by injecting a mixture of plasmids: a plasmid with an antisense segment under the control of a specific promoter and a plasmid with a corresponding sense segment, driven by the same promoter.[17] It is also possible to inject plasmids that have an inverted repeat (IR) of coding sequence under the regulation of a specific promoter.[16] RNA transcribed from such a construct folds back into a hairpin dsRNA. However, cloning of constructs with IRs can be difficult because they are unstable in *E. coli*. To overcome this problem, there are special *E. coli* strains that are more tolerant of IRs. In these bacterial strains, genes are deleted that are involved in DNA repair and recombination (e.g., RecA, RecB, RecJ, uvrC, umuC, and SbcC). The introduction of an unrelated spacer between the IR can also be helpful. The spacer can be from coding sequence of another gene but also of intronic sequence, which is spliced out in the worm, forming a piece of dsRNA without a large single-stranded loop.[21] In principle, it should also be possible to inject plasmids containing a coding sequence flanked by two copies of a promoter in opposite orientation, similar to the vector used to produce dsRNA-expressing bacteria, to obtain transgenes that generate dsRNA.

Promoters driving broad expression can be cloned into the constructs to induce general continuous RNAi. Inducible promoters, such as a heat shock promoter, can be used for stage-specific RNAi. Interestingly, by using plasmids with a tissue-specific promoter fused to an IR, transgene-induced RNAi can be exploited to dissect tissue-specific functions of genes. When dsRNA is introduced by soaking, feeding, or injection into the different parts of the body, it can exert an RNAi effect in tissues that are located distantly from the tissue where the dsRNA was introduced.[6,13,15] This indicates that there is spreading of RNAi throughout the worm (systemic RNAi). In several cases in which hairpin dsRNA is expressed from a transgene, it does not seem to result in spreading of the RNAi effect.[22] However, very strong tissue-specific promoters can induce RNAi outside the cells where they are active.[22]

11.3 METHOD OF DELIVERY

11.3.1 INJECTION OF DSRNA

Initially, RNAi experiments in *C. elegans* were done by injecting dsRNA into the gonads of young adult hermaphrodites and examining their progeny for a phenotype. Some phenotypes can be detected already in the injected hermaphrodite (P0). The injection method is the same as used for DNA transformation. dsRNA can also be injected into other body parts such as the body cavity (pseudocoelom) or intestine.[6]

Injection introduces one dose of dsRNA that causes degradation of mRNA and depletes gene product in the tissues of the mother, including the germline. This results in loss of maternal mRNA and maternal gene product that is deposited into the egg. Some of the dsRNA molecules, or downstream effecter molecules of RNAi, are also delivered via the germline to the progeny, leading to mRNA degradation in the progeny. This method is discussed in greater detail in Chapter 12.

dsRNA injection is very effective for studying gene function in the embryo but appears to work less well for genes expressed later during development compared to soaking or feeding of dsRNA. In addition, RNAi by injection is not useful for experiments that require large quantities of affected animals, such as biochemical experiments. Furthermore, RNAi by injection is not ideally suited to high-throughput RNAi analysis, since these experiments also require many individual animals to be injected. Nevertheless, Gonczy et al.[23] analyzed 96% of the genes on Chromosome III of *C. elegans* to identify genes required for cell division using RNAi by injection and identified 133 genes necessary for distinct cellular processes in early embryos. They injected the dsRNAs in pools of two (larger pool sizes were found to result in significantly lower penetrance of the RNAi effect) to increase speed.[23]

11.3.2 Soaking in dsRNA

To obtain an RNAi effect, worms can also be soaked in solutions of dsRNA. This method was first developed by Tabara et al. and later improved by Maeda et al.[13,24] Soaking can be applied to worms of any developmental stage, which is useful for determining the function of genes at different developmental stages. When young adults are soaked in dsRNA and the F1 generation is analyzed, mainly maternal mRNA is degraded in these animals. In some cases this can result in embryonic RNAi phenotypes that mask later postembryonic phenotypes. When L1 larvae are soaked in dsRNA and analyzed, the RNAi effect starts later in development. Therefore, soaking of L1 larvae can help to reveal the postembryonic functions of genes.[25] This is less feasible with RNAi induced by injection, as injections are fairly difficult in the larvae of *C. elegans*.

To synchronize a population at the L1 stage (adult) worms are treated with hypochlorite solution.[26] The worms will dissolve in this solution, whereas eggs survive longer due to their solid eggshell. When the eggs are kept overnight without food, they will hatch and the worms arrest in the L1 stage due to the lack of food. Introduction of food allows the worms to resume development from the same starting age.

Other advantages of soaking as compared to injecting are that soaking is less labor intensive and does not require injection equipment such as an inverted compound light microscope, needle puller, and pressurized nitrogen cylinder. In addition, multiple RNAi experiments can be performed concurrently. The dsRNA synthesis and worm soaking can be performed in 96-well format.

11.3.3 Feeding of Bacteria Expressing dsRNA

C. elegans is maintained in the laboratory on Petri plates filled with Nematode Growth (NGM) agar and on top of the agar a bacterial lawn of *E. coli*.[1] The *E. coli* are the food source for the animals. To induce RNAi, bacteria expressing dsRNA can be used (Protocol 11.2).[15,18] This method of RNAi induction can be performed with worms of any developmental stage and for many generations. This method is easily scalable; RNAi can be performed on thousands of worms or dsRNA for many genes can be tested in a high-throughput fashion. When analyzing large numbers of individual animals and/or dsRNAs, 12-, 24-, or 96-well plates can be used. If 96-well plates are used, they can be filled with liquid growth medium and bacteria. To obtain sufficient amounts of material for biochemical experiments, a large number of worms can be grown on plates or in liquid culture with bacteria expressing dsRNA.

Compared to injection or soaking, RNAi by feeding has a similar efficiency in generating phenotypes. However, the penetrance for several kinds of phenotypes seems to be more variable. This is not the case for postembryonic phenotypes, which are detected more frequently when RNAi is induced by feeding than with the other methods.[27] This could be due to the fact that both the mothers and their progeny are constantly exposed to the dsRNA, whereas with injection or soaking, only the mothers are subjected to dsRNA.

The Ahringer laboratory has made a genome-wide RNAi feeding library and used this library to identify genes with easily detectable loss-of-function phenotypes such as lethality, sterility, defects in body shape, and animal movement.[8,19] In total, RNAi phenotypes were detected for 1722 genes. This data set provides a starting point for many studies. One can get an overview of which roles genes with a particular RNAi phenotype have in development or analyze the relationship between the function of a gene and its sequence or location in the genome. Another possibility is to study certain genes that give an RNAi phenotype associated with a specific biological process in further detail. Using this library, many more genome-wide RNAi-based screens using more specific assays can be performed. In this way genes involved in all kinds of biological processes will be identified, which will give many new insights and greatly accelerate research. Examples of large-scale screens for specific processes using the RNAi feeding library are a screen for fat-regulatory genes,[7] screens for genes involved in lifespan,[28,29] a screen to identify genes involved in genome stability,[9] and a screen to identify genes involved in transposition.[10]

11.3.4 Transgenic Worms Expressing dsRNA

Transgenic worms are mainly made by injecting DNA into the gonads of young adult hermaphrodites.[20] To induce RNAi, DNA that, when transcribed, forms dsRNA can be injected. DNA encoding a selectable marker has to be co-injected to facilitate the identification of transformants in the progeny. The injected DNA forms an extra chromosomal array. A fraction of the F1 transformants contains heritable arrays. The transformant lines can be maintained for many generations on agar plates, and for long-term storage they can be frozen at –80°C in medium containing glycerol. However, when maintained on agar plates, it is often necessary to pick transformants every generation to prevent loss of the semistable extra chromosomal array. Extra chromosomal arrays can be stabilized by radiation-induced integration into a chromosome. In a stable transgenic line, all offspring inherit the transgene. A disadvantage of DNA transformation by injection is that learning the technique of DNA transformation requires practice; also the making of stable transgenic lines is quite laborious.

Now several labs use micro-particle bombardment as an alternative method for DNA transformation.[30] In this approach, gold beads are coated with DNA and accelerated to high speed, allowing them to penetrate the *C. elegans* cuticle. This results at low frequency in chromosomal insertions. The integrated lines obtained with this approach typically contain only a few copies of the transforming DNA. A selectable cotransformation marker is used to identify the rare transformants with an integrated array within the population of bombarded animals and their descendants. This method of DNA transformation by micro-particle bombardment is a good method to directly obtain integrated lines. However, specific equipment is needed.

Transgene-induced RNAi can be useful for RNAi in neurons. In general, RNAi against neuronally expressed genes is difficult in *C. elegans*.[18,31] However, using a transgene it was shown that RNAi could be effective in the neurons,[16] also indicating that the machinery to perform RNAi is proficient in these cells. Furthermore, using transgene-induced RNAi, large quantities of transgenic worms can be grown for biochemical assays.

11.3.5 Additional Remarks and Tricks

11.3.5.1 Range of Phenotypes

A range of phenotypes is often seen in an RNAi-affected population, from weak to strong loss-of-function phenotypes. This can be useful when the null phenotype is lethality; the weaker phenotypes could give insight into other processes in which the gene is involved. In addition, due to the weaker phenotypes obtained with RNAi, relevant genes with lethal null phenotypes that were missed in a classical forward genetic screen could be identified in an RNAi screen. In principle, to obtain a range of strong and hypomorphic phenotypes, the RNAi effect can also be titrated by reducing the amount of dsRNA.[27]

11.3.5.2 Combinatorial RNAi

Individual members of gene families often have partially redundant functions. In principle, combinatorial RNAi can be used to identify genes that functionally overlap.[32–34] However, combining more than two different dsRNAs is known to reduce the penetrance of the RNAi effect.[23]

11.3.5.3 Zygotic RNAi

As mentioned earlier, RNAi can be induced at different developmental stages to determine the stage-specific roles of genes. Transgenes under the control of inducible or stage-specific promoters can be used for induction of RNAi at a specific developmental stage. Soaking or feeding of L1 larvae is effective to inactivate postembryonic gene functions and dsRNA injection, and soaking or feeding of young adults/L4 larvae inactivates primarily embryonic gene functions in the F1 progeny. Since the eggshell of *C. elegans* is impermeable, dsRNA can only be introduced into eggs via the germline of the mother. To induce RNAi early in embryogenesis without affecting the mother, Herman[34] described a useful protocol: dsRNA is introduced into RNAi-deficient *rde-1* hermaphrodites, which are then mated with wild-type Bristol N2 males and the cross-progeny is analyzed.[35] Due to the mutation in *rde-1*, the mother and, therefore, also the maternal contribution of the target mRNA are not affected by the dsRNA. RNAi becomes active in the *rde-1*/+ progeny when *rde-1* is expressed. Thus, zygotically expressed target mRNA is degraded.

11.3.5.4 Maternal Effects

There are many genetic null mutants that are viable as embryos but do not develop to adults (larval arrest), or develop into sterile adults. Often these phenotypes are due to maternal contribution. Heterozygous mothers contribute mRNA that allows the homozygous progeny to develop into larvae or sterile adults, but zygotic mRNA is necessary for further development. Using conventional methods such as mosaic analysis, it is possible but relatively difficult to determine the maternal role of these genes. RNAi targets both maternal and zygotic mRNA and can, thus, be used to investigate the maternal effect of genes. RNAi experiments for genes with a maternal role often result in embryonic arrest due to loss of the maternal contribution.

11.3.5.5 Hypersensitive Strains

There are many factors that can influence the outcome of an RNAi experiment. These factors include method of delivery, developmental stage at which the dsRNA is delivered, the amount of dsRNA, temperature at which the experiment is conducted, choice of the dsRNA sequence, and stability of the target mRNA and the encoded protein.[27] In addition, RNAi against genes specifically expressed in the neurons of *C. elegans* is inefficient.[18,36]

When studying the family of RNA-directed RNA polymerases of *C. elegans,* it was found that mutations in one of these genes, *rrf-3* (*pk1426*), makes the animals hypersensitive to RNAi.[36,37] In general, when RNAi is induced in *rrf-3* mutant animals, more and stronger RNAi phenotypes are detected compared to wild-type. Importantly, several RNAi phenotypes for neuronal genes that are not detected with wild-type are detected with *rrf-3* animals. Thus, this mutant strain can be useful to study gene function when no phenotype is detected with wild-type, to increase the strength of an RNAi phenotype, or specifically to study neuronal genes. It is important to take into account that this mutant has additional phenotypes that could complicate experiments; *rrf-3* worms have a reduced brood size, produce a low percentage of dead eggs, have a slightly increased frequency of males, are temperature-sensitive sterile at 25°C, and are more sensitive to transgene silencing. In addition, these mutant phenotypes, in combination with RNAi, could in principle result in synthetic phenotypes. Other groups have screened for mutants that have increased RNAi efficiency in neurons making it possible to use such mutants specifically for neuronal RNAi.

11.4 APPLICATION AND READ-OUT

As discussed, RNAi is a rapid way to link sequence and gene function. Selection of the genes that are tested can be based on their sequence motif, expression pattern, or results from many other experiments. The other way around, RNAi can be used to identify the genes involved in a specific biological process. In particular, by using the genome-wide RNAi feeding library,[8,19] high-throughput functional screens can be conducted for genes involved in a wide range of biological processes.

The assays for the different processes can range from detection of lethality or morphology changes to activation of a reporter gene. Morphology features seen under a dissecting microscope include body size and shape and some changes in the positions or sizes of the internal organs. Defects in the vulva and male tail are often visible under the dissecting microscope. More subtle changes in morphology can be seen using the greater resolution of differential-interference contrast microscopy (DIC, Nomarski), including the presence of extra cells or cells that are misplaced. Time-lapse microscopy can be used to follow cell divisions. The use of fluorescence or other cytological tags can help with the identification of changes in cell patterns. Mutant behaviors are examined through the dissecting microscope. Relatively easy to detect are, for instance, changes in the rate or pattern of movement. There are also standard assays for egg laying, defecation, pharyngeal pumping, chemotaxis, chemical avoidance, thermotaxis, mechanosensory responses, dauer formation, food response, and male mating. Besides observing a phenotype directly, there are many other assays possible with dyes and reporters. For instance, 4', 6-diamidino-2-phenylindole (DAPI) staining of the chromosomes in the germline can be used to examine the nuclear organization.[38] Examples of reporters are the green fluorescence protein (GFP) and beta-galactosidase (*lacZ* gene).

In our laboratory we have conducted several large-scale screens using the RNAi feeding library to identify components of different biological processes. One example is the identification genes involved in genome stability.[9] In this screen the bacterial strains of the RNAi feeding library were fed to a *C. elegans* strain, NL3400 (*pkIs1604* [*rol-6 (su1006) hsp16/2::ATG(A)17gfp/LacZ* (out of frame)]), with an integrated transgene containing an out-of-frame LacZ reporter gene. Transgenes that acquire a mutation that brings the reporter gene back in frame can be visualized by histo-chemical staining of beta-galactosidase (Color Figure 11.3).* The bacteria were seeded onto 6-cm Petri plates containing NGM agar with IPTG, incubated overnight at room temperature, and then 2 to 4 worms of the *C. elegans* reporter strain in the L3/L4 developmental stage were put on the plates. The worms were grown at 20°C for 5 days, washed off the plates, and fixed and stained for the presence of beta-galactosidase with X-gal. Sixty-one genes were identified, for which knockdown by RNAi resulted in an increase in the level of spontaneous mutagenesis. These genes are highly conserved and in higher species could be tumor suppressors.

Another screen was designed to identify genes involved in transposition.[10] To monitor transposition, a *C. elegans* strain, NL960 (*unc-22* [*st136::Tc1*]), with a visible twitching phenotype caused by a Tc1 insertion in the muscle gene *unc-22* was used. Nine-centimeter plates with NGM agar with IPTG and ampicillin were seeded with

* Color figures follow page 204.

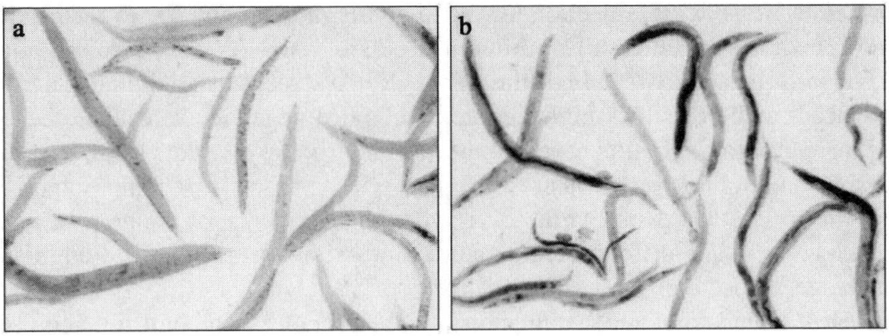

COLOR FIGURE 11.3 RNAi-mediated genome instability. To identify genes involved in genome stability, a *C. elegans* strain containing a heat shock promoter–driven out-of-frame *gfp-lacZ* reporter gene fusion was assayed on bacteria expressing dsRNAs for all genes of the *C. elegans* genome. (a) Control: worms fed on bacteria containing the vector without insert. (b) Example of dsRNA that induces genome instability. The worms were fed on bacteria that express dsRNA homologous to *msh-6*, a gene involved in mismatch repair. These worms have blue patches due to mutations that restore LacZ expression.

the bacterial strains from the genome-wide RNAi feeding library. Between one and three L2–L4 developmental stage animals of the NL960 strain were put on these plates and grown at room temperature until they were starved (4 weeks). The worms were rinsed off the plates and transferred to 15-cm NGM plates that were seeded with regular *E. coli* OP50 (not dsRNA producing). The worms were grown until starvation (two weeks) and then scored for revertants (wild-type moving worms). In these wild-type moving worms, the transposon is no longer in the *unc-22* gene, having jumped out of the *unc-22* gene in the germline of the mother, which restores the function of the *unc-22* gene. In total, 27 genes were identified that, when silenced, cause transposition of the transposon Tc1. Several of these genes suggest a role for chromatin structure in the regulation of transposition. Since most of the identified genes have homologs in other species, and since transposons are found throughout many different species, it seems likely that they also protect the genomes of other species against transposon activity in the germline.

11.5 LIMITATIONS OF RNAi

Not all genes are subject to RNAi; many dsRNAs only lead to a partial reduction of function phenotype or even fail to generate any mutant phenotype. Kamath et al.[27] determined the percentage of missed phenotypes in their genome-wide screen (this screen was for general phenotypes using the RNAi feeding library and wild-type worms) by analyzing the results for 323 loci, of which the null allele has an easily scorable phenotype. They found that for only 63.5% of the genes, phenotypes were observed; for the remaining 36.5%, no phenotype was detected. This demonstrates that a negative result with RNAi is not informative.

Another point of caution is that there is a certain degree of inter-experimental variation when RNAi is applied. Variation between high-throughput RNAi screens preformed using soaking, injection, or feeding[8,39] is most likely due to the use of the different methods which result in different levels of dsRNA, etc. More importantly, there is also variation between high-throughput RNAi screens using the same method.[31] This appears to be due to a high frequency of false negatives in each screen. It is important to take this variation into account when interpreting the data from the different screens. Because of this variation in effectiveness, it can be useful to repeat experiments several times. It is also worthwhile to assay another sequence of the gene of interest, do the experiment at a different temperature, start with worms of a different developmental stage, or use a different delivery method.

Clearly, genetic mutants remain indispensable; for one thing, null-mutants provide the definite information on the loss-of-function phenotype. In addition, genetic mutants are, in general, easier to work with in follow-up experiments, such as suppressor, enhancer, or synthetic screens, and biochemical experiments, to further unravel the pathway in which your gene of interest is involved. Furthermore, there are also alleles that cannot be replicated with RNAi, e.g., gain-of-function alleles or alleles that have mutations in specific domains of the gene that change interactions of the encoded protein with, e.g., DNA or other proteins. Thus, genetic methods and RNAi should be used complementarily.

11.6 REFERENCE LITERATURE FOR *C. ELEGANS* CULTURE TECHNIQUE, MICROSCOPY, AND GENETIC METHODS

1. Sulston, J. and Hodgkin J., in *The Nematode* Caenorhabditis elegans. Wood W.B., Ed., Cold Spring Harbor Laboratory Press, New York, 1988, p. 587–606.
2. Epstein H.F. and Shakes D., Eds. Caenorhabditis elegans: *Modern Biological Analysis of an Organism.* Academic Press, San Diego, 1995.
3. Hope I.A., Ed. *C. elegans*. Oxford University Press, New York, 1999.

PROTOCOLS

Protocol 11.1 dsRNA-Expressing Bacteria

Materials

- DNA fragment of gene of interest
- Timmons and Fire's feeding vector L4440 (Ref. 15) (The 1999 FireLab Vector Kit); also see Carnegie Institution Web site: http://www.ciwemb.edu
- *E. coli* strains: DH5alpha and HT115(DE3) (Caenorhabditis Genetics Center (CGC) Web site: http://www.cbs.umn.edu/CGC/CGChomepage)
- Appropriate antibiotics
- Culture media for *E. coli*

Experimental Procedures

1. Clone a fragment of the gene of interest into L4440 plasmid vector. This is a modified version of Bluescript with a T7 promoter on each side of the MCS.[15] In

our lab we typically clone genomic fragments of 500 bp to 1 kb in length, obtained by PCR, into the vector. Primers should be selected to give a fragment with a maximum portion of coding sequence. The cloning steps should be performed in DH5α bacteria or another standard cloning strain.

2. Transform the vector with insert into HT115(DE3) bacterial cells, plate onto standard LB Agar + tetracycline + ampicillin plates, and grow overnight at 37°C. HT115 (DE3) is an RNase III deficient *E. coli* strain with IPTG inducible T7 polymerase activity.[18] The strain is tetracycline resistant and grows on standard LB + tetracycline. Competent cells can be made using standard techniques.
3. Keep the plate with bacterial colonies at 4°C and make a glycerol stock (-80°C). Bacteria that have been stored on plates at 4°C for a longer period of time often lose effectiveness. Therefore, it is better to streak a new plate from glycerol stocks or retransform when starting RNAi experiments.

Protocol 11.2 RNAi Induced by Feeding

Materials

- dsRNA-expressing bacteria
- Culture media for *E. coli*
- Appropriate antibiotics
- IPTG (e.g., Invitrogen)
- Cholesterol (Sigma)
- NGM Agar (3 g NaCl, 17 g Agar, 2.5 g Bacto-Peptone, H_2O to one liter; sterilize by autoclaving; cool to approximately 60°C; add 1 ml 1 M $CaCl_2$, 1 ml 5 mg/ml cholesterol in ethanol, 1 ml 1 M $MgSO_4$, 25 ml 1 M K_2HPO_4/KH_2PO_4 (pH 6.0))
- M9 buffer (3 g KH_2PO_4, 6 g Na_2HPO_4, 5 g NaCl, H_2O to one liter; sterilize by autoclaving; add 1 ml 1 M $MgSO_4$)
- Sterile Petri plates and other plastic consumables
- 37°C shaker and 15°C / 25°C incubator
- *C. elegans* culture of the desired genotype
- Wormpick and dissecting microscope

Experimental Procedures

1. Pick a single colony and grow bacteria for 6 to 12 h in LB + 50 μg/ml ampicillin at 37°C. Do not add IPTG or tetracycline to the liquid culture, as this can reduce the RNAi effect.
2. Seed the bacteria onto NGM agar plates with Carbenicillin (25 μg/ml final concentration) and IPTG (1 mM final concentration). The ratio of bacteria to worms is important; RNAi will be lost when the worms become starved.
3. Let the plates with bacteria dry overnight at room temperature. The IPTG will induce the production of dsRNA. Dependent on the kind of plates that are used (e.g.,12-well plates dry more slowly than 6-centimeter plates) and what kind of phenotype is analyzed (e.g., movement or lethality; on wet plates subtle differences in the sinusoid movement of *C. elegans* are hard to observe, whereas lethality can

be clear), it is sometimes better to let the plates dry longer. After drying, the plates with bacteria can be stored for 2 to 3 weeks at 4°C without significant loss of potency.

4. Add worms to the plates.

Technical notes. It is important to minimize the amount of non-dsRNA-expressing bacteria that are transferred together with the worms. Therefore, it is best to add worms by handpicking or to wash worms first in M9 buffer and then aliquot them onto the assay plates. Different developmental stages can be put onto the bacteria. When L3/L4-stage hermaphrodites are put onto the plates, it is best to look in the F1 progeny for phenotypes. The P0 adult animals can sometimes show effects of the RNAi, but it is usually at a lower level than in the F1 animals. L3/L4 stage hermaphrodites become adults and start to lay their first eggs after 40 h at 15°C or 20 h at 25°C. The first progeny are not always completely affected. To avoid scoring of these progeny, transfer the adult worms onto new plates seeded with the same dsRNA-expressing bacteria after they laid a few eggs. Allow the worms to lay eggs for 24 h at 20°C, then remove the adults from the plates and score the progeny for phenotypes. It may be worthwhile to test a given gene using 15°C and 25°C, because penetrance of the phenotypes can vary depending on the incubation temperature and gene.

ACKNOWLEDGMENTS

The authors wish to thank Marcel Tijsterman, Robin May, Sylvia Fischer, and Rik Korswagen for their comments on the manuscript.

REFERENCES

1. Wood, W. B., *The Nematode* Caenorhabditis elegans. Cold Spring Harbor Laboratory Press, New York, 1988.
2. Genome sequence of the nematode *C. elegans*: a platform for investigating biology. The *C. elegans* Sequencing Consortium, *Science,* 282, 2012, 1998.
3. Jansen, G. et al., Reverse genetics by chemical mutagenesis in *Caenorhabditis elegans*, *Nat. Genet.,* 17, 119, 1997.
4. Jansen, G. et al., The complete family of genes encoding G proteins of *Caenorhabditis elegans*, *Nat. Genet.,* 21, 414, 1999.
5. Kim, S.K. et al., A gene expression map for *Caenorhabditis elegans*, *Science,* 293, 2087, 2001.
6. Fire, A. et al., Potent and specific genetic interference by double-stranded RNA in *Caenorhabditis elegans*, *Nature,* 391, 806, 1998.
7. Ashrafi, K. et al., Genome-wide RNAi analysis of *Caenorhabditis elegans* fat regulatory genes, *Nature,* 421, 268, 2003.
8. Kamath, R.S. et al., Systematic functional analysis of the *Caenorhabditis elegans* genome using RNAi, *Nature,* 421, 231, 2003.
9. Pothof, J. et al., Identification of genes that protect the *C. elegans* genome against mutations by genome-wide RNAi, *Genes Dev.,* 17, 443, 2003.
10. Vastenhouw, N.L. et al., A genome-wide screen identifies 27 genes involved in transposon silencing in *C. elegans*, *Curr. Biol.,* 13, 1311, 2003.

11. Simmer, F. et al., Genome-wide RNAi of *C. elegans* using the hypersensitive rrf-3 strain reveals novel gene functions, *PLoS. Biol.,* 1, E12, 2003.
12. Montgomery, M.K., Xu, S., and Fire, A., RNA as a target of double-stranded RNA-mediated genetic interference in *Caenorhabditis elegans*, *Proc. Natl. Acad. Sci. U.S.A.,* 95, 15502, 1998.
13. Tabara, H., Grishok, A., and Mello, C.C., RNAi in *C. elegans*: soaking in the genome sequence, *Science,* 282, 430, 1998.
14. Parrish, S. et al., Functional anatomy of a dsRNA trigger. Differential requirement for the two trigger strands in RNA interference, *Mol. Cell,* 6, 1077, 2000.
15. Timmons, L. and Fire, A., Specific interference by ingested dsRNA, *Nature,* 395, 854, 1998.
16. Tavernarakis, N. et al., Heritable and inducible genetic interference by double-stranded RNA encoded by transgenes, *Nat. Genet.,* 24, 180, 2000.
17. Tabara, H. et al., The rde-1 gene, RNA interference, and transposon silencing in *C. elegans*, *Cell,* 99, 123, 1999.
18. Timmons, L., Court, D. L., and Fire, A., Ingestion of bacterially expressed dsRNAs can produce specific and potent genetic interference in *Caenorhabditis elegans*, *Gene,* 263, 103, 2001.
19. Fraser, A.G. et al., Functional genomic analysis of *C. elegans* chromosome I by systematic RNA interference, *Nature,* 408, 325, 2000.
20. Kimble, J. et al., Suppression of an amber mutation by microinjection of suppressor tRNA in *C. elegans*, *Nature,* 299, 456, 1982.
21. Smith, N.A. et al., Total silencing by intron-spliced hairpin RNAs, *Nature,* 407, 319, 2000.
22. Timmons, L. et al., Inducible systemic RNA silencing in *Caenorhabditis elegans*, *Mol. Biol. Cell,* 14, 2972, 2003.
23. Gonczy, P. et al., Functional genomic analysis of cell division in *C. elegans* using RNAi of genes on chromosome III, *Nature,* 408, 331, 2000.
24. Maeda, I. et al., Large-scale analysis of gene function in *Caenorhabditis elegans* by high-throughput RNAi, *Curr. Biol.,* 11, 171, 2001.
25. Kuroyanagi, H. et al., SPK-1, a *C. elegans* SR protein kinase homologue, is essential for embryogenesis and required for germline development, *Mech. Dev.,* 99, 51, 2000.
26. Epstein, H. F. and Shakes, D. C., *Caenorhabditis elegans: Modern Biological Analysis of an Organism.* Academic Press, San Diego, California, 1995.
27. Kamath, R.S. et al., Effectiveness of specific RNA-mediated interference through ingested double-stranded RNA in *Caenorhabditis elegans*, *Genome Biol.,* 2, 2000.
28. Lee, S.S. et al., A systematic RNAi screen identifies a critical role for mitochondria in *C. elegans* longevity, *Nat. Genet.,* 33, 40, 2003.
29. Dillin, A. et al., Rates of behavior and aging specified by mitochondrial function during development, *Science,* 298, 2398, 2002.
30. Praitis, V. et al., Creation of low-copy integrated transgenic lines in *Caenorhabditis elegans*, *Genetics,* 157, 1217, 2001.
31. Simmer, F. et al., Genome-wide RNAi of *C. elegans* using the hypersensitive rrf-3 strain reveals novel gene functions, *PLoS. Biology,* 1, 2003.
32. Kuznicki, K.A. et al., Combinatorial RNA interference indicates GLH-4 can compensate for GLH-1; these two P granule components are critical for fertility in *C. elegans*, *Development,* 127, 2907, 2000.
33. Longman, D., Johnstone, I.L., and Caceres, J.F., Functional characterization of SR and SR-related genes in *Caenorhabditis elegans*, *EMBO J.,* 19, 1625, 2000.

34. Kawano, T., Fujita, M., and Sakamoto, H., Unique and redundant functions of SR proteins, a conserved family of splicing factors, in *Caenorhabditis elegans* development, *Mech. Dev.,* 95, 67, 2000.
35. Herman, M., *C. elegans* POP-1/TCF functions in a canonical Wnt pathway that controls cell migration and in a noncanonical Wnt pathway that controls cell polarity, *Development,* 128, 581, 2001.
36. Simmer, F. et al., Loss of the putative RNA-directed RNA polymerase RRF-3 makes *C. elegans* hypersensitive to RNAi, *Curr. Biol.,* 12, 1317, 2002.
37. Sijen, T. et al., On the role of RNA amplification in dsRNA-triggered gene silencing, *Cell,* 107, 465, 2001.
38. Colaiacovo, M.P. et al., A targeted RNAi screen for genes involved in chromosome morphogenesis and nuclear organization in the *Caenorhabditis elegans* germline, *Genetics,* 162, 113, 2002.
39. Piano, F. et al., Gene clustering based on RNAi phenotypes of ovary-enriched genes in *C. elegans*, *Curr. Biol.,* 12, 1959, 2002.

12 Delivery of RNAi Reagents in *C. elegans* by Microinjection

Peter J. Appleford and Alison Woollard

CONTENTS

12.1 INTRODUCTION

Caenorhabditis elegans was the first metazoan for which the entire genome sequence was made available.[1] This, coupled with the ease of maintenance, favorable genetic properties, and invariant cell lineage, makes *C. elegans* an excellent model organism for biological analysis. With the sequence of all 19,000 or so genes available, it is natural that research groups should seek to use a reverse genetic approach in order to knock out genes of interest. As well as being of great relevance to established worm labs, reverse genetic analysis of *C. elegans* genes has also become an increasingly popular approach in nontraditional worm research environments.

Although several well-established techniques for the generation of *C. elegans* knockout mutants are described in the literature (for a useful introduction to *C. elegans* reverse

0-8493-2141-7/05/$0.00 + $.50

genetics, see the guide by Robert J. Barstead[2]), these are currently comparatively labor-intensive. RNA interference (RNAi) by microinjection is particularly useful where a reverse genetics approach is required for silencing a small number of genes, in that it enables a laboratory to use a directly targeted approach, without recourse to an extended screen. An additional bonus is that genes that share functional redundancy, or that interact genetically in some other way, may be targeted simultaneously by performing interference with more than one species of RNA.

In RNAi by injection, the F1 progeny of injected mothers (~300 individuals in a normal brood) is examined for the phenotypic outcome of gene silencing. Gene silencing is not inherited by subsequent generations, but phenotypic effects can sometimes be identified in the injected mother herself. RNAi affects both embryonic and postembryonic development of F1s and reduces both maternal and zygotic RNAs in embryos. In the case of essential genes, phenotypic analysis by RNAi is often simpler than using a heterozygous mutant strain, in which the maternal contribution of RNA in homozygous mutant embryos may confuse the terminal phenotype. Although gene silencing by injection of double-stranded RNA (dsRNA) has some obvious weaknesses (for example, any interference phenotype is transient and noninherited and some genes are recalcitrant to RNAi), the method is very reproducible, is relatively straightforward, and is the standard RNAi technique used in many laboratories. In this chapter we describe the methods currently used in our own laboratory and many others for the generation of dsRNAs and for delivery by microinjection. We also briefly consider the kinds of phenotypes that might be observed after performing RNAi by injection and the steps a researcher might take to interpret them.

Although wide-ranging genetic screens have been performed using RNAi by injection (in one study, 96% of open reading frames on Chromosome III were targeted in order to identify genes involved in cell division[3]), typically a large-scale RNAi screen would be designed to make use of the bacterial feeding mode of delivery as described in Chapter 11. The microinjection method described here, originally developed for delivery of transgenic DNA constructs into worms, is particularly convenient when a smaller number of genes are to be silenced (e.g., 20 or so) and may be a very useful way of performing functional analysis of a moderately sized gene family.

Early RNA interference experiments on *C. elegans* were attempted using an antisense RNA approach.[4] However, subsequent work has revealed that potent interference is dependent on the presence of dsRNA.[5] Very small amounts of dsRNA are required to produce an interference effect, suggesting that the mechanism of interference is catalytic or amplified in some way. RNA molecules are also able to cross cell membranes, resulting in gene silencing in parts of the worm into which they have not been directly introduced;[5] this appears to be dependent on transport by the trans-membrane protein SID-1.[6,7] The mechanism of RNAi has now been investigated in a number of different experimental systems. On entry into cells, dsRNA is cleaved by an RNase III–like enzyme known as Dicer (*dcr-1* in *C. elegans*)[8] into 21- to 23-nt RNAs known as "small interfering RNAs" (siRNAs). These siRNAs are bound by a large complex known as the RNA-Induced Silencing Complex (RISC).[9] This complex binds the target mRNA that is cleaved, releasing the siRNA, and the endogenous target mRNA is then degraded by endo- and exonucleases. In worms, the siRNA can also act as a template for an RNA-directed RNA polymerase (RdRP),[10] which results in amplification of an antisense strand 5' of the domain targeted by the injected RNA.

12.2 WORM STRAINS AND MAINTENANCE

For those workers not experienced in handling and maintaining *C. elegans* in the laboratory, we would strongly advise that they refer to the various previously published sources on routine worm maintenance.[11,12] For advice on recognizing specific stages of the worm, see the introduction to *C. elegans* biology in *The Nematode* Caenorhabditis elegans.[13]

For silencing of many genes in *C. elegans*, it is normal practice to use the standard wild-type Bristol strain N2. Wild-type populations of *C. elegans* are composed almost entirely of self-fertilizing hermaphrodites, however, so in cases in which gene silencing is anticipated to give rise to a male-specific phenotype, it would be more appropriate to perform RNAi in a background that gives rise to a high proportion of male worms. In such a situation, we perform interference on a strain containing a mutation in *him-8*. This mutation results in meiotic nondisjunction of the X chromosome, with the effect that approximately 37% of the resulting progeny are XO male. For genes that appear to be refractory to RNAi in N2 or *him-8* backgrounds, consideration should be given to using the RNAi-hypersensitive *rrf-3* mutant strain,[14] discussed elsewhere in this book.

12.3 DELIVERY OF DSRNA INTO WORMS

12.3.1 PREPARATION OF DSRNA

In our laboratory RNA preparations are typically transcribed *in vitro* from a PCR template incorporating T3 and T7 promoter tags (Protocols 12.2 and 12.3). Alternatively, where a cDNA or fragment of exon-rich genomic DNA is available in a cloning vector, flanked by T3 and T7 promoters, *in vitro* transcription can be carried out by first linearizing the vector with an appropriate restriction endonuclease, followed by RNA synthesis according to Protocol 12.3. An important consideration in the design of oligonucleotide primers for generating a PCR template is the size and genomic location of the PCR fragment used as an RNA template. In our experience, a template of approximately 500 bp is adequate for silencing, although the primers should preferably be chosen to amplify genomic sequence within a single exon or, if this is not possible, should cover an exon-rich region. Primers should be designed to avoid using regions of DNA that may be highly conserved elsewhere in the genome (e.g., > 90%) to avoid any possibility of cross-interference. The gene-specific parts of both forward and reverse primers are typically chosen to have a melting temperature of approximately 60°C to allow standardized PCR conditions.

12.3.2 Microinjection of dsRNA

Gonad injection of *C. elegans* hermaphrodites is a technique that is widely used in many laboratories as a method of incorporating transgenic DNA constructs such as Green Fluorescent Protein (GFP) reporters as extrachromosomal arrays. Microinjection is also used as a mainstream protocol for the delivery of dsRNA into hermaphrodite worms, with the purpose of gene silencing in the subsequent F1 generation. Although both processes require essentially the same equipment and techniques, we have found that RNAi by injection requires considerably less skill and precision to be successful and has worked well in the hands of undergraduate project students with comparatively little microinjection experience. This is probably because the site of injection is not as critical as when injecting worms with a transgenic DNA reporter construct, when the syncytial gonad *must* be injected. Although we have found RNAi to be effective when other parts of the worm are injected, we do get the optimal and most reproducible results following the standard gonad injection technique.

12.3.2.1 Preparation of Injection Pads

Wild-type *C. elegans* adults are small (approximately 1 mm in length) and very mobile; successful injection requires that the animals are immobilized by adhesion to dried agarose pads. The pads are prepared on glass cover slips using molten agarose, as described in Protocol 12.4. Note that the thickness of the resultant pad is dependent on the temperature of the agarose when the second cover slip is dropped onto it. It is normal to wait a few seconds after dropping agarose onto the first cover slip, or the pad will be too thin. If the agarose has cooled too much and does not spread out immediately after dropping the upper cover slip onto it, the second cover slip may be pressed gently to aid spreading of the agarose.

We dry the freshly made pads for 10 to 15 min on a hot-block set to about 80 to 90°C, although the pads may be dried in a drying oven or by being left at room temperature overnight. Pads can be made in advance in large numbers and stored in a clean plastic Petri dish or in an old cover slip container. Try to keep the pads in a dry place, as worms will not stick to damp pads. It is not a disaster if the pads do become damp, however, as they can be redried if required.

12.3.2.2 Preparation of Microinjection Needles

Needles for microinjection are pulled from borosilicate glass capillary tubes. A variety of different types of microinjection needle pullers exist — the one in our laboratory is a Sutter P-2000 (Sutter Instrument Co., Novato, CA) that works by means of a laser heating element. Alternative designs of needle puller exist, many of which incorporate electrical heating elements. In either type of system, the basic method of pulling the needle is the same: a capillary is heated midway along its length by the heating element while a pulling force is applied to either end of the capillary, typically by means of electromagnets. Correct temperature setting is crucial to pulling a good injection needle — if the temperature of the heating element is too high, the resulting needles are too thin and flexible, whereas a low heat usually results in a stubby needle. A perfect needle is stiff enough to easily penetrate the cuticle of the worms while being thin enough to cause a minimum of damage during injection. A certain amount of trial and error will be required in setting up a new needle puller to pull good needles. A number of needles may be pulled ahead of time and stored in a dust-free container, such as a small plastic box (we rest the needles on strips of modelling clay so as to avoid damage to the tip of the needle).

Needles can be loaded in several ways — one method is to use a drawn-out plastic pipette tip (as used for loading sequencing gels). The end of the pipette tip is inserted into the blunt end of the microinjection needle as far as it will go, and the RNA solution is gently expelled. Note that if the RNA solution is visible to the naked eye once loaded into the injection needle, then there is an adequate quantity for many injections. A simpler but less reliable method of loading the microinjection needle is to simply up-end the needle into the RNA solution (so that the blunt, nondrawn-out end is immersed in the solution). The liquid should fill the needle by capillary action (although an occasional needle will fail to be filled by this method). However the needle is loaded, once the RNA solution is visible at the tip of the needle, it is important to ensure that there are no bubbles present in it. This can be done by viewing the tip of the needle under a dissection microscope; should any bubbles be present at the tip of the needle, they can be removed by holding the loaded needle vertically, point downwards, and flicking it gently with the fingers. In the majority of cases, any bubbles will float upward away from the tip of the needle.

After loading a needle with RNA solution, it is normally necessary to break the tip of the needle (sealed during the pulling process) prior to its use for injection. The preferred method for doing this varies between workers, although the basic technique is similar in every case. Needles may be broken on the edge of the injection pad, against a grain of sand or piece of dirt on the pad, on a short length of drawn-out glass capillary, or on a scratch made on the cover slip using a diamond pen. Note that if the needle is to be broken on a grain of sand, it is necessary to drop a few grains of fine clean sand onto the agarose pad prior to drying on the hot block. This method for breaking the needle assumes that the edge of the agarose pad is being used, although it is a simple matter to apply it to the various other methods. The loaded needle should first be mounted in the needle holder of the micromanipulator. It is a good idea to centralize the tip of the needle in the field of view of a low power (5 ×) objective so that it is easy to find the needle again. On our micromanipulator apparatus, once the needle tip is centered on the field of view, it is possible to press a "home" key that withdraws the needle to a safe distance from which it does not obstruct the stage. A subsequent press of the same key returns the needle to an identical position (note that it is good practice to always withdraw the needle before placing cover slips on the stage or removing them, to avoid accidental damage to the tip of the needle). Select an injection pad (if the needle is to be broken on the pad edge, pick one with a well-defined edge or it will be difficult to locate under high power). Now place a drop or two of injection oil onto the cover slip (we use heavy liquid paraffin) so that the paraffin covers the edge of the agarose pad. Paraffin is necessary to prevent the worms from drying out while on the agarose pad and also aids the visualization of liquid flowing from the broken needle. Place the cover slip on the sliding stage of the microscope so that the edge of the pad is near the center of the field of view on low power. The needle should now be brought back into the center of the field of view (with the "home" key if the micromanipulator has one, or manually otherwise) and lowered so that it is close to the surface of the cover slip. Switch to a high-power objective (40 ×) and try to find the edge of the pad — it should be reasonably easy, assuming that the edge of the pad is well defined. The tip of the needle should then be moved close to the edge of the pad; if the needle tip is not visible, it may be above the plane of the cover slip — this can normally be ascertained by focusing the microscope up and down. If the needle is too high, bring it down carefully so that the needle tip and agarose pad are both in focus simultaneously. The needle should then be pushed against the pad so that the tip is broken off.

To test the successful break of the needle, press the "inject" button (on an automated pump apparatus) or briefly switch on the gas supply manually and the flow of the RNA solution should be clearly visible under the paraffin (the flow will be *up* the outside of the needle). An ideal needle will still appear fine at its tip when viewed under high power yet will give a good flow of RNA solution when the "inject" button is depressed. Note that as with injection itself, some people prefer to move the pad toward the needle (by pushing the sliding stage gently), whereas others leave the stage in place and move the needle into the pad. This is very much a matter of personal preference and should be determined through practice and experience. As mentioned above, a good needle is one of the critical factors in successful injecting — very fine and flexible needles are often hard to break and may block easily, whereas worms are often damaged by injection with thick needles.

12.3.2.3 Gonad Microinjection

Once a good needle has been selected and broken, worms can be mounted on a cover slip for injection. This is one of the most frustrating aspects of microinjection for the beginner, and it is therefore a good idea to practice a little before undertaking actual injections. Select a good pad of a suitable thickness (this comes from experience) and breathe on the cover slip once or twice to moisten the agarose very slightly. (If you are not sure which side the agarose is on, it is normally possible to feel the edge of the injection pad with a fingernail.) If the pad is not moistened sufficiently, the worms may dry out before they can be recovered from the pad. (If the worms will not stick to the pad at this point, select a new pad but do not moisten it by breathing on it.) Worms are then mounted on the pad under a drop of paraffin and injected according to Protocol 12.4. Below we consider in more detail some important aspects of mounting and positioning worms for injection and the microinjection process itself.

Worms can be picked onto the injection pad using a "conventional" worm pick (a piece of platinum wire, often with a flattened end, mounted at the end of a glass Pasteur pipette) or with an eyebrow hair taped to a cocktail stick. When picking the worms, try to transfer as little bacteria to the injection pad as possible, as bacteria will clog the injection needle and also hinder the adhesion of the worm to the pad. (Try to pick worms from a part of the plate which has no bacteria on it.) If a wire pick is to be used, it may be flamed before use (to prevent contamination of the stock plate), and picks of any design can be dipped in paraffin to aid picking of the worms. If the worms to be injected are wild type (N2) or otherwise reasonably healthy (i.e., of a normal size with no serious movement defects or the like), they may be scooped onto the flattened end of the pick and allowed to wriggle off onto the pad. Once part of the worm touches the pad it should stick to it and the pick can be withdrawn, leaving the worm immobilized. It might be necessary to touch the worm gently onto the agarose pad using the pick in order to ensure it is stuck along its entire length (be gentle). A worker experienced in microinjection might be able to mount 10 or more worms in this way, inject, and recover them before they become desiccated, but it is recommended that a beginner should start by attempting to inject only one or two worms at a time.

For RNA-interference purposes, it is not strictly necessary to inject any one part of the hermaphrodite worms, although we prefer to inject RNA into the syncytial gonad, which is most easily done with the worms oriented on their sides. It is possible to stick the worms down this way by allowing them to "swim" in the paraffin prior to sticking them onto the agarose pad. The waves of movement are in the dorso-ventral axis, so if the worm appears to be swimming side to side under the dissecting microscope, when stuck to the pad it will be in the required orientation. It is also possible to roll a worm gently into the required orientation once it is stuck to a pad. With one or two worms attached to the pad, the cover slip should be placed on the sliding stage of the injection microscope. Those with keen eyesight may be able to see the worms on the cover slip with the naked eye and center them relative to the microscope objective — alternatively one may use the lowest-power-objective lens (5 ×). The injection needle should then be moved the worm. By focusing the microscope up and down using a high-power objective (40 ×), the nuclei of the syncytial gonad should be visible. Adjust the plane of focus of the microscope so that the nuclei at the *outer edges* of the gonad are clearly visible (see Figure 12.2 for help in identifying the syncytial gonad). Make fine adjustments to the height of the tip of the injection needle so that the very tip is also in focus. The needle should now be moved to penetrate the worm cuticle — some worm workers move the needle into the worm by means of the micromanipulator, whereas others prefer to leave the needle at the center of the field of view and move the worm onto it by gentle movement of the sliding stage. Either way, the needle should cleanly penetrate the cuticle and be visible within the gonad.

Should the needle not penetrate the cuticle easily but the worms remain stuck to the pad, it is sometimes possible to pierce the cuticle by *gently* tapping the needle holder. If the worms are not adequately stuck to the pad or the injection needle is broken badly (i.e., too thick at its tip), the worm may move away from the needle. It is often easiest to penetrate the cuticle when the worm is turned on the stage so that the needle is at a shallow angle (20 to 30°) from the anterior–posterior axis of the worm. To deliver the RNA into the worm, press the inject button on the injector apparatus (or operate another device, e.g., syringe/gas pressure); a flow of liquid into the worm should be clearly visible as the nuclei of the syncytial gonad move apart slightly. Do not inject the worms for too long, or they will be damaged by too much flow of liquid. Correct injection pressure depends on a number of factors, including the type of injection apparatus and how the needle has been broken. With our Eppendorf apparatus, a range of pressures from 300 to 1500 hPa would be used, depending on the break of the needle alone. With a very fine needle, a high pressure is required in order to get a free flow of RNA solution, whereas a needle broken farther from its tip would require less pressure. It is important not to over-inject the worms — if too high a pressure is used, there is a danger that the worms may be damaged (this may be obvious by observing the number of progeny arising out of an injected worm; damaged worms will lay only a few viable eggs). By trial and error, it is recommended that a pressure be found for the needle such that a *gentle* flow of liquid is observed in the gonad.

Should the needle become blocked, it is possible to increase the injection pressure somewhat to attempt to force the RNA to flow again. If this proves fruitless, the needle may be rebroken carefully (try to avoid breaking too much of the needle off or the needle tip will become too large, resulting in damage to the worms). If this fails to work, replace the needle with a newly loaded one.

After withdrawing the needle following successful injection, it is important to work as quickly as possible to recover injected worms from the pads before they desiccate. To unstick the worms from the pad, drop 1–2μl of recovery buffer onto the worm (we use M9 buffer; other groups add 4% (w/v) glucose to this) and then work using a dissecting microscope. The worms should become detached from the agarose and can be picked to seeded NGM plates with a worm pick.

It is essential that worms are on the dry agarose pad for the minimum time to ensure optimum survival rates; worms are surprisingly robust but will dry out if left for too long. If a high proportion of the worms do not recover after being removed from the pads, the pads could be too dry or too thick. In this case, we would advise ensuring the pads are moistened properly with two or three long breaths prior to overlaying them with paraffin. Beginners to the technique of microinjection should also attempt to inject a maximum of one or two worms at a time.

If the worms do not stick at all or stick for 1 or 2 seconds before wriggling free of the agarose again, it becomes almost impossible to inject them successfully. In this instance it is important to check that the agarose pads have been dried correctly — if they are suspected of being damp, it is possible to redry them. A wet (e.g., recently poured) NGM plate will often lead to worms that are very damp. Such worms may fail to stick adequately to the agarose pad and it is thus recommended that worms to be injected are kept on fairly dry NGM agar plates. We have also experienced difficulties in making the worms adhere where the paraffin has become contaminated with moisture. If this is suspected, simply discard the old paraffin and replace it with a fresh batch.

If worms appear to recover from the injections but lay very few viable embryos, it is most likely that they have been injected too roughly or with too much RNA solution. Reduce the injection pressure and/or injection time, or use a finer needle.

12.4 ANALYSIS OF PHENOTYPES

Following injection, a number of worms may appear to be seriously damaged and sick (normally as a result of rough handling or being on the injection pad for too long). Such worms should be discarded and their progeny not examined for an interference phenotype. After an interval of at least 4 h post-injection, healthy-looking injected worms should be picked to fresh seeded NGM plates and allowed to lay eggs (normally at 20°C). For close study of progeny, it is preferable to pick injected hermaphrodites onto a fresh plate every 12 h or so and to observe the progeny arising to check for any effect of gene silencing. Hypomorphic phenotypes may be observed in the first cohort of progeny following injection of a hermaphrodite mother, and such hypomorphic effects are often phenotypically informative, especially in cases in which the maximal effect is lethality. The time following injection at which the maximal RNAi phenotype is seen in the progeny will vary from gene to gene, depending of the half-life of the relevant protein, among other factors. Where RNAi results in a very gross phenotype, such as arrested embryos, it should be relatively easy to score the penetrance of the phenotype. However, RNAi phenotypes may be very subtle and might be missed without a degree of experience. An initial inspection of the progeny of injected worms under a dissecting microscope may reveal the most obvious gross morphological defects, and these should then be examined more closely by Nomarski microscopy in an attempt to decipher the true nature of the defect. In our experience, for very close analysis of RNAi phenotypes, it is often necessary to repeat injections several times, depending on the effect of gene silencing. A "first-pass" experiment might consist of analysis of the progeny of 10 to 20 injected hermaphrodites.

A wide variety of different phenotypes may be observed following an RNAi experiment, which can be broadly classified as maternal, embryonic, and postembryonic (for a database of RNAi phenotypes, see http://www.wormbase.org/db/searches/rnai_search). Some phenotypes are more easily identified than others; for example, gross morphological phenotypes such as Dumpy (Dpy) (worms are shorter and fatter than wild-type) or Muv (multivulva) are relative easily scored when compared to a subtle behavioral phenotype (such as avoidance of a particular chemical, especially considering the transient nature of interference by injection. Similarly, embryonic arrest is an easily scored phenotype. (An inexperienced investigator should be careful not to confuse arrested embryos with unfertilized oocytes, which may be laid by roughly- handled hermaphrodites.) Of course, assessing the reason for such embryonic arrest is a much more complicated matter, involving a close inspection of embryonic development. As with many other aspects of biological analysis, it is often difficult to decide how penetrant a phenotype should be before it is judged to be significant. It may be that workers would choose to ignore any phenotypes that were observed at a frequency of less than, say, 5 to 10% in the progeny of injected animals, but, of course, there will be cases in which even weakly penetrant phenotypes are informative and significant.

Note that RNAi of a limited number of genes can be used in certain instances as a way of positionally "cloning" genes once a reliable map position is known. To make use of RNAi in this way, one would simply perform silencing on genes in the vicinity of the mutation under investigation to look for that gene that phenocopies the mutant phenotype when subjected to RNAi. The identity of the mutated gene can then be confirmed by targeted transgenic rescue experiments. In some cases this has proved quicker than more conventional approaches involving looking for rescue with cosmids.

12.4.1 If RNAi Does Not Give a Phenotype

One of the shortcomings of reverse genetics by RNAi, especially in the hands of the less-experienced researcher, is that absence of a visible phenotype does not necessarily mean that the gene in question has no important function. In the absence of a phenotype, it may need to be established that silencing of the target gene *has* taken place and that the gene in question is neither fully nor partially redundant. The issue of redundancy is one that may be relatively easily resolved, especially if the gene being silenced has a very close homologue or is one half of an obviously recent duplication (but note that a gene may be functionally redundant with another with which it does not share high homology at the sequence level). In a case in which a candidate gene is known that may share some redundancy with the gene under investigation, it is a simple matter to inject two species of RNA simultaneously and look for a phenotype. In this way, we have performed combinatorial RNAi on pairs of genes believed to have overlapping function and have noted clear and reproducible phenotypes where the individual RNA species gave no observable effect. The precise maximum number of RNAs that can be combined in this fashion is not known. In one study in which pools of eight dsRNAs were used to identify genes involved in RNAi, the authors reported that silencing of a known gene was not inhibited nonspecifically by dilution with other RNA species.[15] However, a maximum pool size of two RNAs was chosen for a study of Chromosome III gene function, as dampening of the RNAi effect was reported with larger pools.[3]

In some circumstances, it may be necessary to confirm that gene expression has been silenced. If a transgenic line carrying a GFP reporter tagged to the gene of interest is available, it is possible to look for reduced or absent expression of the GFP. However, such an analysis will only confirm the presence or absence of the transgene and not the endogenous protein or RNA. We would suggest the use of Western or Northern blotting, or semiquantitative RT-PCR, to confirm the reduction of endogenous protein or RNA as appropriate.

12.5 USEFUL CONTROL EXPERIMENTS

A useful test run for RNAi by injection might be to silence the *unc-22* gene (a target in early RNAi experiments[5]) and check that a "twitching Unc" phenotype is observed. This would confirm not only the efficacy of RNAi but also give an inexperienced researcher some confidence that he is able to recognize a fairly subtle phenotype.

12.6 CAVEATS

RNAi by injection has been shown to be a robust technique that is in use in many laboratories. However, there are some weaknesses with the technique that should be taken into consideration; the most important of these is that some genes, notably those expressed in the nervous system, are recalcitrant to RNAi. Silencing of such genes by "conventional" RNAi experiments may be unsuccessful and, if so, an alternative strategy for analysis is necessary. Conventional chromosomal deletions will need to be generated in these cases or specialized RNAi techniques will need to be used. Such RNAi techniques may involve the use of transgenic inverted-repeat constructs driven *in vivo* by tissue-specific promoters and are described elsewhere in this book. As RNAi techniques evolve, problems associated with RNAi-resistant genes may well become easier to solve in the future.

12.7 CONCLUDING REMARKS

RNAi by injection is a simple and effective means of gene silencing that requires little optimization and gives highly reproducible results. It is not an overstatement to say that RNAi has revolutionized *C. elegans* research. While classical forward genetic strategies still have, and will continue to have, a very important place in worm genetics, the ease of generating targeted knockout phenotypes by RNAi has enabled a very rapid investigation of the biological importance of thousands of genes. Indeed, the bacterial feeding RNAi strategy described elsewhere in this book has enabled genome-wide functional screens to be performed in a way that would have been thought impossible a few years ago. The rate-limiting factor for the future will, of course, lie in the interpretation of phenotypes. It is likely that our standard laboratory examinations and assays will be inadequate for the investigation of many phenotypic effects and will need to be improved as we strive to meet the challenge of deciphering the function of the genome.

PROTOCOLS

Protocol 12.1 Worm Strains and Maintenance

Materials

- Worms (N2 or other strain (see Section 12.2) — note that *C. elegans* strains can be requested from the *Caenorhabditis* Genetics Center (CGC) (http://biosci.cbs.umn.edu/CGC/CGChomepage.htm)
- 1 M KPO_4 (108.3 g KH_2PO_4, 35.6 g K_2HPO_4 to 1 liter, pH 6.0)
- NGM agar (3 g NaCl, 17 g agar, 2.5 g peptone, 1 ml cholesterol (5 mg/ml in ethanol), to 1 liter with H_20; autoclave, then add the following by sterile technique, mixing between additions: 1 ml $CaCl_2$ (1 M), 1 ml $MgSO_4$ (1 M), 25 ml KPO_4 (1 M))
- 5-cm-diameter NGM agar plates seeded with OP50 bacteria (also available from CGC)

- Worm pick (to make a worm pick, melt a short piece of platinum wire into the end of a glass Pasteur pipette — many workers like to flatten the end) and/or scalpel
- Bunsen burner (for heat-sterilizing worm pick and scalpel)
- Dissecting microscope

Experimental Procedures

1. Prepare 50-mm-diameter NGM agar plates seeded with OP50 bacteria.
2. Maintain stocks at 20°C, picking 6 or 7 young adult hermaphrodites to fresh, seeded plates every 3 days or so. Alternatively, passage worms by cutting a small chunk of agar containing worms with a sterile (flamed) scalpel and placing this chunk worm-side down on a NGM plate containing a lawn of OP50 bacteria.

Technical notes. In order to ensure that the worms are healthy and the RNAi results representative, it is essential to perform injections on worms that have been regularly passaged onto fresh NGM plates and not recently (within two generations) subjected to any stresses such as starvation or heat shock. Since the process of injection requires that young adult worms be immobilized on dry agarose pads, it is preferable to use NGM plates that have been kept at room temperature for a week or more before use to maintain worm stocks — newly poured plates are often damp, and worms picked from them tend not to stick to the agarose pads very efficiently. To avoid having to search for worms of a suitable age during the injection session, young L4 hermaphrodite larvae may be picked to seeded NGM plates the day before injecting and allowed to mature overnight into young adults.

Protocol 12.2 Generation of DNA Templates by PCR

Materials

- PCR oligonucleotide primers tagged with T3 and T7 promoter sequences (for design, see below)
- RNase-free microcentrifuge tubes and pipette tips
- Template DNA (typically genomic or cosmid)
- Standard PCR reagents (e.g., Qiagen, Promega, Bioline)
- Thermal cycler

Experimental Procedures

1. Design PCR primers incorporating T3 and T7 promoter tags as follows:
   Forward primer:
   5' AAT TAA CCC TCA CTA AAG Gnn nnn nnn nnn nnn nnn nnn n
   Reverse primer:
   5' TAA TAC GAC TCA CTA TAG Gnn nnn nnn nnn nnn nnn nnn n
   Sequences in CAPITALS represent the T3 and T7 promoters, respectively; "n's" represent gene-specific oligonucleotide sequences.
2. Set up a PCR reaction as follows:

Template DNA	0.1–1.0 µg/25-µl reaction
Forward primer	2.5 µl of a 4-µM working stock

Reverse primer	2.5 μl of a 4-μM working stock
dNTPs	4 μl of a 2.5-μM working stock
Taq DNA polymerase buffer	2.5 μl of a 10 × stock
Taq polymerase	0.5 U/25 μl
dH_2O	to 25 μl

3. Add 1 drop of mineral oil (for thermal cyclers without heated lids).
4. Program the thermal cycler for an appropriate cycle. Cycling parameters may have to be adjusted according to the primer sequences. A typical example used in our laboratory is given here:
   i. 95°C, 3 min
   ii. 94°C, 40 sec; 55°C, 40 sec; 72°C, 1 min (35 cycles)
   iii. 72°C, 3 min
5. Assess the quality of PCR products by electrophoresis of a sample in a 1% (w/v) agarose gel. Further purification may be required prior to *in vitro* transcription if the PCR product gives rise to nonspecific bands.

Protocol 12.3 *In Vitro* Transcription and Preparation of dsRNA for Microinjection

The following procedure is an adaptation from the Fire laboratory protocol.[5]

Materials

- Standard *in vitro* transcription reagents and enzymes (e.g., Ambion, Promega)
- Stop solution (1 M Sodium acetate, 10 mM EDTA, 0.2% (w/v) SDS)
- 20 mg/ml glycogen
- Phenol:chloroform:isoamyl alcohol (25:24:1), pH 5.2
- Chloroform
- Hotblock set at 68°C
- Incubators or water baths at 37°C and 25°C
- Absolute and 70% (v/v) ethanol
- RNase-free TE (1 mM Tris-HCl, pH 8.0, 1 mM EDTA)
- Equipment for agarose gel electrophoresis
- DNase I (if required)
- RNA cleanup spin columns (if preferred to phenol/chloroform extraction)

Experimental Procedures

1. Make up a transcription mix containing:

Template DNA (PCR product or linearized clone)	4 μl (0.05–0.2 μg)
5 × transcription buffer	4 μl
5 × rNTPs (2.5-mM working stock)	4 μl
RNase inhibitor (RNasin)	1 μl
100-mM DTT stock solution	2 μl
RNase-free H_2O	4 μl

2. Mix well and remove 9.5 μl to a sterile, RNase-free 1.5-μl microcentrifuge tube; add 1 μl T3 polymerase. Add 1 μl T7 polymerase to the remaining mix.
3. Incubate the T3 reaction for 1.5 h at 25°C and the T7 reaction for 1.5 h at 37°C.
4. Combine 9.5 μl from each reaction in a new, RNase-free tube (keep the remaining 1 μl of each ssRNA at –20°C for analysis on a gel).
5. Add 380 μl 1 × stop solution and 0.3 μl glycogen (20 mg/ml). Note that the glycogen is not essential for RNA transcription but enables visualization of the RNA pellet during the washing process.
6. Add 200 μl phenol/chloroform/isoamyl alcohol (25:24:1), vortex briefly, and spin in a microcentrifuge at top speed for a few seconds.
7. Transfer the upper (aqueous) phase to a new RNase-free microcentrifuge tube; add 200 μl chloroform. Mix by vortexing, and spin to separate layers as above.
8. Transfer upper phase to a new tube. Incubate RNAs at 68°C for 10 min, followed by 30-min incubation at 37°C to anneal the 2 single strands.
9. Add 1 ml absolute ethanol to precipitate the RNA, mix vigorously, and spin at top speed in a benchtop microcentrifuge for 10 min. Carefully dump the supernatant ***without disturbing the pellet***.
10. Add 1 ml 70% (v/v) ethanol to wash the pellet. Invert gently, and spin at top speed for 10 min.
11. Remove ethanol supernatant and dry the pellet for several minutes by leaving the tube open on the bench. Resuspend pellet in 10 μl RNase-free TE.
12. Check 1 μl dsRNA against the retained single-stranded samples on a 1% (w/v) agarose gel (Figure 12.1). We generally do not attempt to accurately quantify RNA concentrations prior to injection, as experience has shown that concentrations of dsRNA that can be easily visualized by running 1 μl in a 1% (w/v) agarose gel are sufficient to enable gene silencing. As an alternative to the phenol extraction protocol for purification of dsRNA previously outlined, we have also used RNA clean-up kits such as the Qiagen "RNeasy" spin column, that have proved to be a quick and easy method of obtaining good-quality RNAs for injection.

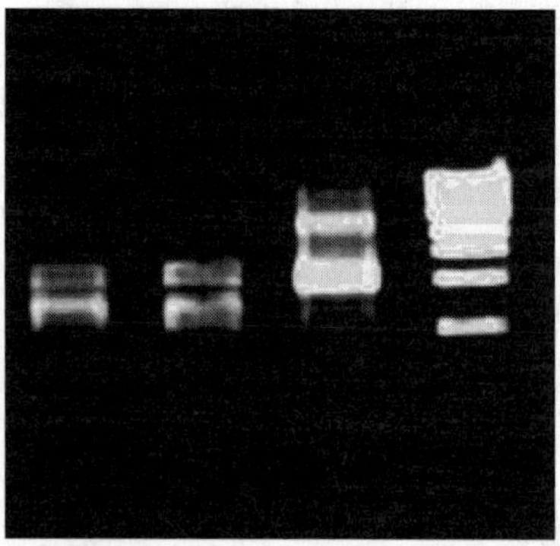

FIGURE 12.1 One percent agarose gel showing transcribed RNAs. The lanes are 1 μl T3 ssRNA, 1μl T7 ssRNA, 1μl annealed dsRNA, and 1kb ladder, respectively. Note the size and intensity difference between single- and double-stranded RNAs.

Technical notes. Removal of template DNA. Although we do not routinely remove template DNA when checking initial RNAi phenotypes, there is a very slight risk that template DNA may affect the observed phenotype as a result of becoming incorporated as an extrachromosomal array. If this is a concern to those carrying out gene silencing, we would suggest that template DNA be removed by the addition of 1 μl DNase I to the transcription reactions after step 3 above, followed by a 1-h incubation at 37°C.

Protocol 12.4 Gonad Microinjection of dsRNA

Prior to injection of dsRNA into worms, we recommend that the RNA solution be centrifuged to remove any particulate matter that might result in a blocked needle (a major cause of frustration while injecting). Fifteen minutes at top speed in a benchtop microcentrifuge is normally adequate for this purpose. Remove the RNA solution carefully to a new, sterile RNase-free microcentrifuge tube.

Materials

- Inverted microscope (with sliding stage) capable of Nomarski/DIC optics, with 5 × and 40 × objective lenses (e.g., Zeiss "Axiovert" S100 inverted microscope)
- Dissecting microscope
- Microscope cover slips (22 × 32 mm or similar size)
- Needle holder and micromanipulator (such as the Eppendorf InjectMan system)
- Microinjector or other suitable pressurized device (e.g., the Eppendorf FemtoJet injection system; many labs use a manually operated injection apparatus pressurized by a CO_2 or nitrogen cylinder)
- Agarose (molecular biology grade)
- Plastic or glass Pasteur pipettes
- Hot block or drying oven
- Bunsen or spirit burner
- Borosilicate glass capillaries (e.g., Harvard capillaries, 1.2 mm x 0.69 mm; Harvard Apparatus Ltd.)
- Needle puller (e.g., Sutter P-2000; Sutter Instrument Co.)
- Liquid paraffin (heavy) (BDH catalog number 294375J)
- M9 buffer (3 g KH_2PO_4, 6 g Na_2HPO_4, 5 g NaCl, 1 ml 1 M $MgSO_4$, H_20 to 1 liter; sterilize by autoclaving)

Experimental Procedures

Preparing injection pads

1. Carefully melt a 2% (w/v) solution of agarose in water in a glass reaction tube over a Bunsen flame. (Agarose solution can be left in a hot block at 60°C to be ready for use whenever required.)
2. To make pads, place several glass microscope cover slips on the bench and, using a disposable plastic Pasteur pipette, drop one or two drops of the molten agarose onto each cover slip (try to avoid getting bubbles in the pads).
3. Drop a second cover slip on top of the molten agarose to create a thin circular pad.
4. Leave for 30 sec and separate the two cover slips by sliding them apart, leaving the pad on one of the two cover slips.

5. Dry pads by placing cover slips pad side up on a hot block set to 80 to 90°C for 10 to 15 min.

Preparing needles

6. Pull several needles using an appropriate needle puller (see Section 12.3.2.2) and store in a dust-free container, such as a small plastic box.
7. Load needles with the dsRNA solution using one of the methods described in Section 12.3.2.2.
8. View the needle under a dissecting microscope to ensure that there are no air bubbles in the loaded RNA solution.
9. Insert the loaded needle into a micromanipulator and break the sealed tip using one of the methods described in Section 12.3.2.2.

Microinjection and recovery of worms

N.B. For a full discussion of microinjection techniques, see Section 12.3.2.3.

10. Drop a small amount of liquid paraffin onto a suitable pad, and spread it to cover the surface.
11. Working under a dissecting microscope, use a worm pick to transfer a young adult hermaphrodite worm to the injection pad and allow it to stick to the agarose on its side.
12. Place the injection pad on the sliding stage of the inverted microscope and locate the worm at the center of the field of view of a 5 × objective lens.
13. Bring the tip of the needle near the worm using the micromanipulator.
14. Switch to high power (40 × objective) and focus the microscope on the nuclei of the syncytial gonad. Adjust the plane of focus of the microscope so that the nuclei at the *outer edges* of the gonad are clearly visible (see Figure 12.2).

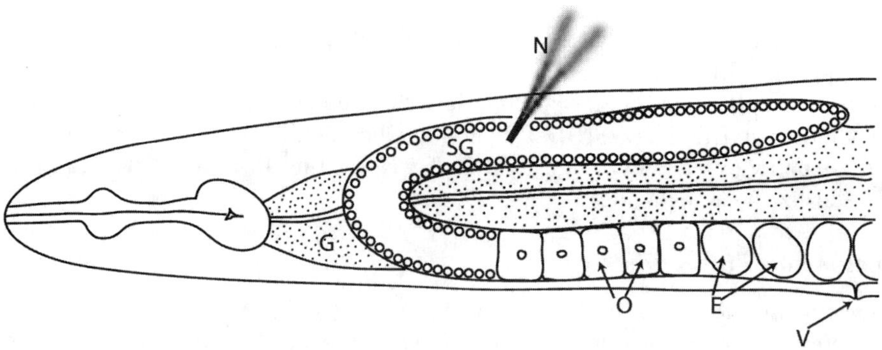

FIGURE 12.2 Injection of the syncytial gonad (5G) of adult hermaphrodite worms. The gut (G) has a granular appearance under Nomarski optics, whereas the syncytial gonad appears clearer and has many circular nuclei at its periphery. The plane of focus should be altered so that the nuclei at the edges (as seen from above) of the gonad are in focus. The tip of the needle (N) should then be brought into exactly the same plane of focus. The needle is pushed through the cuticle and the worm is injected with RNA solution. O = unfertilized oocytes, E = fertilized eggs, V = vulva.

15. Make fine adjustments to the height of the tip of the injection needle so that the very tip is in the same plane of focus.
16. Move the needle to penetrate the worm cuticle by means of the micromanipulator or, alternatively, move the worm onto the needle by gently pushing the sliding stage.
17. Deliver the RNA solution into the gonad by pressing the "inject" button of the microinjector or by operating a suitable manual device.
18. Gently withdraw the needle from the worm. The withdrawn needle can then be removed to a safe distance from the stage using the "home" key if this is available, or manually.
19. Remove the agarose pad from the stage of the injecting microscope and place it underneath the dissecting microscope.
20. To detach the worm from the pad, drop 1–2 μl of recovery buffer onto the worm (simple M9 buffer or M9 with added 4% (w/v) glucose).
21. Pick the detached worm with a worm pick to seeded NGM plates.
22. Transfer injected worms to fresh plates at suitable intervals and observe progeny for phenotypic effects, as described in Section 12.4.

REFERENCES

1. The *C. elegans* Sequencing Consortium, Genome sequence of the nematode *C. elegans*: a platform for investigating biology, *Science*, 282, 2012, 1998.
2. Barstead, R.J., Reverse genetics, in *C. elegans: A Practical Approach*, 1st ed., Hope, I.A., Ed., Oxford University Press, Oxford, 1999, Chapter 6.
3. Gönczy, P. et al., Functional genomic analysis of cell division in *C. elegans* using RNAi of genes on chromosome III, *Nature*, 408, 331, 2000.
4. Fire, A. et al., Production of antisense RNA leads to effective and specific inhibition of gene expression in *C. elegans* muscle, *Development*, 113, 503, 1991.
5. Fire, A. et al, Potent and specific genetic interference by double-stranded RNA in *Caenorhabditis elegans*, *Nature*, 391, 806, 1998.
6. Winston, W.M., Molodowitch, C., and Hunter, C.P., Systemic RNAi in *C. elegans* requires the putative transmembrane protein SID-1, *Science*, 295, 2456, 2002.
7. Feinberg, E.H. and Hunter, C.P., Transport of dsRNA into cells by the transmembrane protein SID-1, *Science,* 301, 1545, 2003.
8. Ketting, R.F. et al., Dicer functions in RNA interference and in synthesis of small RNA involved in developmental timing in *C. elegans*, *Genes Dev.*, 15, 2654, 2001.
9. Hammond, S.M. et al., An RNA-directed nuclease mediates post-transcriptional gene silencing in *Drosophila* cells, *Nature*, 404, 293, 2000.
10. Sijen, T. et al., On the role of RNA amplication in dsRNA-triggered gene silencing, *Cell,* 107, 465, 2001.
11. Sulston, J. and Hodgkin, J., Methods, in *The Nematode* Caenorhabditis elegans, 1st ed., Wood, W.B., Ed., Cold Spring Harbor Laboratory Press, New York, 1988, 587.
12. Stiernagle, T., Maintenance of *C. elegans*, in C. elegans: *A Practical Approach*, 1st ed., Hope, I.A., Ed., Oxford University Press, Oxford, 1999, Chapter 4.
13. Wood, W.B., Introduction to *C. elegans* biology, in *The Nematode* Caenorhabditis elegans, 1st ed., Wood, W.B., Ed., Cold Spring Harbor Laboratory Press, New York, 1988, Chapter 1.

14. Simmer, F. et al., Loss of the putative RNA-directed RNA polymerase RRF-3 makes *C. elegans* hypersensitive to RNAi, *Curr. Biol.,* 12, 1317, 2002.
15. Dudley, N.R., Labbé, J.-C., and Goldstein, B., Using RNA interference to identify genes required for RNA interference, *Proc. Natl. Acad. Sci. U.S.A.,* 99, 4191, 1999.

13 RNAi in *Drosophila* Embryos

Robert W. Williams

CONTENTS

13.1 INTRODUCTION

Research on *Drosophila melanogaster* has provided tremendous insights into basic cellular biology as well as the complex patterns and ultimate outcomes of multicellular development. Now in the postgenomic era, new resources continue to make *Drosophila* a powerful model organism in which to address a variety of biological questions. One recently added and notable resource is RNA interference (RNAi),[1,2] the process whereby double-stranded RNA (dsRNA) silences genes containing identical sequences (see Chapter 1 for a detailed description).

In some model organisms such as *C. elegans*, in which delivery of double-stranded RNA is fast, efficient, and easy,[3] RNAi is being used as a forward genetic tool in a systematic manner to assess gene function (see Chapter 11). RNAi in *Drosophila* requires a greater labor investment than in *C. elegans*. However, due to the wealth of existing genetic/genomic tools, library of knowledge, and history of comparative evolutionary-developmental studies, fruit fly embryos are still an extremely attractive system for more directed RNAi experiments.

The difficulty of delivering dsRNA to the *Drosophila* embryo is rooted in the structure of the embryo itself. Prior to hatching, the embryo is surrounded by a hydrophobic

0-8493-2141-7/05/$0.00 + $.50

vitelline membrane that is an effective barrier to the diffusion or uptake of exogenous molecules. Thus, neither feeding dsRNA to the embryo nor soaking the embryo in a solution containing the dsRNA appears to be effective (unpublished data). So far, the only successful method for delivering exogenous dsRNA to a *Drosophila* embryo is by microinjection.[2] Even with this limitation, scores of *Drosophila* genes affecting embryogenesis have been analyzed using RNAi. Presented here are detailed procedures for investigating gene function by RNAi in *Drosophila* embryos, many of which are largely applicable to other insect embryos, as well.

13.1.1 Preparative Note

Once proficient, these procedures should take 2 to 3 days. I recommend allocating several days for trial runs prior to starting an experiment. Practice embryo collection and alignments first. Once comfortable, perform mock injections using water. This will allow you to evaluate the effect of needle shape and dessication on embryo survival. The first few attempts may be frustrating and not successful. Give yourself time to learn the procedures without wasting RNA and without the expectation that the first experiments will work.

13.2 DESIGNING A dsRNA TRIGGER

In order to investigate a gene's function in the *Drosophila* embryo, that gene must be expressed in the embryo. There are approximately 13,676 genes in the *Drosophila* genome,[4] not all of which are expressed in the embryo. By searching available microarray data sets, it is possible to see if any given gene is expressed, either maternally or zygotically, in the embryo and to get an approximation of its temporal profile of RNA abundance. A systematic approach to studying the spatial pattern of all *Drosophila* genes expressed during embryogenesis using *in situ* hybridization is also well underway.[5] With a little data mining, one can ask if a gene is a good candidate for embryo-based RNAi studies and may be guided to the time and place in the embryo to look for a phenotype (http://www.fruitfly.org/cgi-bin/ex/insitu.pl). Conversely, one may identify all genes expressed in a specific embryonic structure and then assay each gene's role (http://www.fruitfly.org/cgi-bin/ex/insitu_hierarchy.pl).

Once a gene is identified for analysis, one needs to decide the exact sequence of the dsRNA trigger. In general, the longer the dsRNA, the better, with a minimum size about 80 bp.[6,7] The region of the gene to be targeted should be unique and not have regions of identity or strong homology with other genes in the genome. This is particularly important when analyzing a single member of a gene family. Since the long dsRNA is converted to 21-bp short interfering RNA (siRNA),[8,9] even the presence of short regions of identity between two genes may result in a phenotype resulting from the reduction of both gene activities. dsRNA directed to 5' and 3' UTRs are effective at reducing gene activity. Since these noncoding regions tend to be divergent, even between closely related homologue, they can be a good place to look when analyzing one of several very closely related genes. The presence of intron sequences in the dsRNA trigger does not hinder RNAi but should not be relied upon to give a knockdown phenotype. Conversely, if redundancy is a concern, a single well-designed dsRNA trigger, or a combination of several triggers, might knock down a whole gene family.

Long dsRNA is made by *in vitro* transcription of each strand. This can be done in two separate reactions or, more efficiently, both strands can be transcribed in a single tube. The two strands are then annealed. Even if both strands are cotranscribed in a single reaction, boiling and slow cooling is necessary to optimize the amount of double-stranded RNA (Figure 13.1A). The *in vitro* transcription reactions can be done using linearized plasmid DNA as a template. However, I prefer using a PCR product resulting from primers that have T7 promoter sequences added to their 5' ends (Figure 13.1B). This allows for the simultaneous transcription of both strands using a single RNA polymerase (Protocol 13.1). T3 or SP6 promoter sequences could be used, as well.

13.2.1 The siRNA Option

As an alternative to *in vitro* transcription of long dsRNA, several companies offer custom siRNA synthesis (see Chapter 2). This requires merely the identification of a 21-nt region of the mRNA to be targeted. Each siRNA strand is 21 to 22 nt long, resulting in a duplex of 19 to 20 bp with 2 nucleotide 3' overhangs. The siRNA strands are 5' phosphorylated and contain 3' hydroxl groups. After synthesis, the two oligonucleotides need to be deprotected, 5' phosphorylated, and annealed to produce the functional siRNA. Many companies, such as Dharmacon, provide these additional services as well as tools to help design the siRNA sequence. If necessary, the siRNA can be 5' phosphorylated using polynucleotide kinase (New England Biolabs). Different siRNAs may have different efficacies, so several custom siRNA sequences may need to be tried.

A

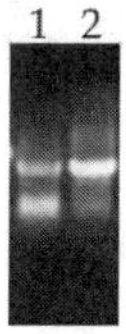

B TAATACGACTCACTATAGGGAGACCAC

FIGURE 13.1 Preparation of *in vitro* transcribed dsRNA. (**A**) Co-transcribed sense and antisense RNA before annealing can still be single stranded (lane 1) but after annealing is largely double stranded (lane 2). (**B**) The T7 sequence placed at the 5' ends of oligonucleotides used as PCR primers to generate gene-specific templates for *in vitro* transcription.

13.3 EMBRYO INJECTIONS

Once a sufficient amount of clean/pure dsRNA at the appropriate concentration is produced, it needs to be loaded into needles of sufficient length, width, and rigidity to cleanly puncture the embryonic vitelline membrane and syncytial plasma membrane (Protocol 13.2). Needles are produced by pulling heated glass capillary tubing. This requires some trial and error (see note below). Just prior to injection, the tip of the capillary tubing will be broken off to produce a functional needle.

13.3.1 Note on Needle Preparation

The proper rigidity and sharpness of the needle are crucial to successful injections. Needles that are too long will tend to bend before puncturing the embryo. Needles that are dull or have wide bores will deform the embryo prior to puncturing and often result in cytoplasmic leakage and death of the embryo. I have included the equipment and settings that work for me, but some trial and error are required to reproducibly get good needles. Once you determine the settings that work for you, keep a reference needle so if the settings get changed or the heating filament gets deformed you can make the necessary adjustments to recreate the appearance of the reference needle.

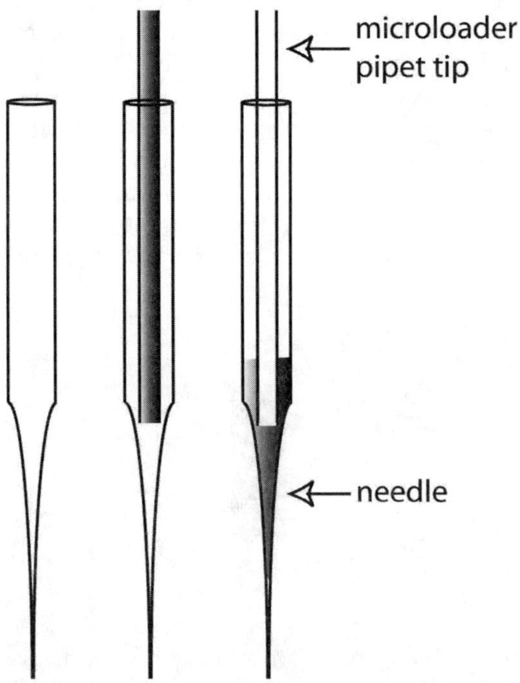

FIGURE 13.2 Diagram of needle loading. A pulled needle (left) is loaded by insertion of a micropipet tip containing dsRNA into the wide bore end (center). The dsRNA solution is carefully deposited into the needle. 2–5 μl of liquid should fill the needle past the shoulders of the tip (right).

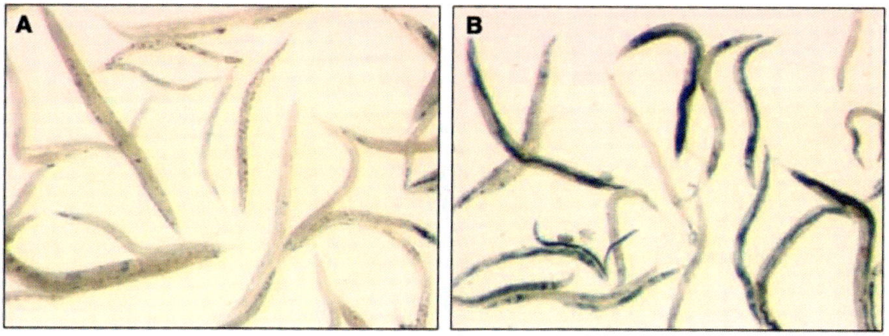

COLOR FIGURE 11.3 RNAi mediated genome instability. To identify genes involved in genome stability, a *C. elegans* strain containing a heat shock promoter-driven out-of-frame *gfp-lacZ* reporter gene fusion was assayed on bacteria expressing dsRNAs for all genes of the *C. elegans* genome. A. Control: worms fed on bacteria containing the vector without insert. B. Example of dsRNA that induces genome instability. The worms were fed on bacteria that express dsRNA homologous to *msh-6*, a gene involved in mismatch repair. These worms have blue patches due to mutations that restore LacZ expression.

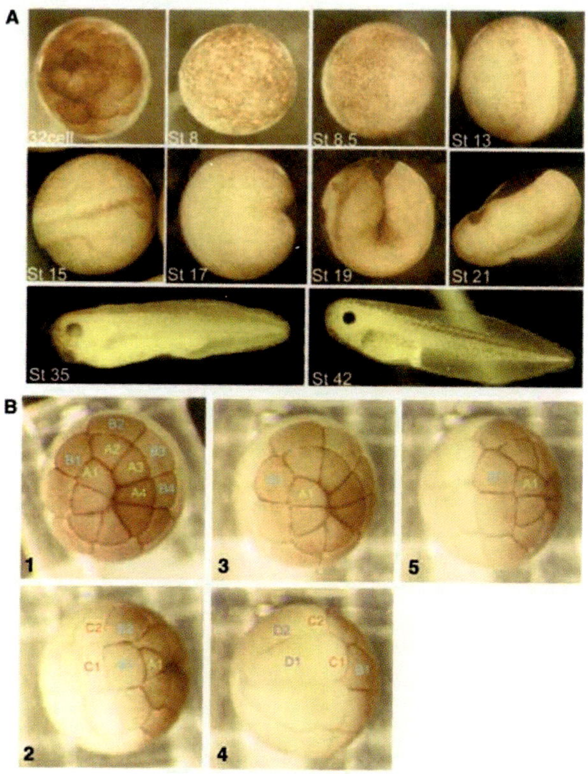

COLOR FIGURE 15.1 Embryonic development of *Xenopus laevis*. A. Different stages of *Xenopus* embryos from 32-cell to tadpole stage. B. Nomenclature of blastomeres at 32-cell stage. 1 and 2: animal pole view. 3, 4, and 5: left side view.

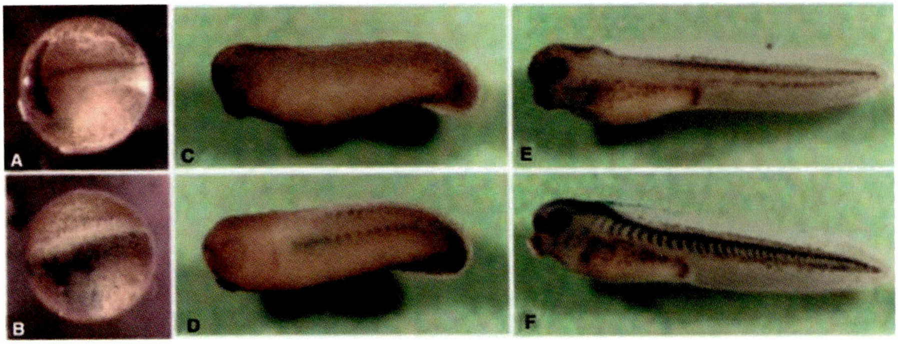

COLOR FIGURE 15.2 Injection of LacZ mRNA into blastomere B1 on the left side of a *Xenopus* embryo at 32-cell stage. A, C, and E: uninjected embryos. B, D, and F: injected embryos. A and B: stage 15. C and D: tailbud stage. E and F: tadpole stage.

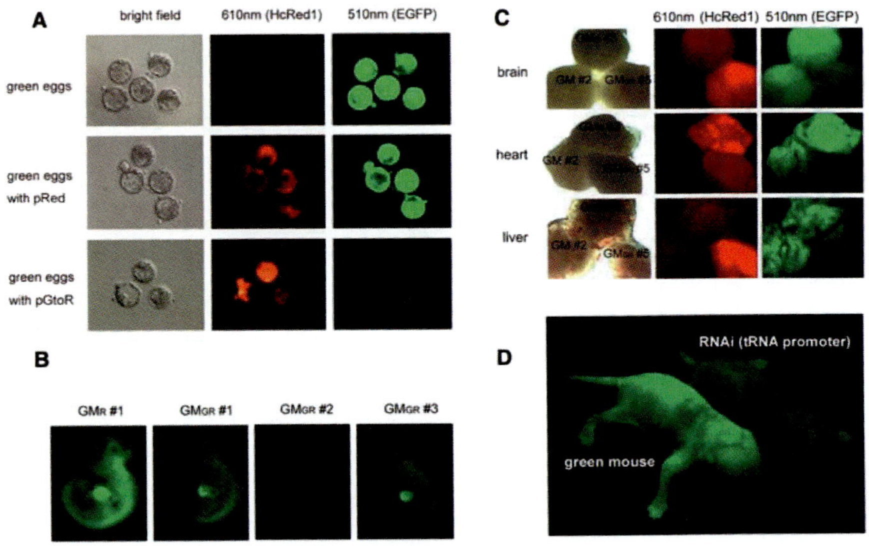

COLOR FIGURE 19.3 A. Green eggs obtained from wild-type females mated with homozygous "green" males were injected with pGtoR, cultured in kSOM for 3 days, and observed for HcRed1 (middle column) and EGFP (right column) expression by fluorescence microscopy. "Green" blastocysts with no additional transgene (top row), "green" blastocysts with the pRed transgene (middle row), and "silenced" blastocysts with the pGtoR transgene (bottom row). B. Day 10.5 embryos were recovered and their fluorescence observed (from left, "green" embryo with pRed GM_R #1, with $pGto_R$ GM_{GR} #1, 2, and 3, respectively). C. Various organs were removed from newborn founder mice (GM #2, GM_R #3, GM_{GR} #5) containing the genetic backgrounds indicated above and photographed under normal lighting (left), 510 nm (right), and 610 nm (center) band pass filter. The exposure time for each panel was adjusted to obtain optimal images. D. A newborn "green" mouse and a "silenced green" mouse with the ptRNA/$siRNA_{EGFP}$ transgene.

13.3.2 Embryo Preparation

While only 50 to 100 embryos may be needed for a given experiment, the embryos need to be near synchronous and collected during a short time window, thus necessitating a large fly population. Embryos to be injected with dsRNA should all be at the syncytial blastoderm stage. This allows for the uptake of dsRNA into the cytoplasm of individual cells during cellularization. Although cellularized embryos are still competent to respond to microinjected dsRNA, they are easily damaged by the injection needle. Furthermore, if the gene to be silenced is active during very early embryogenesis, by waiting until after cellularization to inject, you may miss the critical period to induce a phenotype.

Prior to injection, the embryos are aligned on an agar pad (Figure 13.3B and C) and then transferred to a glass slide with double-sided tape (Figure 13.3D). The anterior

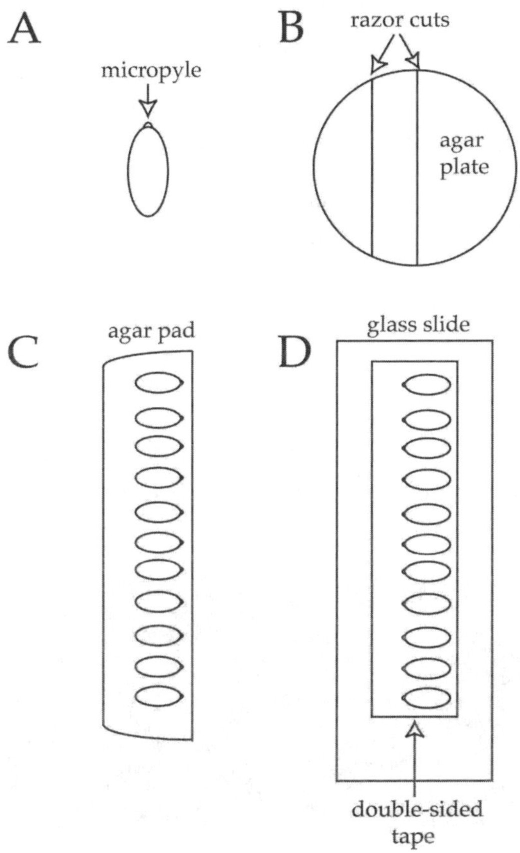

FIGURE 13.3 Diagram of embryo alignment. (**A**) The micropyle marks the anterior end of the embryo. (**B**) An agar pad is created by using a razor blade to cut an approximately ½-inch strip from a 60-mm agar plate. The ends of the pad can be cut off if a meniscus lip is prominent. (**C**) Embryos aligned on the agar pad with their anterior ends in the same orientation. (**D**) Embryos after transfer to a double-sided tape strip on a glass slide.

end of the embryo contains a visible protrusion called the micropyle (Figure 13.3A). The micropyle is used to orient the embryos so they are all facing in the same direction (Figure 13.3C). The procedure presented here describes injecting the posterior region of the embryos. The injection site may be changed to any other region of the embryo by changing how the embryos are aligned and/or presented on microscope stage. Since the effects of RNAi may be localized (see Section 13.6), changing the injection site may be necessary, depending on the gene being studied.

13.4 PHENOTYPIC ANALYSIS

The techniques used to characterize an RNAi-based embryonic phenotype will be dictated by the function of the gene being silenced. To determine the efficacy of RNAi, you can perform RT-PCR or Western analysis on your target gene. For Westerns, 50 to 100 injected embryos will be enough material for a couple of lanes on a gel. Fewer embryos are needed for RT-PCR. Bear in mind that not all injected embryos will show a concomitant phenotype, and those that do will have a range of phenotypic strengths. As such, you may only see a reduction of gene expression, not a complete absence. To collect tissue, follow Protocol 13.6, steps 1 and 2.

For a more detailed description of an RNAi-induced phenotype, it may be desirable to analyze the spatial patterns of gene expression by RNA *in situ* hybridization or immuno-localization. The embryos need to be removed from halocarbon oil, and removed from the vitelline membrane. Follow Protocol 13.6, steps 1 to 2, then proceed with any of the available and routine procedures.

13.5 EXPECTED RESULTS

The number of embryos you inject is a function of your skill level, the type of phenotypic analysis to be done, and the expressivity of the induced phenotype. Many embryos do not survive the injection process. Once competent, it is reasonable to have < 25%

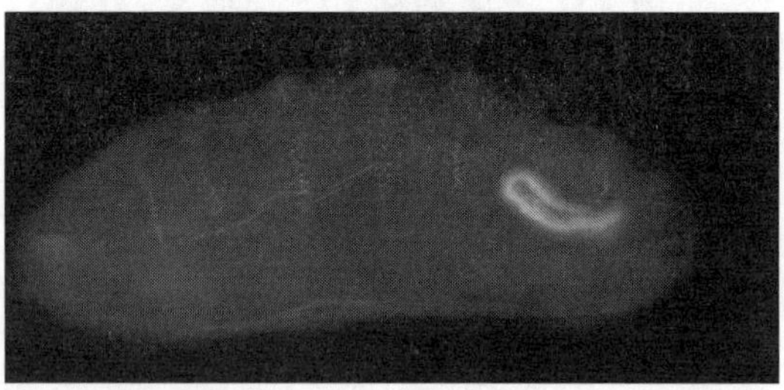

FIGURE 13.4 Cuticle prep. The ventral denticle belts, seen as white stripes along the top of the embryo, are visualized as outlined in Protocol 13.6. A wild-type embryo has eight denticle belts. This embryo was injected with dsRNA corresponding to the *even-skipped* gene, which results in fewer belts (in this case, six).

lethality as a result of injection alone. As a first pass, align 50 to 100 embryos per slide and inject 1 or 2 slides per dsRNA trigger.

The number of surviving embryos that express the RNAi-induced phenotype will also vary. Reported percentages are commonly in the range of 30% to 75%.[2,10] The expressivity may be gene specific. Increasing the concentration of dsRNA up to 5 μM may increase the percentage of embryos exhibiting a phenotype.[11] RNAi is a saturable process, however.[6] Large amounts of dsRNA may induce nonspecific phenotypes by overloading the RNAi machinery. For careful analysis, test several concentrations of dsRNA to achieve the optimum expressivity of the gene-specific phenotype.

13.6 PROBLEMS AND PITFALLS

The dsRNA trigger appears to be spatially restricted in the *Drosophila* embryo. Frequently, a strong phenotype is seen only near the injection site (Figure 13.5). To minimize this, it may be advisable to inject the embryo near the region of endogenous gene expression. A thorough knowledge of a target gene, including its expression pattern, is therefore invaluable. Localized phenotypes may also be overcome by increasing the concentration of injected dsRNA to increase diffusion throughout the embryo.

The dsRNA trigger also appears to be temporally restricted in the *Drosophila* embryo. In other words, dsRNA injected into embryos will generally not give a larval or adult phenotype.[12] This is in contrast to *C. elegans*, for which RNAi-induced phenotypes can persist for several generations.[13] As such, embryo injection-based RNAi experiments must be limited to assaying embryonic phenotypes.

RNAi in general, however, is not limited to the embryo in *Drosophila*. Silencing in larvae and adults is commonly the result of transgene-based inverted repeat constructs.[14] While affective and stable, a transgenic approach does not lend itself to high throughput assessment of gene function. Recently, injection of dsRNA into the abdomen of adult flies was reported to systemically induce gene silencing.[15] There is also a report of RNAi being achieved in larvae by adding dsRNA to their food.[16] These reports may lead to more efficient uses of RNAi as a tool for gene silencing in *Drosophila*.

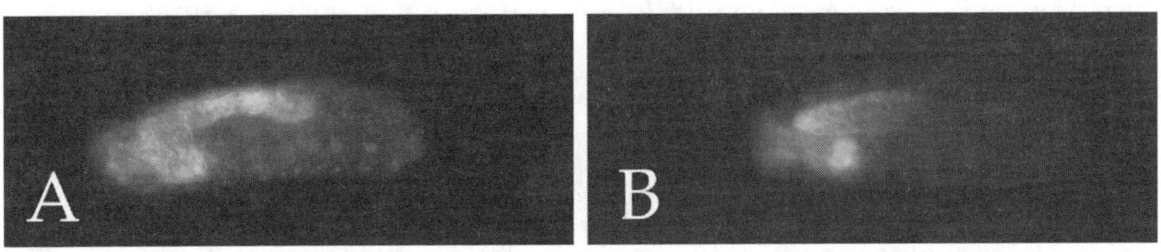

FIGURE 13.5 Posterior localized silencing of *GFP* in *ELAV-GFP* fly embryos. (**A**) Mock injected control, showing GFP fluorescence in the central and peripheral nervous system throughout the length of the embryo. (**B**) Representative embryo injected with dsRNA corresponding to *GFP* coding sequence into the posterior end, showing reduction of GFP fluorescence primarily in the posterior end of the embryo.

13.7 TRANSFERABILITY

The procedures presented here have focused on *Drosophila*. However, they may be generally applicable. Microinjection-based delivery of dsRNA to embryos has successfully achieved RNAi phenotypes in a wide range of insects. Some examples include Tribolium,[17] house flies,[18] and silkworms.[19] One particularly interesting example took advantage of evolutionary-developmental comparisons to *Drosophila*. Schröder[20] showed that the *orthodenticle-1* and *hunchback* genes functionally replace *bicoid* during beetle embryogenesis. Clear *bicoid* orthologs are not found in insects other than higher diptera, suggesting that *Drosophila's* mode of anterior patterning may be lineage specific. By using RNAi in the beetle, an insect with few developed genetic tools, Schröder was able to use information gained from *Drosophila* as a guide to elucidate a more general program of embryo development.[20]

PROTOCOLS

Protocol 13.1 *In Vitro* Production of Double-Stranded RNA

Materials

- Taq DNA polymerase (Amersham, #27-0799-01)
- Qiaquick PCR Purification Kit (Qiagen, #28104)
- T7 MegaScript Transcription Kit (Ambion, #1334)
- Agarose gel (Cambrex Bio Science, #54903)
- 10 × TBE electrophoresis buffer
- Horizontal gel electrophoresis system (BioRad)
- Ethidium bromide, 10 mg/ml (Sigma)
- 6 × bromphenol blue/xylene cyanol tracking dye
- Eagle Eye Gel Imaging System (Stratgene)

Experimental Procedures

1. Perform a standard 50-µl PCR reaction using 100 ng each of the two gene-specific oligos containing 5' T7 sequences (Figure 13.1B).
2. Clean up the product using a PCR purification kit and elute in a 50-µl volume.
3. Verify size and approximate quantity by agarose gel electrophoresis.
4. Set up a 20-µl *in vitro* transcription reaction, following the manufacturer's guidelines, using approximately 1 µg of the PCR product as template.
5. Incubate reaction at 37°C for 2 h, then digest template DNA with DNase I.
6. Boil reaction for approximately 5 min, then remove heat and let water bath with reactions cool to room temperature for several hours or overnight.
7. Dilute reactions to 200 µl with TE and ethanol precipitate.
8. Resuspend pellet in 25 µl of TE, spin in microfuge at full speed for 10 min, carefully avoid pellet, and remove RNA solution to new tube. This step removes particulate matter commonly present in these reactions. Particulates can clog the microinjection needle in later steps (Protocol 13.5, step 12).
9. Quantify RNA concentration by measuring A_{260}. Adjust concentration to 1 µM (approximately 1 µg/µl of 1.4 kb of dsRNA).

10. Verify size, quantity, and double-stranded nature of RNA by agarose gel electrophoresis (see Figure 13.1A).
11. Store the dsRNA in the freezer. It is stable for at least several months.

Protocol 13.2 Preparing Needles

Materials

- Glass capillary tubing (Drummond #9-000-1171)
- Microelectrode puller (Narishige #PN-3)
- Microloader pipette tip (Brinkmann #930001007)

Experimental Procedures

1. Pull capillary tubing using the following settings:
   Heater 7
   Magnet 6.5
   Main Magnet 9.5
2. Discard the needle end held near the filament. Keep the other end, being careful not to break the tip.
3. Spin RNA samples at full speed for 10 min and withdraw 5–10 μl of RNA into a microloader pipet. Carefully avoid the bottom of the tube while removing the RNA. This minimizes particulate matter that can clog the narrow tip of the needle during injections (see Protocol 13.5, step 12).
4. Insert into the back (or wide) end of the needle and load 1–2μl of RNA as close to the tip as possible (Figure 13.2). Avoid introducing air bubbles into loaded needle.
5. Loaded needles should be used within a few hours.

Protocol 13.3 Embryo Collections

Materials

- Stock bottles (Applied Scientific #AS-359 and AS-277)
- *Drosophila* food (Applied Scientific #AS-153)
- Dried yeast (Fleischmann's Yeast #2139)
- Embryo collection cages (Doc Frugal #59-101)
- 100 × 15 mm Petri dishes (Doc Frugal #32-100)

Experimental Procedures

1. Prepare fly food and dispense into bottle as described by manufacturer. Maintain bottles of *Drosophila* stocks at 25°C. Adults are most fertile 3 to 7 days after eclosion.
2. Transfer all the adults from five bottles to one large embryo collection cage fitted with a Petri dish containing *Drosophila* food and a dab of freshly made yeast paste. The collection chambers and all subsequent work should be done at 18°C.
3. Change egg collection plates 1 to 2 h prior to starting embryo collections. Fertilization usually occurs near the time of egg deposition. However, females

occasionally retain developing embryos for several hours prior to deposition. Fresh food stimulates the females to lay all eggs, including those that have begun development, so this first prelaying collection results in a synchronous population of embryos in subsequent collections.

Protocol 13.4 Aligning Embryos

Materials

- Bleach
- Cellector Tissue Sieve Kit (VWR Cat #62399-872)
- 150-mesh screen (VWR Cat #62399-883)
- SuperFrost Glass Microslides (VWR #48311-703)
- ¼" double-sided tape (3M)
- Drierite (W.A. Hammond Drierite Co.)
- Halocarbon Oil Series 700 (Halocarbon Products Corporation)

Experimental Procedures

1. After "pre-laying," collect for 30 to 45 min at 18°C. Cellularization takes about 2 h at 18°C, so most of the embryos during this timed collection are still syncytial. For the remainder of the injection procedure, work quickly because the embryos will start cellularizing over the next one and a half to two hours.
2. Flood the embryo collection plates with several milliliters of 50% bleach. This removes the outer chorion of the embryo, exposing the vitelline membrane, detaches the embryos from the surface of the agar plate, and dissolves any unused yeast paste.
3. If needed, the embryos can be gently dislodged using a small paint brush.
4. Collect embryos in a Cellector with a 150-mesh screen.
5. Thoroughly rinse the embryos several times with H_2O while still in the Cellector. Any residual bleach on the embryos will result in increased lethality after injection.
6. Make an agar pad by cutting a strip from a 60-mm molasses plate (Figure 13.3B).
7. Using a fine paint brush or forceps, align 50 to 100 embryos along the edge of the agar pad with the anterior end/micropyle facing out (Figure 13.3A and C).
8. Prepare a glass slide by placing a strip of double-sided tape long enough to pick up all embryos on the agar pad.
9. Remove tape cover and gently press the glass slide to the edge of the agar pad to pick up the aligned embryos (Figure 13.3D).
10. Dessicate embryos by placing the slide in a sealed jar half filled with Drierite. The exact time will vary and should be optimized. For fresh Drierite, it will be around 5 minutes. For older Drierite, I have used up to 9 minutes. If over-dessicated, the embryos will pull away from the vitelline membrane, have a "saggy" appearance, and deform during injection. Under-dessicated embryos will easily leak and/or pop during injection (Table 13.1).

TABLE 13.1
Common Observations during Needle Insertion

Embryo Deformation	Cytoplasmic Leakage	Likely Reason	Action
Excessive > 10% EL	None to little	Embryos overly dessicated	Reduce time Protocol 12.4, step 10
Excessive > 10% EL	A lot/Embryo pops	Needle tip too wide	Break and try a new needle
OK < 10% EL	A lot	Embryos under-dessicated	Increase time Protocol 12.4, step 10
OK <10% EL	None to little	Perfect	Keep going

11. When proficient, you can align another slide of embryos while the previous one is being dessicated.
12. After dessication, generously cover the embryos with halocarbon oil.

Protocol 13.5 Microinjecting

Materials

- Eppendorf transjector (model #5246)
- Eppendorf micromanipulator (#5171)
- Zeiss inverted microscope (IM35)
- Polystyrene humidity boxes (Ted Pella, cat. #139-206)
- Blotting paper (Schleicher and Schuell #GB004)

Experimental Procedures

1. Insert a loaded, unbroken needle into the transjector arm.
2. Tilt micromanipulator arm down 10 degrees; this is necessary to provide space between the needle barrel and the glass slide.
3. Put glass slide on microscope stage such that the posterior ends of the embryos are facing the needle. If you want to inject the anterior end of the embryo (see Section 13.6), merely turn the slide around.
4. Focus on an embryo, then slowly and carefully bring the needle tip into the same plane. Be careful not to break needle or bring needle into contact with the tape on the slide.
5. Move stage and refocus on the vertical edge of the glass slide.
6. Using the micromanipulator Z-axis control, bring tip of needle into focus just in front of the glass slide.
7. Slowly, move the stage until the needle tip touches the glass slide. You should see the end of the needle quiver a little and the tip should be slightly blunted. The tip should now be broken, allowing the delivery of the RNA solution from the needle.

8. Using the micromanipulator Z-axis control, lift the needle out of the way of the glass slide.
9. Move slide to bring embryos back into view, and focus. When focused on the middle of the embryo, the posterior vitelline membrane should be clear and distinct.
10. While focused on the embryos, carefully lower the needle back into the halocarbon oil and move needle until the tip is aligned with, and in the same plane of focus as, the embryos.
11. Turn on transjector and set Pi to 5 (injection pressure), Pc to 2.5 (constant/back pressue), and t to 0.8 s (injection time).
12. Push "Clean" button until all air in the needle escapes and liquid begins to flow. If no flow is seen, then either the needle tip is not broken or particulate matter is clogging the needle. Try rebreaking the needle. If that does not work, try a new needle. With a properly broken needle tip, there should be very subtle but visible liquid flow even when not pressing the "Clean" button. If no liquid flow is seen at all, try increasing Pc and Pi. Although subjective, "excess" constant flow may be a sign that the needle bore is too wide.
13. (Optional) Do a mock injection into the halocarbon oil by either pressing the "Inject" button or using a foot pedal. If a photomicrometer is available, estimate the volume of delivered liquid by measuring the diameter of the droplet. Using $V = 4/3\mu r^3$, a diameter of approximately 30 μm equals 100 pL. 100 pL of a 1-μM solution delivers 0.1 fmol of dsRNA to the embryo. For comparison, embryos are approximately 450 μm in length.
14. With the needle tip and embryo now in the same plane of focus, move the microscope stage until the needle punctures the embryo. The embryo will deform a little prior to needle insertion, and a little cytoplasmic leaking may occur. If the embryo indents more than about 10% of the length of the embryo (EL) before needle insertion and/or does not quickly reform immediately after needle insertion, the embryos may be overdessicated, resulting in poor viability. Discard slide and shorten time in Protocol 13.4 step 10. Conversely, if the embryo does not deform but does leak, the embryos may be underdessicated. Increase dessication time. If embryos deform excessively and then leak or pop after needle insertion, the needle may not be sharp enough or may be too wide (Table 13.1).
15. Once inserted into the embryo, dispense RNA by pressing the "Inject" button. Visible clearing of the cytoplasm around the needle tip indicates good transfer of solution.
16. Using the microscope stage, move embryo away from needle. A little cytoplasmic leakage may come from the injection site. Excess leakage (i.e., greater than the injection volume) may result in lethality. If questionable, kill the embryo by reinserting the needle and rip out the side.
17. Move to next embryo and inject. If an embryo appears old (i.e., cellularization is visible), kill it as described in step 16.
18. Store slide of injected embryos in a plastic humidity box with moist blotting paper taped to the lid.

19. Let embryos recover and age at 18°C prior to phenotypic analysis. The length of time spent at 18°C will depend on the stage of embryogenesis when a phenotype is to be scored. Larvae will begin to hatch 36 to 48 h after injection.

Protocol 13.6 Cuticle Preparation

Materials

- Heptane (Sigma-Aldrich #H219-8)
- Bleach
- PBT
- Lactic acid (85%; Sigma-Aldrich #W26,111-4)

Experimental Procedures

1. To remove halocarbon oil, place slide in a glass Petri dish and repeatedly wash embryos with heptane. Some embryos may come off the slide at this step. They should be washed into the dish and can still be devitellinized (next step).
2. Using forceps and a needle, manually dissect embryos from the vitelline membrane in PBT.
3. On a glass slide, put two small cover slips at the ends to function as "supports."
4. Place embryos on slide between the two "supports" and remove as much PBT as possible by wicking with a chemwipe.
5. Cover embryos with 85% (v/v) lactic acid and place another cover slip on top of the embryos and two "supports." Let sit either overnight at room temperature or for 3 h at 65°C. The lactic acid dissolves the internal soft tissue of the embryo but leaves the cuticle and external morphology intact.
6. Take phase contrast pictures immediately, as the cuticles will turn yellow over time. Depending on the degree of lactic acid penetration, not all internal tissue may be removed (Figure 13.5).

ACKNOWLEDGMENTS

A special thanks goes to Gerry Rubin, in whose lab these procedures were implemented and optimized. Clare Bolduc advised on microinjection procedures and Todd Laverty helped with equipment, fly maintenance, and overall knowledge. Eric Lai generously provided the ELAV-GFP fly stocks. The preparation of this chapter was supported by the Howard Hughes Medical Institute and by a fellowship from the Helen Hay Whitney Foundation.

REFERENCES

1. Fire, A. et al., Potent and specific genetic interference by double-stranded RNA in *Caenorhabditis elegans*, *Nature*, 391, 806, 1998.

2. Kennerdell, J.R. and Carthew, R.W., Use of dsRNA-mediated genetic interference to demonstrate that *frizzled* and *frizzled2* act in the *wingless* pathway, *Cell*, 95, 1017, 1998.
3. Timmons, L. and Fire, A., Specific interference by ingested dsRNA, *Nature*, 395, 854, 1998.
4. Misra, S. et al., Annotation of the *Drosophila melanogaster* euchromatic genome: a systematic review, *Genome Biol.*, 3, 2002.
5. Tomancak, P. et al., Systematic determination of patterns of gene expression during *Drosophila* embryogenesis, *Genome Biol.*, 3, 2002.
6. Yang, D., Lu, H. and Erickson, J.W., Evidence that processed small dsRNAs may mediate sequence-specific mRNA degradation during RNAi in *Drosophila* embryos, *Curr Biol.*, 10, 191, 2000.
7. Elbashir, S.M., Lendeckel, W., and Tuschl, T., RNA interference is mediated by 21- and 22-nucleotide RNAs, *Genes Dev.*, 15, 188, 2001.
8. Zamore, P.D. et al., RNAi: double-stranded RNA directs the ATP-dependent cleavage of mRNA at 21 to 23 nucleotide intervals, *Cell*, 31, 25, 2000.
9. Hammond, S.M. et al., An RNA-directed nuclease mediates post-transcriptional gene silencing in Drosophila cells, *Nature*, 404, 293, 2000.
10. Williams, R.W. and Rubin, G.M., *ARGONAUTE1* is required for efficient RNA interference in *Drosophila* embryos, *Proc. Natl. Acad. Sci. U.S.A.,* 14, 6889, 2002.
11. Schmid, A., Schindelholz, B., and Zinn, K., Combinatorial RNAi: a method for evaluating the functions of gene families in *Drosophila*, *Trends Neurosci.*, 25, 71, 2002.
12. Misquitta, L. and Paterson, B.M., Targeted disruption of gene function in *Drosophila* by RNA interference (RNA-i): a role for *nautilus* in embryonic somatic muscle formation, *Proc. Natl. Acad. Sci. U.S.A.*, 16, 1451, 1999.
13. Grishok, A., Tabara, H., and Mello, C.C., Genetic requirements for inheritance of RNAi in *C. elegans*, *Science*, 287, 2431, 2000.
14. Kennerdell, J.R. and Carthew, R.W., Heritable gene silencing in *Drosophila* using double-stranded RNA, *Nat. Biotechnol.*, 18, 896, 2000.
15. Dzitoyeva, S. et al., Intra-abdominal injection of double-stranded RNA into anesthetized adult *Drosophila* triggers RNA interference in the central nervous system, *Mol. Psychiatry*, 6, 665, 2001.
16. Eaton, B.A., Fetter, R.D., and Davis, G.W., *Dynactin* is necessary for synapse stabilization, *Neuron*, 30, 729, 2002.
17. Brown, S.J. et al., Using RNAi to investigate orthologous homeotic gene function during development of distantly related insects, *Evol. Dev.*, 1, 11, 1999.
18. McGregor, A.P. et al., Rapid restructuring of bicoid-dependent hunchback promoters within and between Dipteran species: implications for molecular coevolution, *Evol. Dev.*, 3, 397, 2001.
19. Quan, G.X., Kanda, T., and Tamura, T., Induction of the white egg 3 mutant phenotype by injection of the double-stranded RNA of the silkworm white gene, *Insect Mol. Biol.*, 11, 217, 2002.
20. Schröder, R., The genes *orthodenticle* and *hunchback* substitute for *bicoid* in the beetle *Tribolium*, *Nature*, 422, 621, 2003.

14 RNAi in *Xenopus Laevis* Oocytes

Earl Gordon and Geoffrey W. Abbott

CONTENTS

14.1 INTRODUCTION

Xenopus laevis, the South African clawed frog (Figure 14.1A), is named for its country of origin and distinctive long claws on the three inside toes of either hind foot, which it uses to tear apart larger food items.[1] Although the primarily aquatic existence of *X. laevis* is characteristic of true frogs, it belongs to neither the *Rana* (true frogs) nor *Bufo* (toads) families and is classified in the family *Pipidae*. *X. laevis* has no ears or eyelids and, unlike *Rana* frogs, lacks a tongue suited for the trapping of prey. Though endogenous to southern African countries, *X. laevis* can now be found widely distributed in many parts of the world. *X. laevis* was used for human pregnancy tests in the 1940s, as chorionic gonadotrophin present in the urine of pregnant women was shown to induce the laying of eggs when injected into adult female frogs.

The use of frog oocytes in experimental biology dates back to 1952, when Briggs and King performed their early work on nuclear transplantation.[2] Initially, oocytes from the leopard frog *Rana pipiens* were utilized, but later experiments were performed with oocytes from *X. laevis*. The oocytes of *X. laevis* (Figure 14.1B) have been widely used since then as tools for protein expression and analysis because injection of complementary RNA (cRNA) for specific genes results in highly efficient protein expression, even of mammalian proteins and membrane proteins. In particular, *X. laevis* oocytes are commonly used for electrophysiological analysis of the function of mammalian channels and transporters. The first reports of injection of ion channel transcripts into oocytes emerged in the early 1980s, when RNA from particular tissues was injected into oocytes and functional assays were performed to deduce the function of the

0-8493-2141-7/05/$0.00 + $.50

translation products.[3] *X. laevis* oocytes are ideal for use in electrophysiology because of their robustness and relatively large size (1 to 1.2 mm), which ensure relative ease of handling, injection, and recording.

Oocytes are precursors to mature egg cells and are stored in the lobes of the ovaries of the adult female frog. Oocytes are classified in stages, from stage I to stage VI; stage I is the smallest and most immature and stage VI is the largest and most mature. At any given time oocytes from all stages can be found in the ovary of an adult frog, and it is estimated that it takes approximately 8 months for oocytes to progress from stage I to stage VI.[4] In most electrophysiological experiments, stage V and stage VI oocytes are used. Oocytes are harvested from frogs using one of two methods: survival surgery or following euthanasia of the frog.[5] Large numbers of oocytes can be obtained using either approach. It was recently proposed that eggs from a different species, *Xenopus tropicalis*, might have distinct advantages as a model in some experimental settings,[6] but this is outside the scope of this work, and further references to oocytes in this chapter will be taken to mean those from *X. laevis*.

14.2 RNAi OF GENES ENCODING ENDOGENOUS *XENOPUS* OOCYTE ION CHANNEL β SUBUNITS

In the past several decades, *X. laevis* oocytes have been widely used as a heterologous expression system for the study of mammalian ion channels. Oocytes are particularly suited for study of voltage-gated ion channels because they can be clamped at particular voltages using the two-electrode voltage clamp (TEVC) system and then whole-cell currents can be recorded to measure the current–voltage relationship for the ion channel studied. In addition, pharmacological agents can be passed over the oocyte via a bath system, thus allowing the pharmacological properties of the expressed channels to be assessed. Further, membrane patches can be pulled from the oocyte and electrical currents recorded using patch-clamp electrophysiology; thus, if expression levels are suitable, the currents through single channel proteins can be resolved.

In several studies involving heterologous expression of mammalian voltage-gated potassium (Kv) channels, the data suggested that factors intrinsic to the expression system might be altering the properties of the overexpressed channel α subunit (pore-forming subunit); results varied depending upon expression system (oocyte vs. mammalian cell) or even expression level within the same system.[7–10] We hypothesized that *X. laevis* oocytes might contain endogenous *Xenopus* β subunits, relatives of the mammalian proteins that co-assemble with Kv channels to alter their functional properties. To test this hypothesis, we BLAST-searched the *X. laevis* genome for sequences of one class of Kv channel β subunits — the MinK-related peptides (MiRPs) — and found several *X. laevis MiRP* genes.[11] MiRPs are short (~100 to 170 residues), single membrane-spanning proteins that can radically alter Kv channel properties. In human cardiac tissue, MiRPs such as MinK and MiRP1 are thought to modulate Kv channels such as KCNQ1 and HERG.[12,13] Both KCNQ1 and HERG are essential for human

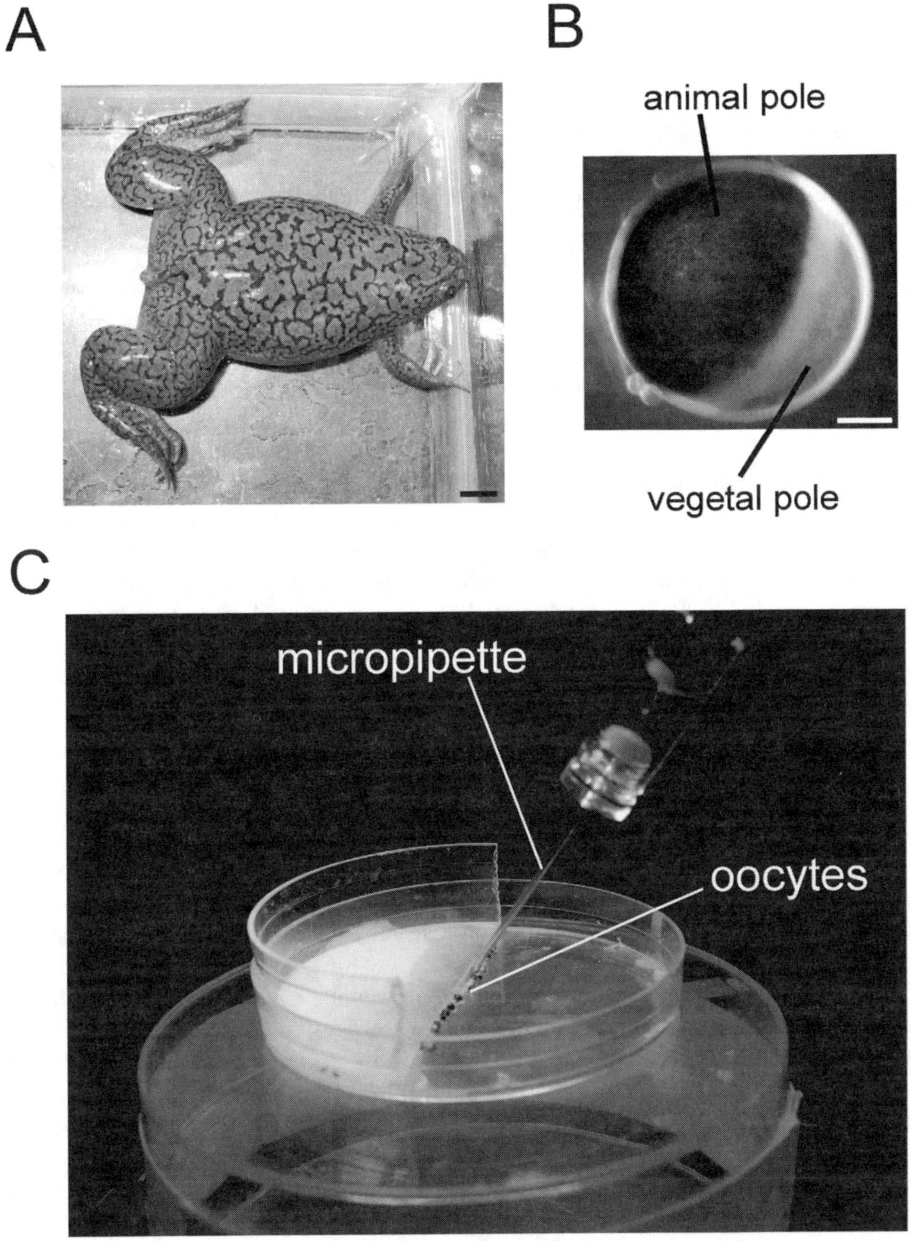

FIGURE 14.1 The *Xenopus laevis* oocyte expression system. (A) A mature female *Xenopus laevis* (scale bar, 1 cm). (B) A stage VI *Xenopus laevis* oocyte with animal and vegetal poles labeled (scale bar, 0.2 mm). (C) Injection micropipette and oocytes queued up on a modified 60-mm plate for injection.

cardiac repolarization and have been widely studied by overexpression and TEVC in *X. laevis* oocytes.

xMiRP2 was cloned from cDNA isolated by RT-PCR from stage V and VI *Xenopus laevis* oocyte mRNA using forward primer 5'CAGTTTGATTGGAGAGTGGGATTC and reverse primer 5'TAGACCCCTGGGGCCTCGTC. xMiRP2 and other MiRP genes were identified by a BLAST search of EST databases with mammalian MiRP gene sequences using the tblastn algorithm. All genes showed a canonical initiation ATG site with G at – 3, A at + 3, and no intervening methionines 3' of the nearest 5' stop codon. For functional expression, xMiRP2 was ligated into pBluescriptRN3 and linearized with *Sfi* I for transcription of cRNA for injection into oocytes. For studies in oocytes, cRNA transcripts for xMiRP2 and HERG (in pSP64, Promega) were produced as before.[12]

Reverse genetics, in which the function of a gene is elucidated by gene-manipulation techniques, has revolutionized the analysis of gene function. One such technique is the use of RNA interference (RNAi); although the specific mechanism of action of double-stranded RNAi oligonucleotides is unknown, it is believed to involve targeting of specific mRNAs for degradation.[14,15] This action results in sequence-specific silencing of expression of the targeted gene. We thus applied RNAi techniques to suppress endogenous oocyte MiRP expression to assess the contribution of endogenous *X. laevis* MiRPs in heterologous Kv channel expression studies performed in oocytes. Results showed that RNAi gene silencing of endogenous xMiRP2 altered heterologously expressed HERG currents, suggesting that xMiRPs in general may play a significant role in determining the properties of Kv channels studied in *X. laevis* oocytes.[11]

14.3 CONCLUSION

Xenopus oocytes provide a versatile system for robust heterologous expression of mammalian proteins. For electrophysiologists, the ease of handling and recording from oocytes has facilitated decades of research into ion channel structure and function. The cloning of a great many ion channel genes together with improved transfection techniques and commercially available patch-clamp rigs optimized for recording from mammalian cells, have encouraged a partial shift in recent years to ion channel heterologous expression studies being carried out in relatively electrically silent mammalian cells such as Chinese Hamster Ovary (CHO) cells and Human Embryonic Kidney (HEK) cells. *Xenopus* oocytes, however, are still widely used by electrophysiologists in both academia and industry. The successful application of RNAi techniques to *Xenopus* oocytes facilitates the elimination of heterologous expression artifacts introduced by variable contribution of endogenous ion channel modifier subunits but, by the same token, provides a means to exploit the oocyte as a model system to study the many roles and actions of channel β subunits in their own right.

PROTOCOLS

PROTOCOL 14.1 OOCYTE HARVESTING

Materials

- OR-2 buffer (82.5 mM NaCl, 2.5 mM KCl, 1.0 mM $MgCl_2$, 5.0 mM HEPES, pH to 7.4 with NaOH)
- ND-96 buffer (96.0 mM NaCl, 2.0 mM KCl, 1.8 mM $CaCl_2$, 1.0 mM $MgCl_2$, 5.0 mM HEPES, penicillin/streptomycin 10,000 U/ml, pH to 7.6 with NaOH)
- Type I collagenase, lyophilized (Gibco)
- Anesthetic (Tricaine methane sulphate)
- Dissection instruments (disposable scalpel, scissors, two pairs of blunt forceps, and suture thread)
- Oscillating platform
- 60-mm tissue culture plates and Pasteur pipettes for oocyte agitation
- 15°C incubator and 24-well plates for oocyte storage

Experimental Procedures

1. Anesthetize female frogs in water containing 0.5–2 g/L Tricaine methane sulphate, and place on their backs.
2. Make a small incision in the lower abdominal wall such that both the outer and inner skin, are breached.
3. Using blunt forceps, remove 1 to 2 lobes of the ovary while leaving the remainder of the ovaries intact to speed recovery.
4. Place the lobes in OR-2 buffer (pH 7.4), and suture the wound of the frog.
5. Separate oocytes (which will number several hundred per harvest) into groups of < 5 using surgical forceps.
6. Defollicate the oocytes by agitation at room temperature in Ca^{2+}-free OR-2 solution containing 2 mg/ml collagenase for 45 to 120 min. Choice of collagenase is of great importance, as quality and specific action vary widely between manufacturers and even batch to batch; we favor Gibco Type I collagenase. Once the particular condition required for a specific batch of collagenase is found, it is prudent to order more of that specific batch to avoid further titration of conditions for future treatments.
7. Once collagenase treatment has resulted in complete separation of oocytes from one another, use gentle aspiration through a Pasteur pipette (wide enough not to place shear stress on the oocyte) to remove weakly attached follicular layer.
8. Place separated stage V and stage VI oocytes in 60-mm-diameter tissue culture plates.
9. Incubate the oocytes in ND-96 buffer (pH 7.6) and store in an incubator maintained at 15°C prior to selection for microinjection.

Oocyte selection

10. Select oocytes by inspection under a dissecting microscope. Discard unhealthy oocytes with anomalies in shape or pigmentation of the animal pole.

11. Store selected oocytes in an incubator maintained at 15°C.

Technical notes. In mature, healthy oocytes two distinct regions can be easily identified — the animal pole is dark brown in color and the vegetal pole is yellow (Figure 14.1B). A layer of follicle cells surrounds the oocytes; this is removed by collagenase treatment before injection, as it tends to blunt or clog injection electrodes very quickly. Furthermore, it is essential that this layer be removed before electrophysiological recording or quantification of RNAi, as it represents a separate population of cells that will interfere with interpretation of experimental findings. Selection of oocytes for injection must be performed with care since the success of experiments will depend on the suitability of oocytes for injection. In most cases, an unhealthy oocyte can be identified by anomalies in the shape or pigmentation of the animal pole. Since oocytes can be damaged mechanically by injection or subsequently by the material injected into the oocyte, it is usually necessary to inject more oocytes than that which will be required for the proposed experiments.

Protocol 14.2 Oocyte Injection

Oocytes are injected with appropriate amounts of cRNA (usually in the range of 0.1 to 20 ng per oocyte) or short interfering RNA (siRNA) oligonucleotides (we typically use 500 pg per oocyte) using a micropipette injection system (Figure 14.1 C).

Materials

- Borosilicate capillaries with filament (Sutter Instrument Co., Novato, CA)
- Mineral oil for use in microinjection (Fluka)
- RNase-free water
- Microelectrode puller (e.g., PP-830 model, Narishige, Japan)
- Manual microinjector (Sutter)
- Dissection microscope with lamp
- Micromanipulator (Sutter)

Experimental Procedures

1. Prepare glass micropipettes using a vertical micropipette puller from capillaries of borosilicate glass (also see Chapters 12 and 13).
2. Use an oil-filled borosilicate micropipette with a tip diameter of ~15 micrometers to inject each oocyte, typically with 25 to 50 nl of sample.
3. Once injected, keep oocytes in small groups (e.g., in 24-well plates, with 2 to 3 oocytes per well) because some eggs are damaged during injection and tend to subsequently release enzymes that can cause the death of other oocytes with which they are stored.
4. To improve viability of oocytes postinjection, change daily the media in which they are incubated; this applies more to those that are used for electrophysiological recordings, which are in some cases performed up to 2 weeks after injection.
5. Lyse oocytes for molecular biology experiments typically 1 to 4 days after injection, at which time the oocytes are generally demonstrating robust expression of injected cRNAs and during which time injected siRNA oligonucleotides suppress endogenous gene expression.[11]

PROTOCOL 14.3 siRNA OLIGONUCLEOTIDE DESIGN AND SYNTHESIS

Materials

- siACE-RNAi siRNA oligonucelotide duplexes (Dharmacon)

Experimental Procedures

The following guidelines were used for oligonucelotide design:

1. Count 75 bases downstream from the start site.
2. Locate the first AA dinucleotide.
3. Count the next 19 bases after the AA; check that the GC richness is ~50 and no more than 70.
4. Conduct a BLAST search using the NCBI *X. laevis* EST database to check for similar sequences to reduce the possibility that the siRNA oligonucleotides will target other genes.
5. If the conditions in 3 and 4 are not met, move downstream to the next AA.
6. Based on the 21-nt sequence that is selected synthesize two siRNA oligonucleotides and anneal to construct a double-stranded RNAi oligonucleotide (Figure 14.2).

Technical notes. For RNAi of endogenous xMiRP2 in *X. laevis* oocytes, 500 pg of double-stranded siRNA 21 mer oligonucleotide (Dharmacon) targeted to bases 104 to 124 of xMiRP2 (5'AAGGGAACCACACGGACGCCA) (Figure 14.2A) was injected into oocytes immediately after injection of Kv channel α subunit cRNA.[11]

One of the controls used was injection of an siRNA oligonucelotide predicted not to target any *X. laevis* genes. For xMiRP2 studies, we used an siRNA oligo targeted to the rat MiRP2 gene (Figure 14.2 B). This oligonucleotide had no effect on xMiRP2 gene expression in oocytes, indicating specificity of effects, but efficiently silenced endogenous rat MiRP2 in rat pheochromocytoma (PC12) cells (Figure 14.3). Conversely, xMiRP2 siRNA oligonucleotides had no effects on rat MiRP2 in PC12 cells when used as a reciprocal control in the mammalian system.[16] For PC12 cell studies, siRNA oligonucelotides (100 ng per well of a 6-well plate) were transfected using Lipofactamine 2000 (Invitrogen, San Diego, CA).[11,16]

PROTOCOL 14.4 ASSESSMENT OF RNAi IN OOCYTES AT THE mRNA LEVEL USING REVERSE TRANSCRIPTASE POLYMERASE CHAIN REACTION (RT-PCR)

Polymerase chain reaction (PCR) combined with reverse transcription (RT) is a useful tool for the determination of the knockdown of a targeted gene with RNAi as follows:

Materials

- Trizol (InVitrogen)
- SuperScript First Strand Synthesis System for RT-PCR (InVitrogen)
- PCR Platinum Supermix (InVitrogen)
- RNase-free water, ethanol, and isopropanol
- Thermal cycler for polymerase chain reaction

A

xMiRP2 cDNA sequence

1 50

ATGGAGACGGCCCGACCAATGGAAAAGCTGCTGGAGAACCTGCACAATGT

51 100

GCTGCTGGCCATAAACAAGACGCTGAACAGTTTGCCGTGCCAGCCGGACC

101 150

CAC**AAGGGAACCACACGGACGCCA**AAGCGAATAAGGAGCACCGACCCGAC

151 200

AACGCCTATATATTCATCCTCTTTGTTCTGTTTCTCTTTGCGGCCACCGT

201 250

GGGGAGTTTGATACTGGGCTACACCAGGTCTAAGAAAGTGGACAAGCGCA

251 300

GTGACCCGTACCACGTTTACATAAAAAACAACCGCATTTCTGTGATATGA

xMiRP2 siRNA duplex:

GGG.AAC.CAC.ACG.GAC.GCC.AdTdT

dTdT.CCC.UUG.GUG.UGC.CUG.CGG.U

B

rMiRP2 cDNA sequence

1 50

ATGGAGACTTCCAACGGGACTGAGACCTGGTACAAGAGCCTCCACGCTGT

51 100

GCTAAAGGCTCTGAACACAACCCTTCACAGTCACTTGCTCTGCCGGCCTG

101 150

GGCCAGGGCCAGGGTCAGGAACAGGGCCAGACAATCA**AACAGAGGATCAT**

151 200

CGGGCTAGCCTTCCCGGCCGTAATGACAACTCCTACATGTATATTCTCTT

201 250

TGTCATGTTCCTATTCGCTGTCACTGTGGGCAGTCTCATCCTGGGATATA

251 300

CCCGTTCACGCAAAGTGGACAAACGTAGTGACCCCTATCATGTGTACATC

301

AAAAACCGTGTGTCTATGATCTGA

rMiRP2 siRNA duplex:

CAG.AGG.AUC.AUC.GGG.CUA.GdTdT

dTdT.GUC.UCC.UAG.UAG.CCC.GAU.C

FIGURE 14.2 RNAi oligonucleotide design. (A) Sequence of *Xenopus laevis* MiRP2 (xMiRP2) highlighting the RNAi target sequence (bold) and the sequence of the double-stranded siRNA oligonucleotide. (B) Sequence of *Rattus rattus* MiRP2 (rMiRP2), highlighting the RNAi target sequence (bold) and the sequence of the double-stranded siRNA oligonucleotide.

- Vortex mixer
- BioRad Fluor-S MultiImager (BioRad)

Experimental Procedures

Extraction of RNA from oocytes

1. Select 10 oocytes from each group to be analyzed and pool. A sample group (i.e., siRNA oligonucleotide injected) and several control groups can be compared to assess the efficiency of RNAi gene silencing. Controls include uninjected oocytes, oocytes overexpressing the targeted gene (i.e., cRNA-injected) and oocytes injected with a scrambled siRNA oligonucleotide (a 21 mer not having sequence similarity with known *X. laevis* genes).

2. Treat oocytes with Trizol® (1 ml) for the extraction of RNA, and allow the suspension to stand at room temperature for 5 min.
3. Add 200 μl chloroform to the suspension and shake the mixture by hand for 15 sec; incubate at room temperature for 3 min.
4. Centrifuge the suspension at 12,000 × g at 4°C for 15 min.
5. Remove the colorless upper layer, which contains the RNA and mix with 500 μl isopropanol. Incubate the suspension at room temperature for 10 min.
6. Spin the suspension at 12,000 × g at 4°C for ten min; this should result in precipitation of the RNA, which is identifiable as a pellet at the base of the Eppendorf tube.
7. Discard the supernatant and wash the pellet with 75% (v/v) ethanol (1 ml). Mix the resultant solution by vortexing and centrifuge at 7500 × g at 4°C for 5 min.
8. Discard the supernatant and air-dry the precipitated RNA at room temperature for 5 to 10 min. Dissolve the pellet in RNase-free water and store at –80°C.

Reverse transcription

Since PCR involves the amplification of a targeted DNA sequence, it is necessary to transcribe the extracted RNA into cDNA; this is achieved by the process of reverse transcription (RT).

9. Incubate sample RNA (~5 μg) with the following reagents:

dNTP mix (10 mM)	1 μl
Oligo (dT) 12–18 (0.5 μg/μl)	1 μl
DEPC–water	up to10 μl

10. Incubate the reaction mixture at 65°C for 5 min and then place on ice for 1 min.
11. Add the following reagents to the mix:

10 × RT buffer	2 μl
$MgCl_2$ (25 mM)	4 μl
DTT (0.1 M)	2 μl
RNase OUT	1 μl

12. Spin the reaction mix and incubate at 42°C for 2 min.
13. Add 1 μl SuperScript II Reverse Transcriptase, mix, and incubate at 42°C for 50 min.
14. To terminate the reaction, incubate at 70°C for 15 min, followed by chilling on ice.
15. Following brief centrifugation, add RNase H (1 μl) to the mix and incubate at 37°C for 20 min. The oocyte cDNA is now ready for PCR amplification.

Polymerase chain reaction

Since the effect of knockdown is being compared between several groups, it is necessary to ensure that an equal amount of total cDNA from each is being used. This is achieved by normalization with an oocyte "housekeeping" gene. We utilized *X. laevis* β actin, which can be amplified using the following primers:[11]
Forward, 5'AAGGAGACAGTCTGTGTGCGTCCA
Reverse, 5'CAACATGATTTCTGCAAGAGCTCC

16. Initially, perform serial dilutions of the samples to establish the point at which the amplification is not saturated, thus providing a suitable point for comparison.

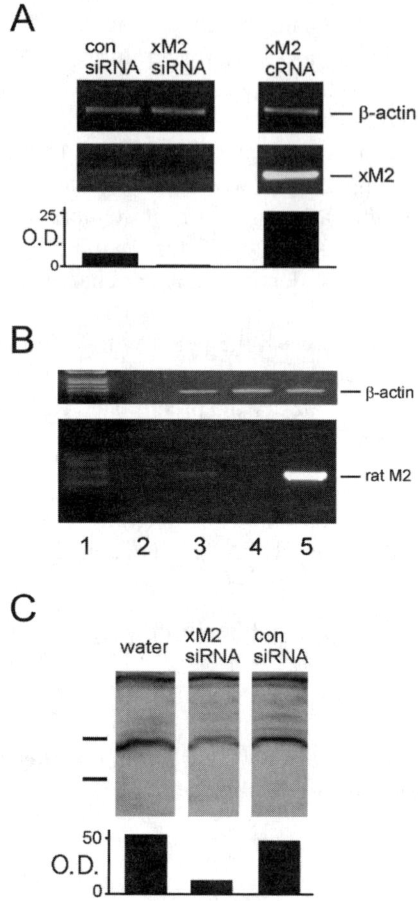

FIGURE 14.3 Quantification of RNAi of endogenous MiRP2 (A) Semiquantitative PCR of cDNA generated by RT-PCR of mRNA isolated from *X. laevis* oocytes. Upper panel: *X. laevis* mRNA normalized to endogenous β actin gene: (left to right) oocytes injected with 500 pg scrambled control siRNA, 500 pg xMiRP2 siRNA, 3 ng positive control xMiRP2 cRNA. Bands were generated by PCR using primers specific for *X. laevis* β actin and indicate equal amounts of mRNA-generated cDNA loaded in each lane. Central panel: PCR from cDNA generated from *X. laevis* oocyte mRNA by RT-PCR and normalized to *X. laevis* β actin (see above): (left to right) oocytes injected with 500 pg control siRNA (targets rat MiRP2, see panel B, but not *X. laevis* MiRP2), 500 pg xMiRP2 siRNA, 3 ng positive control xMiRP2 cRNA. Bands were generated by PCR using primers specific for *X. laevis* MiRP2 and indicate specific post-transcriptional gene silencing of xMiRP2 by xMiRP2 siRNA. Lower panel: Densitometry of xMiRP2 bands indicating 8-fold reduction of xMiRP2 cDNA band intensity using xMiRP2 siRNA and 30-fold increase using MiRP2 cRNA. (B) RNAi gene silencing of endogenous rat MiRP2 in PC12 cells assessed by semiquantitative PCR of products from RT-PCR of PC12 cell mRNA. Upper gel: normalization of RT-PCR products by titration of amplified β actin; lanes show approximately equal β actin; band intensities. Lower gel: RT-PCR for rMiRP2 using normalized samples. Upper and lower gels: Lane 1, marker; lane 2, water control; lane 3, nontransfected PC12 cell RT-PCR product; lane 4, rat MiRP2 siRNA-transfected PC12 cell RT-PCR product; lane 5, MiRP2 cDNA-transfected PC12 cell RT-PCR product. Expected

17. Measure optical densities of samples run on a 1% (w/v) agarose gel and stained with ethidium bromide using, for example, a Fluor-S MultiImager (BioRad).
18. Determine the sample concentration that results in equal β actin amplicon band intensity of each group.
19. Perform PCR on these normalized samples with primers that will amplify the siRNA-targeted gene. Use the following reaction mix for PCR amplification in oocytes:

Platinum Supermix (Invitrogen)	21 μl
Gene specific primer forward	1 μl
Gene specific primer reverse	1 μl
cDNA	2 μl

20. View the results of PCR amplification from normalized samples using target gene-specific primers by running a 10-μl aliquot of the reaction mix on a 1% (w/v) agarose gel and measure the band intensity by spectrophotometry (Figure 14.3A).

Technical notes. Results from RNAi of xMiRP2 showed gene knockdown with xMiRP2 siRNA oligos but not with rat MiRP2 siRNA oligos as expected because of sequence differences (Figure 14.3A). The rat MiRP2 siRNA oligonucleotide did, however, efficiently silence expression of endogenous rat MiRP2 in PC12 cells, indicating it also showed specific function (Figure 16.3B).[11,16]

Protocol 14.5 Assessment of RNAi in Oocytes at the Protein Level Using Western Blot Analysis

Western blotting can be used as a direct measure of the effectiveness of RNAi in oocytes at the protein level.

Materials

- Lysis buffer (150 mM NaCl, 0.2 mM PMSF, 20 mM NaF, 10 mM Na_3VO_4, 50 mM Tris, pH 7.4, with detergents, 1% (v/v) NP-40, 1% (v/v) CHAPS, 1% (v/v) Triton X-100, and 0.5% (w/v) SDS)
- Bradford assay kit (Sigma)
- SDS-PAGE loading buffer
- SDS-PAGE gel apparatus and power supply (BioRad)
- Western transfer apparatus (BioRad)
- PVDF membrane (Millipore)
- BioRad Fluor-S MultiImager (BioRad)

FIGURE 14.3 continued
product sizes for gene fragments, indicated by labeled lines, were 450 bp for actin and 596 bp for rat MiRP2. (C) Upper panel: Western blot using MiRP2-specific antibodies from lysates of *X. laevis* oocytes injected with (left to right) water, 500 pg x MiRP2 siRNA, 500 pg scrambled control siRNA. Protein loaded per lane was normalized to total protein concentration (20 mg per lane) using the Bradford assay. Lines indicate 18 kD (upper) and 6.4 kD (lower).
Lower panel: Densitometry of xMiRP2 bands indicating 5-fold reduction of xMiRP2 protein concentration using xMiRP2 siRNA and no significant reduction with scrambled control siRNA.

- Target protein-specific primary antibodies
- Horseradish peroxidase conjugated secondary antibodies (BioRad or Pierce)

Experimental Procedures

1. To extract proteins, select 10 oocytes from each group. These can either be frozen at –80°C for long-term storage (without buffer) or immediately treated with lysis buffer.
2. Clarify the samples by two separate centrifugations at 10,000 × g for 20 min. At this point the supernatants can be stored at –80°C or continue with quantification.
3. Determine the protein content of each sample using the Bradford assay as recommended by the supplier.
4. Mix aliquots of each lysate containing 20 μg of protein with standard SDS-PAGE loading buffer and separate by SDS-PAGE (in the case of MinK-related peptides, 15% (w/v) or 4 to 20% (w/v) gradient gels are used).
5. Transfer the separated proteins to PVDF membrane.
6. Perform Western blotting with appropriate primary antibodies and peroxidase-coupled secondary antibodies (for xMiRP2, an in-house primary antibody was utilized, raised by injection into rabbits of a mammalian expression vector expressing full-length hMiRP2; detection was via goat antirabbit horseradish peroxidase-coupled secondary antibodies (BioRad) for fluorography).
7. Assess the intensity of the bands by densitometry, using, for example, a FLUOR-S MultiImager (BioRad) (Figure 14.3C). Detection and quantification can also be performed digitally on systems such as the STORM imaging system (Molecular Dynamics).

Protocol 14.6 Assessment of RNAi in Oocytes at the Functional Level Using Two-Electrode Voltage Clamp (TEVC)

X. laevis oocytes are amenable to rapid electrophysiological studies using the two-electrode voltage clamp (TEVC) system. TEVC facilitates the measurement and analysis of currents across the cell membrane at a given membrane potential.[17] In the two-electrode voltage clamp system, as the name suggests, there are two electrodes, a voltage electrode, and a current electrode. The functional readout of voltage clamp experiments using TEVC is the amount of current that is injected by the current electrode to counteract currents flowing though ion channels in the plasma membrane and, thus, keep membrane potential constant. Current–voltage (I/V) plots are the standard method of visualizing and quantifying the currents passed by voltage-gated ion channels in oocytes (Figure 14.4). In the case of analysis of xMiRP2 RNAi in oocytes, TEVC is a robust bioassay for the remaining xMiRP2 function, because overexpressed xMiRP2 (by injection of xMiRP2 cRNA into oocytes) strongly inhibits currents expressed when the mammalian cardiac Kv channel HERG is overexpressed in oocytes (Figures 14.4A and B).[11] When a range of HERG cRNA concentrations are injected into groups of oocytes in the presence or absence of simultaneously injected xMiRP siRNA, however, an increase in HERG current amplitude is observed up to ~5 ng HERG cRNA above which the effects of endogenous xMiRP2 are sufficiently diluted

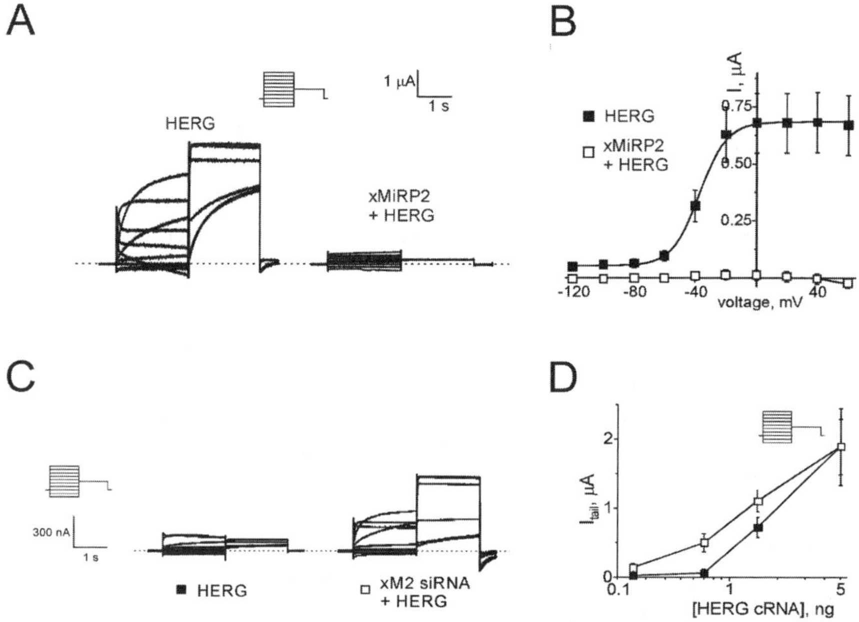

FIGURE 14.4 TEVC — a functional assay of endogenous oocyte xMiRP2 RNAi gene silencing. (A) Exemplar raw current traces generated from TEVC of oocytes injected with 5 ng HERG or co-injected with 3 ng xMiRP2 and 5 ng HERG cRNA. Oocytes were held at –80 mV and stepped to voltages between –120 and +60 mV, with a –30 mV tail pulse (protocol inset). Dashed lines indicate zero current level. (B) Mean tail current–voltage relationships from traces as in panel A for HERG (solid squares) or xMiRP2/HERG (open squares) channels in oocytes, n = 8. Error bars indicate SEM. (C) Exemplar current families recorded between –120 mV and 60 mV from oocytes injected with 0.7 ng HERG cRNA or co-injected with 0.7 ng HERG cRNA and 500 pg xMiRP2 siRNA. Dashed lines indicate zero current level. Oocytes were held at –80 mV, then stepped to voltages between –120 mV and +60 mV with a –30 mV tail pulse (protocol inset). (D) Effect of xMiRP2 siRNA on HERG cRNA dose response. Peak tail currents from oocytes injected with various concentrations of HERG cRNA with or without 500 pg xMiRP2 siRNA using the protocol in panel A (inset), n = 10 to 17 oocytes per point.

that no difference is observed between the two groups (Figures 14.4C and D). Thus, highlighting the utility of RNAi in the *X. laevis* oocyte system, even simultaneous injection of channel α subunit cRNA and target gene siRNA results in efficient target gene knockdown.[11]

Materials

- cDNA template for channel gene
- mMessage mMachine *in vitro* transcription kit (Ambion)
- Model PP-830 microelectrode puller (Narishige, Japan)
- Borosilicate glass capillaries (Sutter)
- TEVC workstation

- Micromanipulators with magnetic stands (Sutter)
- OC-725C amplifier, headstage, electrodes, and oocyte bath (Warner Instruments)
- Dissection microscope with lamp
- Vacuum source for suction of waste
- Vibration isolation table and Faraday cage (Technical Manufacturing Corporation)
- Digidata 1322A low-noise digitizer (Axon Instruments)
- Computer (at least 1-GHz processor) with pCLAMP software (Axon Instruments)

Experimental Procedures

1. Place a defolliculated oocyte, 1 to 14 days following injection of appropriate cRNA and/or siRNA, in a perfusion chamber mounted under a light microscope.
2. Impale the cell with two microelectrodes and clamp at a holding potential of –80 mV. Pull microelectrodes used for recording from borosilicate glass using a vertical microelectrode puller. Fill the microelectrodes with 3 M KCl and, a silver chloride coated silver wire held by electrode holders that are mounted on micromanipulators and connected to an amplifier suitable for oocyte recording. The electrical resistance of the microelectrodes can be measured using the TEVC amplifier and should be 1–5 MΩ.
3. Use a holding voltage of –80 to –100 mV for voltage-gated potassium (Kv) channels, at which most Kv channels are closed.
4. Depending on the protocol used, step up the membrane potential in 10-mV or 20-mV increments from the given holding voltage; at voltages as positive as +60, most Kv channels are at least beginning to saturate. At voltages more positive than +60, contamination from endogenous oocyte ion channels may be problematical.
5. Acquire the data files using suitable software such as pCLAMP9 (Axon Instruments), and convert from analog to digital using a digitizer such as the Digidata 1322A (Axon Instruments).
6. Analyze the digital files using the recording software and measure parameters such as current density, voltage dependence of activation, and channel-gating kinetics.

ACKNOWLEDGMENTS

G.W.A. is supported by a Greenberg Atrial Fibrillation Grant and a Scientist Development Grant from the American Heart Association. We thank Arun Anantharam, Gianina Panaghie, and Drs. Anthony Lewis, Daniel J. Lerner, and Zoe A. McCrossan for their contributions to the oocyte RNAi studies described here, and we thank Dr. Anthony D. O'Connell for the oocyte digital image.

REFERENCES

1. Avila, V.L. and Frye, P. G., Feeding behavior of the African Clawed Frog: effect of prey type, *J. Herpetology,* 12, 1978.
2. Briggs, R. and King, T.J., Transplantation of lining nuclei from blastula cells into enucleated frogs' eggs, *Proc. Nat. Acad. Sci. U.S.A.,* 38, 455, 1952.
3. Miledi, R., Parker, I., and Sumikawa, K., Synthesis of chick brain receptors by frog oocytes, *Proc. Royal Soc. Lond. B Biol. Sci.,* 216, 509, 1982.
4. Kay, B.K. and Peng, H.B., Xenopus laevis: *Practical Uses in Cell and Molecular Biology,* Academic Press, San Diego, 1991.
5. Guille, M., *Molecular Methods in Developmental Biology:* Xenopus *and Zebrafish,* Humana Press, Totowa, NJ, 1999.
6. Marchant, J.S. and Parker, I., *Xenopus tropicalis* oocytes as an advantageous model system for the study of intracellular Ca^{2+} signaling, *Br. J. Pharmacol.,* 132, 1396, 2001.
7. Weber, W.M. et al., Influence of extracellular Ca^{2+} on endogenous Cl-channels in *Xenopus* oocytes, *Pflugers Arch.,* 429, 820, 1995.
8. Weber, W., Ion currents of *Xenopus laevis* oocytes: state of the art, *Biochim. Biophys. Acta.,* 1421, 213–33, 1999.
9. Weber, W.M., Endogenous ion channels in oocytes of *Xenopus laevis*: recent developments, *J. Membr. Biol.,* 170, 1, 1999.
10. Seebohm, G. et al., Dependence of I(Ks) biophysical properties on the expression system, *Pflugers Arch.,* 442, 891, 2001.
11. Anantharam, A. et al., RNA interference reveals that endogenous *Xenopus* MinK-related peptides govern mammalian K^+ channel function in oocyte expression studies, *J. Biol. Chem.,* 278, 11739, 2003.
12. Abbott, G.W. et al., MiRP1 forms IKr potassium channels with HERG and is associated with cardiac arrhythmia, *Cell,* 97, 175, 1999.
13. Abbott, G.W. and Goldstein, S.A., A superfamily of small potassium channel subunits: form and function of the MinK-related peptides (MiRPs), *Q. Rev. Biophys.,* 31, 357, 1998.
14. McManus, M.T. and Sharp, P.A., Gene silencing in mammals by small interfering RNAs, *Nature Rev. Genet.,* 3, 737, 2002.
15. Fire, A. et al., Potent and specific genetic interference by double-stranded RNA in *Caenorhabditis elegans*, *Nature,* 391, 806, 1998.
16. McCrossan, Z.A. et al., MinK-Related Peptide 2 modulates Kv2.1 and Kv3.1 potassium channels in mammalian brain, *J. Neuroscience,* 23, 8077, 2003.
17. Hille, B., *Ion Channels of Excitable Membranes,* Sinauer Associates, Sunderland, MA, 2001.

15 RNAi in *Xenopus* Embryos

Kin Hang Kok and Dong-Yan Jin

CONTENTS

15.1 INTRODUCTION

The African clawed frog *Xenopus laevis* is an invaluable model organism that has contributed substantially to the study of cellular and developmental biology.[1] Thus, studies in this system are seminal to the fields of mitosis and cell cycle control, chromatin assembly, and cytoskeleton, as well as cell signaling. In addition, pioneering work in *Xenopus* has provided grounds for understanding early embryonic development, morphogenesis, and organogenesis.

Xenopus has a unique combination of features that renders it advantageous for deciphering the molecular and cellular mechanisms of important biological processes.[1] *Xenopus laevis* is allo-tetraploid and contains 18 chromosomes. Generation time of *Xenopus laevis* is around one and a half years. Remarkably, the early *Xenopus* embryo takes only three days to develop from fertilized egg to tadpole externally in a Petri dish. The large size of the eggs and embryos facilitates microsurgical manipulation and microinjection. In practice, female *Xenopus laevis* can be induced year round to produce hundreds of eggs in the laboratory, after injecting human chorionic gonadotropin. Large amounts of mature oocytes can be fertilized *in vitro* by incubating with testis extract within half an hour. In an early embryo, each blastomere can be clearly identified under a microscope from the 2-cell to 32-cell stage. The dorsal and ventral sides of the embryo can be distinguished by a change in pigmentation of the animal pole due to cortical rotation after fertilization. The fate map of each blastomere in a 32-cell embryo has been determined,[2] and it paves the way for new molecular approaches in *Xenopus* embryology. For instance, the "animal cap" experiment exploits the property that an explant of animal cap tissue has an ectodermal fate. The response of the explant to extrinsic factors can then be studied. In light of the fate map, specific grafts or molecules can also be introduced into predictable positions, and the effects be monitored through lineage tracing. Additionally, excellent molecular markers are available, and they can be correlated with fate and specification maps. More recently, *Xenopus tropicalis* has emerged as a new model amphibian complementary to the *X.*

0–8493–2141–7/05/$0.00 + $.50

laevis system.[3] *X. tropicalis* has a much shorter generation time and a true diploid genome, which makes it much more amenable to genetic manipulations. Most genes in the two frogs are highly homologous, or almost identical, and they are functionally interchangeable.

While *Xenopus* oocytes and egg extracts are useful for studying chromatin and nuclear assembly, cell cycle, transcription, and protein synthesis, as well as other biochemical and biological processes (see Chapter 14), *Xenopus* embryos provide an ideal model for the study of embryonic development, cell cycle, and cell signaling. Functional studies in *Xenopus* embryos use gain-of-function and loss-of-function approaches. In addition, transgenic technology has been developed recently for *Xenopus*.[4] Loss-of-function approaches in *Xenopus* rely primarily on dominant negative receptors or ligands,[5] antibody depletion, antisense RNA,[6] and ribozyme.[7] However, it is understood that these approaches may sometimes yield nonspecific results that are unexpected and even unexplainable.[1] In addition, the efficacy of antisense RNA is inconsistent from experiment to experiment. Thus, a more reproducible and reliable method for specific loss-of-function analysis in *Xenopus* is desirable.

Double-stranded RNAs (dsRNAs) have been used to induce gene-specific silencing in plants, *Caenorhabditis elegans*, *Drosophila*, mouse embryonic stem cells, and many other systems. However, several reports have documented that the injection of long dsRNA leads to sequence-nonspecific effects in zebrafish and *Xenopus* embryos.[8–10] This is plausibly mediated through dsRNA-dependent protein kinase PKR, which shuts down translation by phosphorylating initiation factor eIF2α and induces apoptosis.[11,12] Exogenously introduced long dsRNAs may also activate 2',5'-oligoadenylate synthetase, leading to nonspecific mRNA degradation by 2',5'-oligoadenylate-activated RNase L.[13] We have demonstrated that small interfering RNAs (siRNAs) suppress endogenous and exogenous gene expression in *Xenopus* embryos in a sequence-specific manner and at a posttranscriptional step, apparently without inducing nonspecific effects.[14] We have shown that siRNAs efficiently reduce the levels of both mRNA and protein of the cognate gene. When co-injected into *Xenopus* embryos, siRNAs significantly knock down the activity of *in vitro* transcribed mRNA. Our work paves the way for the use of siRNA in *Xenopus* embryos.

In this chapter, we will provide a protocol for microinjection of siRNAs into *Xenopus* embryos. We will also briefly describe the procedures for semiquantitative RT-PCR and western blot analysis of the silencing effects.

PROTOCOLS

PROTOCOL 15.1. MICROINJECTION OF siRNAs INTO *XENOPUS* EMBRYOS

Materials

- Stereomicroscope Stemi 2000 (Carl Zeiss)
- Auto-nanoliter injector Nanoject II (Drummond Sci. Co.)
- 30-µl Drummond micropipette needles (Drummond Sci. Co.)
- Stainless steel forceps Dumont 5 (Fine Science Tools)
- 800-µm mesh Nitex screen (Tetko) (placed in a 35-mm Petri dish)

- *Xenopus laevis* male adult and female adult (Xenopus Express, Inc., Cape Town, South Africa)
- Human chorionic gonadotropin (hCG) (Sigma)
- 1 × Marc's Modified Ringers' solution (MMR):[15] 0.1 M NaCl, 2.0 mM KCl, 1 mM $MgSO_4$, 2 mM $CaCl_2$, 5 mM HEPES, pH 7.4
- Dejellying solution: 1 × MMR with 2% (w/v) cysteine at pH 8.0
- 4% (w/v) Ficoll (Sigma) in 30% (v/v) MMR
- Mineral oil (Sigma)
- RNA annealing buffer (100 mM potassium acetate, 30 mM HEPES-KOH at pH 7.4, 2 mM $MgCl_2$)
- SiRNA duplex (Proligo, France; Qiagen; or enzymatically produced, see Chapters 4 and 5): Purify RNA oligos through polyacrylamide gel electrophoresis, reconstitute in DEPC-treated H_2O to 200 µM, and store in aliquots at –20°C. To anneal siRNAs, incubate single-stranded RNAs (40 µM) in annealing buffer (see above) for 1 min at 90°C, followed by 1 h at 37°C. Annealed duplex siRNAs are stored in 1-µl aliquots at –20°C.
- *In vitro* transcribed reporter or tracer mRNA (optional): RNA is *in vitro* transcribed with SP6 RNA polymerase and capped (Ambion). Then purify RNA using RNeasy Mini kit (Qiagen) and store in 1-µl aliquots at –20°C.
- Luciferase, LacZ, or GFP plasmids (CLONTECH or Promega)
- X-gal stock: 20 mg/ml in *N,N*-dimethylformamide (DMF); store in the dark at –20°C
- X-gal stain: 5 mM $K_3Fe(CN)_6$, 5 mM $K_4Fe(CN)_6$, 1 mg/ml X-gal, 2 mM $MgCl_2$ in 1 × PBS

Experimental Procedures

Induction of ovulation

1. To induce full ovulation, inject 500 to 800 U of hCG into dorsal lymph sac of a female adult frog using a 1-ml syringe with a 26-gauge fine needle. Eggs will be laid about 10 h after injection at about 23°C or about 14 h after injection at about 15 °C.
2. Gently hold the frog and massage the belly. Eggs should be laid naturally and be collected in a 60-mm Petri dish containing 1 × MMR.
3. Wash the eggs twice with 30% (v/v) MMR using a plastic pipet. At this stage, eggs are protected by their jelly coats. Separate the eggs by gently pipetting up and down.
4. Discard as much MMR solution as possible prior to *in vitro* fertilization.

In vitro *fertilization*

5. Sacrifice and dissect a male frog on the ice.
6. Use forceps to remove the yellow fat body attached to the testis.
7. Rinse the testis in 1 × MMR and discard the supernatant.
8. Crush part of the testis and add 5 to 10 drops of 5% (v/v) MMR. Mix gently.
9. Keep the remaining testis in 1 × MMR at 4°C. It can be used within 24 h or 48 h.
10. Add the crushed testis into the washed eggs immediately. Remember to mix it well for better contact between crushed testis and eggs.

11. After 20 to 30 min, the pigmented animal hemisphere of fertilized eggs should contract and face upward. Good-quality eggs and testis should give over 95% fertilization within 25 min after mixing together.

Dejellying embryos

12. About 30 min after mixing testis and eggs, transfer the fertilized eggs into a 100-ml beaker containing 10 ml dejellying solution.
13. Swirl the embryos gently and continuously until they start to pack together, normally in less than 5 min. Long incubation may damage the embryos.
14. Quickly discard the dejellying solution and wash the embryos four times in 30% (v/v) MMR.
15. Place the washed embryos in a 60-mm Petri dish with 4% (w/v) Ficoll in 30% (v/v) MMR.

Microinjection of 2-cell stage embryos

16. Line up 10 to 20 embryos in a mesh in a Petri dish containing 4% (w/v) Ficoll in 30% (v/v) MMR. Embryos begin to divide into 2 cells about 1 h 15 min after *in vitro* fertilization.
17. Fill as much mineral oil into the injection needle as possible by using 1-ml syringe with a metal fine needle. Do not trap air bubbles.
18. Fit the oil-filled injection needle into Drummond injector according to the supplier's guidelines.
19. Add 1-μl RNA (0.4 ng/nl for siRNA duplex) or 1-μl DNA (30 pg/nl for reporter plasmid) sample on an empty Petri dish.
20. Suck up the RNA or DNA sample into the oil-filled injection needle.
21. Check whether each injection gives the same size droplet under microscope.
22. (Optional) Calibrate the injection volume by counting how many injections are required to empty 0.5 μl of sample.
23. Inject the sample (5 nl of RNA or DNA) into the animal pole of each blastomere in 2-cell stage embryos. Check whether the injection needle is blocked after every other injection. The injection should be completed within 30 min for the same batch of embryos.
24. Remove 4% (w/v) Ficoll in 30% (v/v) MMR and add 30% (v/v) MMR until gastrulation (stage 7) starts.

Technical note: The dosage of mRNA and DNA sample should not exceed 5 ng and 100 pg, respectively. The injection volume should be less than 10 nl for each embryo.

Microinjection of 32-cell stage embryos

25. Select embryos that have been divided into 8 cells symmetrically. The asymmetric dorsoventral pigmentation should be clearly identified. Approximately 50% of good-quality embryos may show symmetrical cell division and distinguishable asymmetric dorsoventral pigmentation.
26. Line up the selected embryos and wait till 32-cell stage (Color Figure 15.1).*
27. Identify the blastomere to be injected. Nomenclature of blastomeres (A1–A4; B1–B4; C1–C4; D1–D4) at 32-cell stage (Figure 15.1B) has been described.[2] Practically, lighter-pigmented blastomere on the animal hemisphere is the dorsal

* Color figures follow page 204.

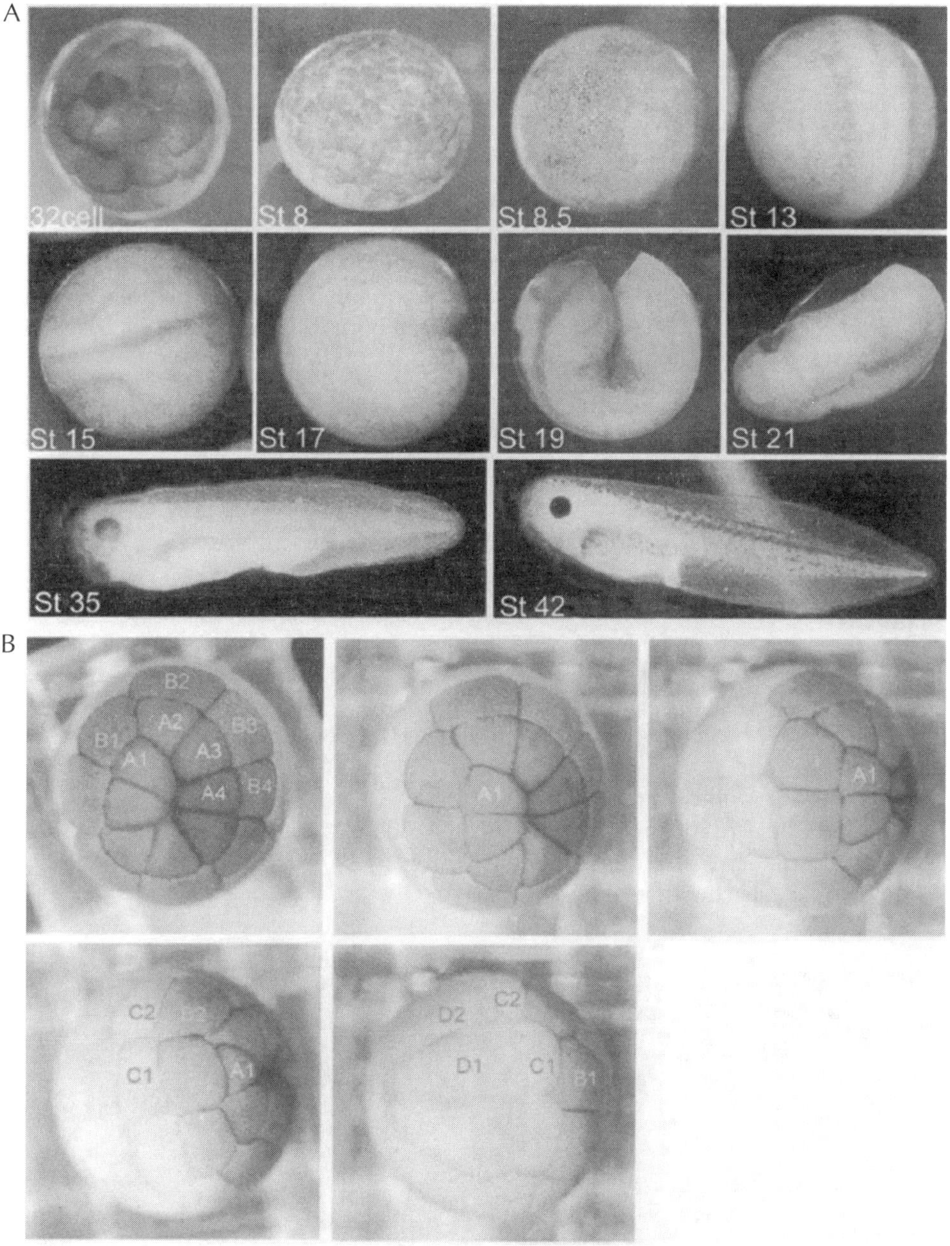

COLOR FIGURE 15.1 Embryonic development of *Xenopus laevis*. A. Different stages of *Xenopus* embryos from 32-cell to tadpole stage. B. Nomenclature of blastomeres at 32-cell stage. 1 and 2: animal pole view. 3, 4 and 5: left side view.

side (e.g., Figure 15.1B, panel 1, A1 blastomere) and darker-pigmented blastomere on the animal hemisphere is the ventral side (e.g., Figure 15.1B, panel 1, A4 blastomere). Likewise, B1 blastomere is right downward of A1 blastomere and C1 blastomere is right downward of B1 blastomere, and so on (Figure 15.1B, panel 2, 3, 4, and 5).

28. Inject the sample (less than 3 nl of 0.4 ng/nl siRNA duplex) with tracer (e.g., LacZ or GFP mRNA, 300 to 600 pg per blastomere).
29. Postinjection path of the sample can be located by X-gal staining or fluorescence microscopy. For X-gal staining, incubate the embryos in X-gal stain at 30°C for 1 to 2 h or longer. For example, injection of LacZ mRNA into B1 blastomere of a *Xenopus* embryo at 32-cell stage showed deep blue color on the somite tissue in later stages after staining (Color Figure 15.2).

Technical note: The injection volume is critical in 32-cell stage injection experiment. It is better to keep the injection volume as small as 1 nl per blastomere. Large volume of injection may result in artifacts in the development.

Protocol 15.2 Semiquantitative RT-PCR Analysis of mRNA Transcripts in *Xenopus* Embryos

Materials

- Trizol (Invitrogen)
- RNase-free DNase I (Sigma, Ambion, Promega)
- 10 × DNase I reaction buffer (200 mM Tris-HCl, pH 8.4, 500 mM KCl, 20 mM $MgCl_2$)
- Oligo(dT) (Ambion, Promega)
- 10 × RT buffer (200 mM Tris-HCl, pH 8.4, 500 mM KCl, 20 mM $MgCl_2$)
- SuperScript II reverse transcriptase (Invitrogen)
- PCR reagents (Invitrogen, CLONTECH)

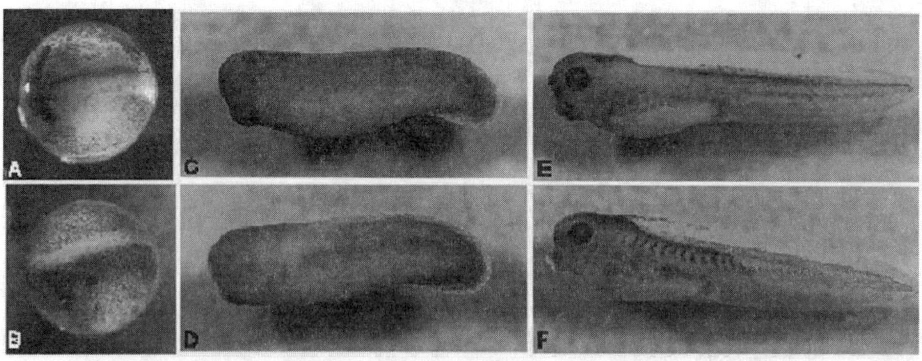

COLOR FIGURE 15.2 Injection of LacZ mRNA into blastomere B1 on the left side of a *Xenopus* embryo at 32-cell stage. A, C, and E: uninjected embryos. B, D, and F: injected embryos. A and B: stage 15. C and D: tailbud stage. E and F: tadpole stage.

Experimental Procedures

RNA extraction

1. Wash the embryos twice with 1 × PBS. Injected *Xenopus* embryos are collected at different stages. Embryos were staged as described[2] (Color Figure 15.1).
2. Dissolve two embryos in 500 μl Trizol in 1.5-ml Eppendorf tube. Vortex for 1 min. Note that late-stage embryos will take longer time to dissolve.
3. Add 0.1 ml chloroform; vortex for 40 sec.
4. Centrifuge at 14,000 × g for 10 min.
5. Transfer the supernatant to a new tube.
6. Add 0.25 ml isopropanol and mix gently for 20 sec.
7. Centrifuge at 14,000 × g for 10 min.
8. Discard the supernatant carefully.
9. Wash pellet with 0.5 ml 75% (v/v) chilled ethanol.
10. Centrifuge at 14,000 × g for 5 min.
11. Remove as much supernatant as possible. Air-dry the pellet for 10 min.
12. Add 15 μl DEPC-treated H_2O to dissolve the RNA pellet and store at –80°C.

DNase I treatment

If injected sample contains expression plasmid that may be amplified by RT-PCR, RNA sample should be treated by RNase-free DNase I before reverse transcription.

13. Mix 2 μg of extracted total RNA, 1 μl of 10 × DNase I reaction buffer, and 1 μl DNase I.
14. Add DEPC-treated H_2O up to 10 μl.
15. Incubate at room temperature for 30 min.
16. Add 1 μl of 25 mM EDTA and incubate for 15 min at 65°C. Store at –20°C.

RT-PCR

17. Mix 2 μg of extracted total RNA or DNase I-treated total RNA with 1 μl of oligo(dT) primer (10 μM) and add DEPC-treated H_2O up to 12 μl.
18. Preincubate at 95°C for 30 sec and then at 70°C for 10 min.
19. Prepare RT master mix by adding 2 μl of dithiothreitol, 2 μl of 10 × RT buffer, 1 μl SuperScript II reverse transcriptase, and 1 μl 10 mM dNTP.
20. Mix 8 μl of RT master mix with 12 μl RNA-primer mix.
21. Incubate at 42°C for 1 h and then at 95°C for 5 min. Store at –20°C.
22. Typically, 1 μl of the RT product is used for PCR.

Protocol 15.3 Western Blot Analysis of Proteins in *Xenopus* Embryos

Materials

- Whole-cell extraction buffer (25 mM Tris-HCl, pH 7.5, 70 mM KCl, 1 mM EDTA, 20% (v/v) glycerol, 5 mM dithiothreitol, 1 μg/μl leupeptin, 1 μg/μl pepstatin)
- 5 × Protein loading buffer (250 mM Tris-Cl, pH 6.8, 10% (w/v) SDS, 0.5% (w/v) bromophenol blue, 50% (v/v) glycerol, 500 mM β-mercaptoethanol
- SDS-PAGE system (Amersham, BioRad)
- Semi-dry blotting apparatus SemiPhor (Hoefer, Amersham)

- Immobilon-P membrane (Millipore)
- Enhanced chemiluminescence (ECL) Western blotting detection reagent (Amersham)

Experimental Procedures

1. Wash the embryos twice with 1 × PBS.
2. Homogenize two embryos in 100 μl whole cell extraction buffer. Store at –80°C.
3. Mix 16 μl of embryo lysate in whole cell extraction buffer with 4 μl of 5 × protein loading buffer.
4. Boil the samples for 3 min, and leave them on ice.
5. Load the boiled samples onto a SDS-PAGE gel.
6. Transfer the proteins onto an Immobilon-P membrane using the SemiPhor apparatus.
7. Incubate the membrane with suitable primary and secondary antibodies.
8. Visualize the proteins by enhanced chemiluminescence.

REFERENCES

1. Sive, H.L., Grainger, R.M., and Harland, R.M., *Early Development of Xenopus laevis: A Laboratory Manual,* 2nd ed., Cold Spring Harbor Laboratory Press, New York, 1999.
2. Dale, L. and Slack, J.M., Fate map for the 32-cell stage of *Xenopus laevis*, *Development,* 99, 527, 1987.
3. Hirsch, N., Zimmerman, L.B., and Grainger, R.M., *Xenopus*, the next generation: *X. tropicalis* genetics and genomics, *Dev. Dyn*., 225, 422, 2002.
4. Amaya, E. and Kroll, K.L., A method for generating transgenic frog embryos, *Methods Mol. Biol.,* 97, 393, 1999.
5. Amaya, E., Musci, T.J,. and Kirschner, M.W., Expression of a dominant negative mutant of the FGF receptor disrupts mesoderm formation in *Xenopus* embryos, *Cell,* 66, 257, 1991.
6. Steinbeisser, H., Fainsod, A., Niehrs, C., Sasai, Y., and De Robertis, E.M., The role of gsc and BMP-4 in dorsal-ventral patterning of the marginal zone in *Xenopus*: A loss-of-function study using antisense RNA, *EMBO J.,* 14, 5230, 1995.
7. Steinbach, O.C., Wolffe, A.P., and Rupp, R.A., Somatic linker histones cause loss of mesodermal competence in *Xenopus*, *Nature,* 389, 395, 1997.
8. Nakano, H., Amemiya, S., Shiokawa, K., and Taira, M. RNA interference for the organizer-specific gene *Xlim-1* in *Xenopus* embryos. *Biochem. Biophys. Res. Commun.*, 274, 434, 2000.
9. Oates, A.C., Bruce, A.E.E. and Ho, R.K. Too much interference: injection of double-stranded RNA has non-specific effects in the zebrafish embryos, *Dev. Biol.*, 224, 20, 2000.
10. Zhao, Z., Cao, Y., Li, M., and Meng, A., Double-stranded RNA injection produces non-specific defects in zebrafish, *Dev. Biol.*, 229, 215, 2001.
11. Der, S.D., Yang, Y.L., Weissmann, C., and Williams B.R.G., A double-stranded RNA-activated protein kinase-dependent pathway mediating stress-induced apoptosis, *Proc. Natl. Acad. Sci. USA*, 94, 3279, 1997.
12. Samuel, C.E., The eIF-2α protein kinases, regulators of translation in eukaryotes from yeast to humans, *J. Biol. Chem.*, 268, 7603, 1993.

13. Kumar, S., Mitnik, C., Valente, G., and Floyd-Smith, G., Expansion and molecular evolution of the interferon-induced 2'-5' oligoadenylate synthetase gene family, *Mol. Biol. Evol.*, 17, 738, 2000.
14. Zhou, Y., Ching, Y.P., Kok, K.H., Kung, H.F., and Jin, D.Y., Post-transcriptional suppression of gene expression in *Xenopus* embryos by small interfering RNA, *Nucleic Acids Res.*, 30, 1664, 2002.
15. Ubbels, G.A., Hara, K., Koster, C.H., and Kirschner, M.W., Evidence for a functional role of the cytoskeleton in determination of the dorsoventral axis in *Xenopus laevis* eggs, *J. Embryol. Exp. Morphol.*, 77, 15, 1983.

16 An RNAi-Based Genomic Library for Forward Genetics in the African Trypanosome

James C. Morris, Zefeng Wang, Shawn A. Motyka, Mark E. Drew, and Paul T. Englund

CONTENTS

16.1 INTRODUCTION

The African trypanosome, *Trypanosoma brucei*, is an important cause of human suffering in sub-Saharan Africa, infecting an estimated 500,000 people. The disease is fatal if untreated and is currently considered a category 1 (emerging or uncontrolled) disease by the Special Programme for Research and Training in Tropical Diseases (TDR) (World Health Organization, www.who.int). The parasite also infects livestock, greatly impacting both nutrition and agriculture in endemic areas.

T. brucei, which is transmitted by the bite of an infected tsetse fly, grows rapidly in the mammalian blood as bloodstream form (BSF) trypanosomes. Some parasites become preadapted for life in the insect vector and, if they are taken up by a feeding tsetse fly, will differentiate into procyclic form (PF) trypanosomes. Parasites then replicate within the fly, migrate to the salivary glands, and are introduced back into a mammalian host during fly feeding to complete the lifecycle.[1]

In addition to its importance as a disease-causing organism, fundamental aspects of eukaryote biology have been discovered using trypanosomes as a model system. For example, research on trypanosome surface molecules led to the elucidation of the first structure of a glycosyl phosphatidylinositol (GPI) anchor as well as the resolution of the GPI anchor biosynthetic pathway.[2,3] Additionally, RNA editing[4] and trans

0-8493-2141-7/05/$0.00 + $.50

splicing[5] were initially identified in trypanosomes but have since been found in other organisms, emphasizing that discoveries in trypanosomes can yield insight into basic eukaryote biology.

There are several reasons why *T. brucei* has been a useful model organism. First, parasites can be manipulated safely in the laboratory because the subspecies *T. brucei brucei* is not pathogenic to humans. Second, it can be grown in large numbers in semi-defined media as either BSF or PF parasites, allowing generation of a sufficient number of parasites for biochemical studies. Third, *T. brucei* is suitable for genetic research. The parasite genome is compact, with ~10,000 genes (a number that may change slightly as the genome database is completed) and few introns.[6] *T. brucei* is amenable to genetic manipulation, with homologous recombination allowing for generation of gene-specific knockouts in these diploid organisms.

Gene knockout technology has advanced our understanding of parasite biology but is of limited use in some cases. *T. brucei* is diploid and cannot be easily crossed sexually in the laboratory, requiring two rounds of deletion by homologous recombination for a complete gene knockout. This usually requires two different selectable markers and is a time-consuming process. Additionally, knocking out essential genes requires the introduction of a regulated ectopic copy of the targeted gene to provide expression while the genomic alleles are replaced.[7] Finally, *T. brucei* contains a number of gene families consisting of nearly identical genes arranged in tandem,[8–11] which makes targeted disruption difficult.

The discovery that RNA interference functions in *T. brucei* provides another powerful tool for genetic manipulation of trypanosomes, allowing one to rapidly generate functional knockdowns of genes, even if they are essential or occur in multiple copies in the genome. The Ullu and Gull labs first demonstrated that PF *T. brucei* was amenable to RNAi,[12,13] with the Ullu group developing methods using either *in vitro* synthesized dsRNA or transient transfection of a vector that expresses a stem-loop RNA for RNAi. Here we describe the use of the vector pZJMα for simplification of RNAi in both PF and BSF *T. brucei* (a similar vector, p2T7^{Ti}, which is almost identical to pZJMα, was developed in the Donelson laboratory).[14] Further, pZJMα has been modified for the generation of an RNAi-based genomic library, making it possible to perform forward genetics in these primitive eukaryotes.

16.2 THE PLASMID pZJMα — A VECTOR FOR RNAi IN *T. BRUCEI*

The plasmid pZJMα (Figure 16.1A) was constructed by modifying pLew100[7] with a cloning site flanked by opposing T7 promoters under the control of tetracycline operators. In order to minimize read-through transcription, pZJMα contains dual T7 terminators that flank the T7 promoters. This cassette replaces the luciferase gene in pLew100.[7]

Plasmid pZJMα is maintained with a 660-bp fragment of the *T. brucei* α-tubulin gene within the *Xho* I/*Hind* III cloning site in order to prevent cruciform formation during plasmid replication in *Escherichia coli*. Prior to inserting a fragment of a gene of interest, the α-tubulin fragment is removed by *Xho* I and *Hind* III digestion.

Fragments of interest are amplified from the *T. brucei* genome using forward and reverse primers (Protocol 16.1). The forward primer adds the required *Xho* I site

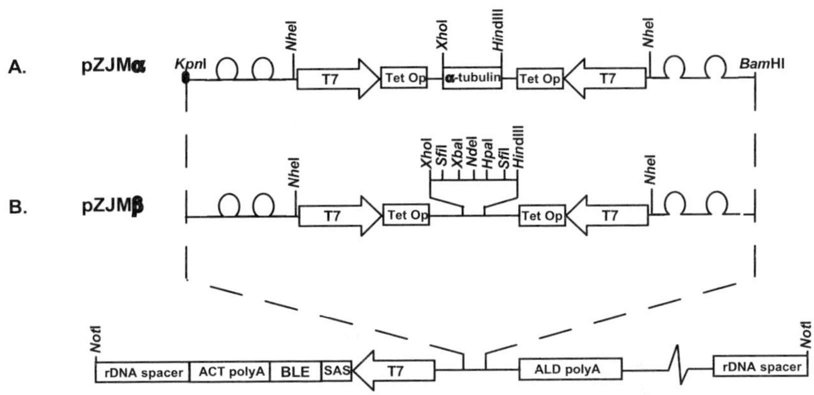

FIGURE 16.1 The pZJM RNAi vectors. These vectors are based on previous constructs devised in the Ullu lab,[25] the Cross lab,[7] and the Fire lab.[26] The tetracycline operator (*Tet Op*), dual T7 terminators (ΩΩ), tetracycline-inducible T7 promoters (*T7 arrows*), ribosomal DNA spacer (*rDNA*), actin poly(A) addition sequence (*ACT polyA*), phleomycin resistance gene (*BLE*), splice acceptor site (*SAS*), and aldolase poly(A) addition sequence (*ALD polyA*) are indicated. The plasmids are shown in linearized form, after cleavage in the rDNA spacer, and are not drawn to scale.

(CTCGAG) plus sufficient extra nucleotides to allow direct restriction enzyme digestion of the PCR product, and the reverse primer adds the *Hin*d III restriction endonuclease site (AAGCTT) required for cloning into pZJMα. The fragment of the gene of interest should be (ideally) between 100 and 1000 bp (we usually use ~500 bp) and should not contain *Xho* I, *Hin*d III, or *Not* I restriction endonuclease sites.

We have found variation in the effectiveness of RNAi, with some mRNAs being less sensitive than others to silencing. Additionally, targeting different regions of a gene can lead to different levels of interference. It has also been observed that RNAi of essential genes can select for cells that no longer express dsRNA but are still resistant to the selectable marker introduced with the RNAi vector.[15] In our experience, if RNAi with pZJM is ineffective, we can often still achieve effective gene silencing with the stem-loop vector.[16]

"Leaky" expression of dsRNA in uninduced parasites, possibly as a result of trace tetracycline in the FBS or as a result of deregulation of the T7 promoters, can complicate interpretation of gene-silencing experiments.[16] Cloning parasites can help to overcome this problem, as clones tend to exhibit tighter regulation of RNAi (perhaps because "leaky" clones are selected against). Alternatively, comparing induced parasites to the parental, untransformed line can be informative. In many examples, leaky expression of dsRNA has no effect on cell growth.

16.3 FORWARD GENETICS USING AN RNAi-BASED GENOMIC LIBRARY

The *T. brucei* genome-sequencing project is nearing completion. Soon we will know the entire gene complement of *T. brucei*, but this is just the first step in resolving major

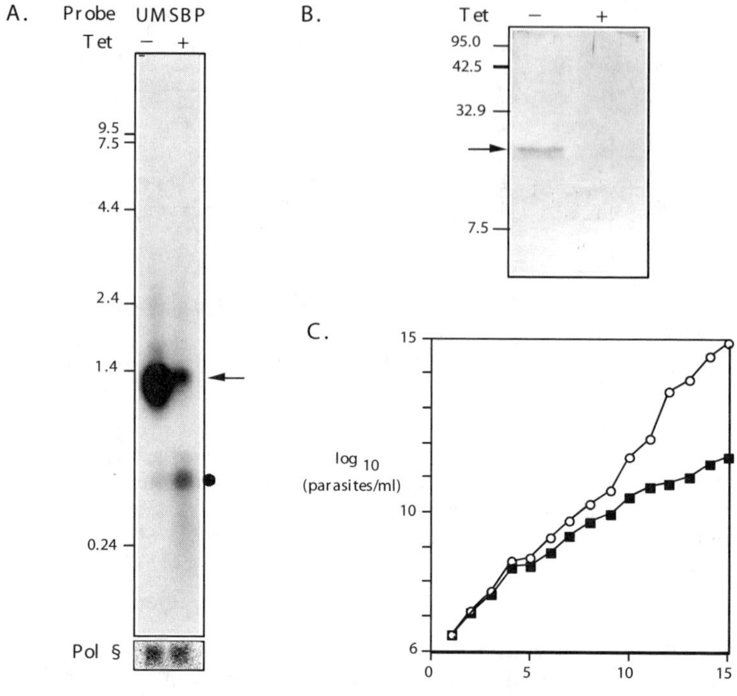

FIGURE 16.2 RNA interference of the universal minicircle sequence-binding protein (UMSBP)[27] using the pZJMα vector. (A) Northern analysis of total RNA from uninduced (–) and induced (+) parasites harboring pZJMα(UMSBP). Northern analysis was performed on RNA from cells induced with 1-μg/ml tetracycline for 40 h using an UMSBP-specific probe. As a loading control, the same blot was stripped and probed for polymerase β mRNA. UMSBP mRNA (arrow) and dsRNA (asterisk) are indicated. (B) Western analysis of whole cell lysates of parasites harboring pZJMα(UMSBP). Proteins from 2.5×10^6 cells, either induced (+ tet, 10 days) or uninduced (–), were separated by SDS-polyacrylamide gel electrophoresis and analyzed by Western blot using an antibody to UMSBP from *Crithidia fasciculata* (a gift from Dr. Joseph Shlomai, Hebrew University). The arrow indicates the cross-reacting species of the predicted molecular weight for *T. brucei* UMSBP (14.6 kDa). (C) Cell growth after silencing of UMSBP. Parasites were grown in the absence (○) or presence (■) of 1-μg/ml tetracycline, and cell density was plotted against days of growth. (Figure is used with permission from *The Journal of Biological Chemistry* and was originally published in Vol. 275, No. 51, pg. 40,177.)

questions regarding the biology of the parasite. Postgenomic analysis, which will order genes (and their products) into their "biological positions" within pathways, will be one of the next major areas of research in the trypanosome field. To dissect these pathways, a comprehensive understanding of gene function will be required. Forward genetics, a powerful approach in which mutant cells are screened for a phenotype of interest followed by identification of the genetic lesion that is responsible for the phenotype, would enhance our ability to place genes in pathways. To that end, we have constructed an RNAi-based genomic library in pZJM to use for forward genetics in PF trypanosomes (Figure 16.3).

Since the trypanosome genome is small (~3.5×10^7 bp[17]) and compact (with ~10,000 genes, most without introns), the library has been developed using genomic

DNA. Instead of a fragment of a gene of interest inserted into the pZJM vector, random fragments of genomic DNA are inserted between the opposing T7 promoters. After transfection and induction of RNAi, the population of parasites theoretically silences (in a random fashion) all genes in the genome. This is the first RNAi-based genomic library for forward genetics (to our knowledge) developed in any system, and, to date, we have used it successfully in two different screens.[18,19]

One of the most difficult aspects of using the RNAi library is the development of a screen for a phenotype of interest. In our first screen, we used killing by a lectin as a selection scheme. Unfortunately, this approach led to the isolation of spontaneously arising mutants, a possibility when screening for survivors of a lethal selection. This observation has led us to develop nonlethal screens. For example, selection of cells by fluorescence-activated cell sorting (FACS) has proven to be useful for the isolation of cells with a phenotype of interest.[18] Additionally, since phenotypes do not develop at the same rate, the same screen may need to be applied to cultures that have been induced for different lengths of time.

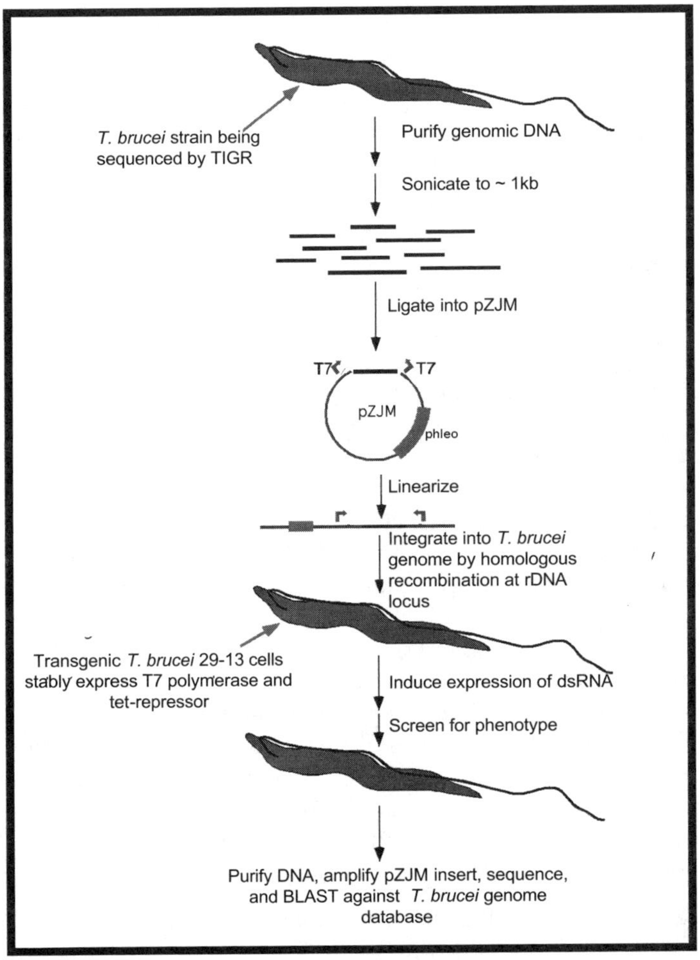

FIGURE 16.3 The scheme for generation of an RNAi-based genomic library in *T. brucei*.

PROTOCOLS

PROTOCOL 16.1 CLONING INTO pZJMα

Materials

- Standard PCR reagents (Promega)
- Thermal cycler (MJ Research)
- Forward primer; 5'GATCCTCGAG$(N)_{20}$ (where N corresponds to at least 20 nucleotides of the 5' end of the fragment to be amplified)
- Reverse primer; 5'GATCAAGCTT$(N)_{20}$ (where N corresponds to the reverse complement of at least 20 nucleotides of the 3' end of the target gene fragment)
- Qiaquick PCR Purification Kit (Qiagen)
- *Xho* I and *Hin*d III (Promega)
- T4 DNA ligase and ligase buffer (Promega)
- Competent *E. coli* DH5α (Invitrogen)
- Ampicillin (Fisher Biotech)

Experimental Procedures

1. Amplify a portion of a gene of interest by PCR using forward and reverse primers, adding 5' *Xho* I and 3' *Hin*d III restriction sites to the ends of the product.

Reaction buffer	1 ×
T. brucei genomic DNA	50 ng
Forward primer	0.25 µM
Reverse primer	0.25 µM
dNTPs	250 µM
$MgCl_2$	1.5 MM
Taq DNA polymerase	2.5 U

2. Perform PCR for 30 cycles, with the annealing temperature based on the calculated T_m of the primers and the extension times appropriate for the insert size (1 min/kb of DNA to be amplified).
3. Confirm by agarose gel electrophoresis that the PCR yielded a single product and purify the fragment by Qiaquick PCR Purification Kit.
4. Digest 2 µg of the resulting DNA with *Xho* I and *Hin*d III for 6 h at 37°C.
5. Mix the digested fragment with *Xho* I/*Hin*d III-digested pZJMα at a 5:1 molar ratio, respectively.
6. Ligate using T4 DNA ligase (3 U) in T4 DNA ligase buffer at 16°C overnight.
7. Transform into competent DH5α according to the supplier's information, and plate on LB medium containing ampicillin at 50 µg/ml.
8. To confirm that the pZJMα insert is correct, analyze plasmid DNA prepared from a few colonies by restriction endonuclease digestion. Plasmids containing inserts of the appropriate size may be sequenced using the primers PhleoF (GGAACGGCACTGGTCAAC) and AldoR (CTCCAAAGCAAACATGCAGA), which correspond to sequences that flank the cloning site.

Technical note. Alternatively, PCR products can be initially cloned into pGEM-T Easy Vector (Promega) and then sub-cloned into pZJMα.

Protocol 16.2 Preparation of Plasmid DNA for Transfection

Transfection efficiencies of PF and BSF trypanosomes are different, with our group achieving efficiencies of ~1×10^{-4} and ~2×10^{-8} for PF and BSF parasites, respectively.[16,20] There is some variation in efficiencies reported among laboratories for reasons that we do not completely understand. Nevertheless, current protocols require that large amounts of linearized pZJMα be used for successful transformation and stable integration in *T. brucei*.

Materials

- *E. coli* with pZJMα clones
- Plasmid Midi Kit (Qiagen)
- Qiaquick PCR Purification Kit (Qiagen)
- *Not I*

Experimental Procedures

1. To prepare DNA for transfection, grow *E. coli* harboring pZJMα under ampicillin selection and isolate plasmid DNA by Qiagen Plasmid Midi Kit.
2. For integration into the nontranscribed rDNA spacer region of the trypanosome genome,[21] linearize pZJMα (10 μg/PF transfection, 100 μg/BSF transfection) with *Not* I for at least 2 h at 37°C.
3. Purify linearized plasmid using Qiaquick PCR Purification Kit, following the manufacturer's instructions, for PCR product purification. Care must be taken to maintain the sterility of the linearized plasmid, including the use of sterile water for elution from the Qiagen column in the final step of purification. Alternatively, linearized pZJMα can be purified from the restriction enzyme digestion components by ethanol precipitation, though in our hands this method yields reduced transfection efficiencies.
4. To ensure that the *Not* I digestion was complete and that recovery following column purification was efficient, compare a small amount of purified linearized pZJMα to undigested plasmid by agarose gel electrophoresis.

Protocol 16.3 Transfection of PF Trypanosomes and Selection for Stable Integration of pZJMα

Materials

- PF strain 29-13 (harbors integrated genes for T7 RNA polymerase and tetracycline repressor[7])
- SDM-79 (see Appendix 16.1)
- FBS (ICN)
- G418 (G) (Fisher Biotech)
- Hygromycin (H) (Alexis Corporation)
- Cytomix[22] (120 mM KCl, 0.15 mM $CaCl_2$, 10 mM K_2HPO_4, 25 mM HEPES, 2 mM EDTA, 5 mM $MgCl_2$)

- Phleomycin (P) (ICN)
- 4-mm gapped cuvettes using a BTX Electro Cell Manipulator 630 or BTX ECM 830 Electro Square Porator (Gentronics)

Experimental Procedures

1. Grow strain PF 29-13 in SDM-79 supplemented with 10% (v/v) FBS (SDM-79/FBS$_{10}$). Include G418 (15 μg/ml) and hygromycin (50 μg/ml) in the medium to maintain the T7 RNA polymerase and the tetracycline repressor constructs, respectively (SDM-79/FBS$_{10}$/G/H).[7]
2. For each transfection, harvest 1×10^8 cells during mid-logarithmic growth (~5×10^6/ml) by centrifugation (3000 × g, 10 min, RT), wash once in 5 ml of cytomix, and resuspend in 0.5-ml of cytomix containing 10 μg of purified, *Not* I linearized pZJMα[16] (Protocol 16.2).
3. Carry out transfections in 4-mm gapped cuvettes using a BTX Electro Cell Manipulator 630 with peak discharge at 1.65 kV in resistance-timing mode R2 (24 ohms). Alternatively, transfection can be performed using a BTX ECM 830 Electro Square Porator with the following settings: 3 pulses, 100-msec pulse length, 200-μsec interval, 1700 volts.
4. Immediately following transfection, transfer cells into 10 ml SDM-79/FBS$_{10}$/G/H.

Selection of stable transformants

5. One day after electroporation, collect the parasites by centrifugation (3000 × g, 10 min, RT) and replace the media with SDM-79/FBS$_{10}$/G/H supplemented with phleomycin (2.5 μg/ml) (SDM-79/FBS$_{10}$/G/H/P). Selection for stable integration is the most difficult aspect of using pZJMα, in part because selection must be balanced carefully against culture growth.

Technical notes. The selection for phleomycin-resistant parasites requires 7 to 10 days, during which the culture continues to grow. In order to prevent overgrowth (and subsequent death), the cell density must be carefully monitored. During selection, when parasite density reaches ~1×10^7/ml, trypanosomes are passed by a 1:5 dilution into SDM-79/FBS$_{10}$/G/H/P. Greater dilutions should be avoided and careful monitoring of culture density continued until stable growth is observed. Severe dilution is detrimental to PF trypanosomes and can complicate selection because killing of untransformed parasites effectively dilutes the culture. Therefore, one needs to be conservative in passing the cultures during selection in order to avoid unintentional overdilution.

Protocol 16.4 Cloning and Storage of PF Trypanosomes

Materials

- SDM-79/FBS$_{15}$/G/H/P (SDM-79 with 15% (v/v) FBS, G418 (15 μg/ml), hygromycin (50 μg/ml), phleomycin (2.5 μg/ml)
- SDM-79/FBS$_{10}$/G/H/P (as above, but with 10% (v/v) FBS)
- Freezing medium (50% (v/v) glycerol in cytomix; filter-sterilize)
- Modified trypanosome freezing carton (Styrofoam racks)

Experimental Procedures

Cloning

Following the selection of stably transfected trypanosomes, it is often useful to generate clonal parasite lines. For example, a population of cells might exhibit a variable RNAi phenotype, which can be made more homogeneous by studying clonal cell lines.

1. Count cells in mid-log phase that have undergone selection and dilute to a concentration of ~5 cells/ml in culture medium supplemented with 15% (v/v) FBS (SDM-79/FBS_{15}/G/H/P); we have cloned cells in SDM-79/FBS_{10}/G/H/P but have found that increasing the serum concentration improves parasite survival.
2. Mix thoroughly and transfer the culture to a 96-well flat-bottomed, untreated tissue culture plate in 100-µl aliquots, yielding ~0.5 cell/well.
3. Seal the plate with parafilm and grow the clones at 27°C in 5% CO_2 until the wells are turbid, usually 14 to 20 days after plating (this cloning method is a personal communication from Dr. W. Gibson, University of Bristol, UK). If plating efficiency is ~30 to 40% (as in our hands), clonal growth should produce 15 to 20 positive wells.
4. Once the cells in a well have become turbid, transfer the entire contents of the well for routine culturing into a flask containing 2.5 ml of SDM-79/FBS_{10}/G/H/P.

Freezing-transformed PF trypanosomes

Once stable parasite lines have been established, a portion of the culture should be stored for future use. Storage serves two purposes. First, if cultures become contaminated or die, frozen stabilates can be used as a replacement. Second, we (and others) have found that long-term, continuous culturing can lead (for reasons not fully understood) to a loss in the effectiveness of RNAi.[15] Therefore, frozen stabilates should be used periodically to eliminate this concern.

5. To freeze PF trypanosomes, add 0.2 ml freezing medium to 0.8 ml of culture containing 1–5 × 10^6 cells/ml in a liquid nitrogen–compatible cryotube.
6. Mix the cells thoroughly by inversion and then freeze at –80°C overnight in a modified trypanosome-freezing carton. This carton is made of two Styrofoam racks (in which 15-ml conical centrifuge tubes are shipped) taped together; samples placed in the carton freeze more slowly than exposed tubes.
7. The next day, transfer the cryotubes to liquid nitrogen.
8. To thaw PF trypanosomes, warm frozen cryotubes to room temperature, transfer the contents of a single tube to 9 ml SDM-79/FBS_{10}/G/H/P, and grow at 27°C in 5% CO_2.

Protocol 16.5 Reverse Genetics Using pZJMα in PF Trypanosomes

Materials

- SDM-79/FBS_{10}/G/H/P
- Tetracycline
- Cytomix (Protocol 16.2)

- Purescript RNA Isolation Kit (Gentra systems, Minneapolis)
- Random Primers DNA Labeling System (Invitrogen)
- Other standard materials for northern hybridization analysis

Experimental Procedures

To determine the effectiveness of RNAi, we routinely analyze cell growth rate, the steady-state amount of targeted mRNA after silencing, and the amount of protein produced from the silenced gene. There are other methods for monitoring an RNAi effect, which may be dependent on the gene under investigation.

Cell phenotypes

1. To initiate RNAi studies in PF parasites, split cells into two flasks at a density between 5×10^5 and 5×10^6 cells/ml.
2. Add tetracycline (tet, 1 μg/ml) to one flask and count both cultures daily to determine the impact of silencing on cell growth. Maintain the cultures at 5×10^5 to 5×10^6 cells/ml and pass into fresh SDM-79/FBS$_{10}$/G/H/P (with tet, if induced). Growth phenotypes have been observed in as few as 4 days and as many as 12 days. See Figure 16.2C for an example, in which growth over several logs is plotted as total cell number (cell density × dilution) through time.

Northern hybridization analysis

Northern blots are useful for assessing the effectiveness of RNAi (Figure 16.2A).

3. Split PF parasites (40 ml of ~ 5.0×10^5 cells/ml) equally into two flasks and add tetracycline (1 μg/ml) to one flask.
4. Following 48 h of growth, count the parasites and centrifuge ~5×10^7 cells (3000 × g, 10 min).
5. Wash once in 10 ml cytomix and centrifuge again. Cell pellets can be stored at –80°C until RNA is prepared.
6. Prepare RNA using Purescript RNA Isolation Kit according to the supplier's guidelines.
7. Resolve total RNA from 1.25×10^7 cell equivalents on a formaldehyde 1.5% (w/v) agarose gel and estimate ribosomal RNA levels by ethidium bromide staining to ensure equal loading of RNA.
8. Transfer the RNA to a nylon support and probe using radiolabeled oligonucleotide probes generated by random hexamer priming, using as template the insert from pZJMα.

Technical notes. To rehybridize, the same blot can be stripped (by boiling in 0.1% (w/v) SDS for 10 min) and hybridized with probes to genes that are homologous to the target (if any) to ensure specificity of the RNAi. Additionally, hybridizing the blot with an α-tubulin probe is useful for confirming equal loading.

Western analysis of the protein product of the targeted mRNA can also be used to determine the impact of RNAi (Figure 16.2B). This should be performed throughout the induction period because phenotypes develop at different rates depending on transcript, protein abundance, and stability. Also, the development of growth phenotypes may not directly coincide with the loss of protein, since the amount of protein required for normal growth may be lower than the threshold of protein detection.

To ensure that the observed phenotype is not the result of unanticipated interference of a gene that shares homology with the target, a second unique fragment of the targeted gene should be inserted into pZJMα. If both constructs target the same transcript, RNAi using either fragment yields the same phenotype. Additionally, northern analysis should confirm that both constructs affect the same transcript. (Alternatively, consider at the outset using two different fragments targeting the same gene in order to save time in the confirmation process.)

Protocol 16.6 RNAi in BSF Trypanosomes

Materials

- HMI-9 (Appendix 16.1)
- HMI-9/FBS_{10}/SP_{10} (HMI-9 supplemented with 10% (v/v) fetal bovine serum and 10% (v/v) Serum Plus (JRH Biosciences, Lenexa, KS))
- G418
- Hygromycin
- Cytomix (Protocol 16.3)
- CO_2 incubator (37°C)
- 4-mm gapped cuvettes using a BTX Electro Cell Manipulator 630 or BTX ECM 830 Electro Square Porator

Experimental Procedures

Cell line growth

RNAi using pZJMα in BSF trypanosomes employs a transgenic BSF cell line, 90–13, that expresses the T7 RNA polymerase and the tetracycline repressor.[7] The T7 RNA polymerase and the tetracycline repressor are maintained by culturing the parasites in the presence of 2.5 μg/ml G418 and 5 μg/ml hygromycin.

1. Grow the parasites in HMI-9/FBS_{10}/SP_{10} medium (warmed to 37°C and equilibrated to 5% CO_2 for 1 h in the growth incubator),[23] with cell densities maintained between 1×10^4 and 1×10^6 cell/ml. Maintaining the cultures between these densities is essential, as overgrowth is lethal to the BSF parasites.

Transfection and selection

2. For transfection, grow the BSF parasites to mid-log phase (~5×10^5 cells/ml).
3. Centrifuge ~200 ml of culture (~1×10^8 cells) at 3000 × g, 10 min, in a rotor warmed to RT.[20]
4. Wash the parasites in 1 ml of prewarmed cytomix and resuspend in 0.5 ml of cytomix containing 100 μg of *Not* I-linearized plasmid.
5. Perform electroporations in 4-mm gapped cuvettes using a BTX electroporator as described for PF trypanosomes (Protocol 16.3), but instead of transferring to flasks for growth, dilute the cells into 24 ml of prewarmed HMI-9/FBS_{10}/SP_{10}/G/H[20] and transfer 1 ml into untreated 24-well tissue culture plates.

6. Grow the cells at 37°C and 5% CO_2.
7. After one day, apply selection by the addition of 1 ml of medium supplemented with HMI-9/FBS$_{10}$/SP$_{10}$/G/H and 5 μg/ml phleomycin (this yields a final concentration of 2.5 μg/ml phleomycin).

Technical notes. Stable clones become evident after 7 to 10 days of growth at 37°C in 5% CO_2. The tissue culture plate well may not be turbid after 7 to 10 days of growth, but clusters of growing cells will be visible by microscopy. These cells should be transferred to a flask (after mixing the well with a pipette) for continued culturing. Using this approach, transfection efficiency is ~2×10^{-8}. Cells that grow up from the 24-well plates are considered clonal because of the low transfection efficiency.

Freezing BSF trypanosomes
BSF parasites are frozen and stored much like PF parasites.

8. Add 0.2 ml of 50% (v/v) filter-sterilized glycerol in cytomix to 0.8 ml of culture containing 5×10^5 to 1×10^6 cells/ml in a liquid nitrogen–compatible cryotube.
9. Freeze the sample overnight at –80°C in the Styrofoam freezing carton before transfer to liquid nitrogen storage.
10. To thaw frozen BSF trypanosomes, thaw cryotubes to room temperature and transfer the contents to 9 ml warmed and equilibrated (37°C, 5% CO_2) HMI-9/FBS$_{10}$/SP$_{10}$/G/H/P.

Protocol 16.7 The RNAi Library

Materials

- Appropriate plasmid vector (e.g., pZJMβ, see below)
- Various restriction endonucleases (Promega, New England Biolabs)
- *T. brucei* strain 927 DNA (prepared using a Puregene DNA Isolation Kit (Gentra Systems))
- Electrocompetent ElectroMAX DH5α-E cells (Invitrogen)
- Qiagen Maxi Plasmid Kit (Qiagen)
- Standard LB/ampicillin (50 μg/ml) agar plates
- SDM-79/FBS$_{10}$/G/H/P (Appendix 16.1)
- SDM-79/FBS$_{10}$/G/H (Appendix 16.1)
- Electroporator

Experimental Procedures

Library construction
To construct the RNAi library, we modified pZJMα by inserting a multicloning site that contains the restriction sites *Xho* I, *Xba* I, *Nde* I, *Hpa* I, two *Sfi* I sites, and *Hind* III into the *Xho* I and *Hind* III sites of pZJMα, yielding pZJMβ (Figure 16.1B). The α-tubulin insert is no longer required, because the 56-bp insert is sufficient to allow replication in *E. coli*.

1. Generate the genomic DNA inserts for the RNAi library by sonication of *T. brucei* strain 927 DNA, followed by incubation with BAL-31 nuclease and T4 DNA polymerase to generate fragments with blunt ends.[18]

2. Anneal and ligate adapter oligonucleotides to the genomic fragments. Then ligate into the *Sfi*I sites of pZJMβ and use the plasmids for electroporation into *E. coli* ElectroMAX DH5α-E cells.
3. To assess the quality of the library, plate *E. coli* and analyze by restriction enzyme digestion using *Xho* I and *Hin*d III plasmids purified from individual colonies (inserts are found in 92% of plasmids with an average size of ~660 bp).
4. Given a genome size of 3.5×10^7 bp, plate a total of ~5×10^5 colonies (~11-fold genome coverage) on 150-mm LB/ampicillin (50 μg/ml) plates (which can accommodate ~20,000 colonies/plate) and grow overnight.
5. Wash the colonies from the plates with 10 ml of liquid LB per plate, using a gloved finger to dislodge the bacteria.
6. Extract plasmid DNA with Qiagen Maxi Plasmid Kit according to the supplier's information.

Transfection of the library into T. brucei

To cover the parasite genome five times with the pZJM library (average insert size 660 bp), one needs to generate 2.7×10^5 transfected parasites. Unfortunately, the transfection efficiency of BSF parasites is too low (at this time) to readily introduce the RNAi library into BSF parasites. Therefore, work to date has been limited to studies in PF trypanosomes.

7. Determine PF trypanosome transfection efficiency using pZJM by electroporating 1×10^8 parasites with 10 μg of *Not* I linearized library.
8. Following a 24-h recovery period in SDM-79/FBS_{10}/G/H, centrifuge cells, resuspend in SDM-79/FBS_{10}/G/H/P, and plate in the first row of a 96-well plate at 5×10^6 parasites/ml. Seed subsequent rows of the plates with 1:1 dilutions of the preceding row.
9. Use the greatest dilution yielding viable wells (along with an observed plating efficiency of ~30%) to calculate a transfection efficiency of ~1×10^{-4}. Transfection of the library into trypanosomes is similar to transfection of pZJMα. Given the transfection efficiency of PF trypanosomes, 50 transfections performed as described for pZJMα (with each containing 10 μg of *Not* I linearized pZJM library and 1×10^8 cells) will yield the required number of transfectants for > 5-fold coverage.
10. Following electroporation, pool the parasites (at 5×10^6/ml) in 1000 ml of SDM-79/FBS_{10}/G/H and allow to recover overnight.
11. Centrifuge cells (3000 × g, 10 min) and resuspend in SDM-79/FBS_{10}/G/H/P. Again, as selection is occurring, cells must be passed to avoid overgrowth and the detrimental effects of large amounts of cell debris. Dividing cells should begin to appear after 7 to 10 days of selection.

Expansion and care of the trypanosome library

The library has been maintained in PF trypanosomes as frozen stabilates of cells (~ 5×10^6/ml). Although expansion of the library can potentially cause a reduction in coverage, we have expanded it both for our own use and to supply it to other laboratories.

12. To expand the frozen library, thaw 5 ml of stabilate and immediately dilute into 45 ml of SDM-79/FBS_{10}/G/H/P.

13. Maintain the culture and pass for 2 to 3 days to dilute out the glycerol required for freezing and also to remove (by passage) dead cells and debris.
14. At this point, make new stabilates and freeze as described above.
15. Induce the remaining culture with tetracycline (1 μg/ml), which is maintained in mid-logarithmic growth (20 ml, between 5×10^5 and 5×10^6 cells/ml).

Technical notes. The library has also been propagated in *E. coli* and stored as purified plasmid DNA, which allows for ease of shipping and long-term storage. To use this form of the library, plasmid DNA is expanded by transforming *E. coli*, which are then plated on selective media and the resulting colonies pooled for plasmid purification. In order to cover the genome at least five times with a 660-bp library, 2.6×10^5 colonies are required. Commercially available electrocompetent *E. coli* (ElectroMAX DH5α-E cells, Invitrogen Life Technologies, with an estimated transfection efficiency of ~1×10^{10} transformants/μg of plasmid DNA) are transformed with 10 ng of pZJM library, and the number of transformants/electroporation is determined by plating serial dilutions of the mix on LB/ampicillin (50 μg/ml) plates.

After calculating the *E. coli* transfection efficiency, enough transfection mix to generate sufficient colonies is plated on 150-mm LB/ampicillin (50 μg/ml) plates and incubated overnight at 37°C. Colonies are washed off the plate in 10 ml of LB, using a gloved finger. The resulting liquid is collected from all of the plates and plasmid DNA is isolated by Qiagen Plasmid Maxi kit (Qiagen).

Protocol 16.8 Identification of the Gene Responsible for a Phenotype of Interest

Materials

- Puregene DNA Isolation Kit (Gentra Systems)
- Primers PhleoF (5'GGAACGGCACTGGTCAAC) and AldoR (5'CTCCAAAGCAAACATGCAGA)
- Primers XlinkF (5'CTCGAGGGCCAGTGAGGC) and BetaLow (5'GATCTAGCCCCTGCAGGAAT)
- Standard PCR reagents (e.g., Promega, Bioline)
- Thermal cycler

Experimental Procedures

1. Following sorting and cloning, isolate genomic DNA (from a total of ~1×10^7 cells) using a Puregene DNA Isolation Kit (Gentra Systems) according to the supplier's information. Resuspend in 20 μl of H_20.
2. Amplify the pZJM insert using a nested PCR approach. Carry out the first round of PCR using the outer primers (PhleoF and AldoR) (95°C, 30 sec; 55°C, 30 sec; 72°C, 2.5 min; 30 cycles).
3. Take 1 μl of the above PCR reaction as template and amplify (95°C, 30 sec, 60°C, 30 sec, 72°C, 1.5 min, 30 cycles) using internal primers (XlinkF and BetaLow). Please note that these primers will not function using pZJMα as template.
4. Either directly sequence the resulting PCR product or first clone into pGEM-T Easy Vector (Promega) and then sequence.
5. Identify the candidate gene by a BLAST search of the *T. brucei* sequence database.

APPENDIX 16.1

Growth Media

1. *SDM-79 for PF trypanosome growth* — slightly modified from [24].

For 1 liter, mix the following dry ingredients: Grace's Insect Cell Culture Media with L-glutamine and without sodium bicarbonate (2.0 g), glucose (1.0 g), HEPES (8.0 g), MOPS (5.0 g), $NaHCO_3$ (2.0 g), pyruvic acid (100 mg), L-alanine (200 mg), L-arginine (100 mg), L-glutamine (300 mg), L-methionine (70 mg), L-phenylalanine (80 mg), L-proline (600 mg), L-serine (60 mg), L-taurine (160 mg), L-threonine (350 mg), L-tyrosine (100 mg), adenosine (10 mg), guanosine (10 mg), D-glucosamine-HCl (50 mg), folic acid (4 mg), *p*-aminobenzoic acid (2 mg), and biotin (0.2 mg).

Dissolve the dry solids (20.2 g) in the following: Minimum Essential Medium (MEM) with L-glutamine (600 ml), MEM amino acid solution 50X without L-glutamine (8 ml), MEM nonessential amino acids 100X (6 ml), bovine hemin (2 mg/ml in 50 mM NaOH) (3.75 ml), Basal Medium Eagle vitamin solution 100 × (10 ml), penicillin/streptomycin solution 100 × (10 ml).

Adjust the pH to 7.3 with 10 N NaOH, and increase the volume to 900 ml with H_20. Filter-sterilize with a 0.22-mm filter. Add heat-inactivated (56°C, 30 min) fetal bovine serum (100 ml) to the filtered solution — do not filter the serum. Store at 4°C for up to 6 months.

2. *HMI-9 for BSF growth* [23]

For 1 liter, mix the following dry ingredients: Hypoxanthine (0.136 g), bathocuproine sulfonate (28 mg), cysteine (182 mg), pyruvic acid (110 mg), thymidine (39 mg)

Add the dry ingredients to the following liquids: 2-mercaptoethanol (14 µl), Iscove's Modified Dulbeccos's Medium with L-glutamine and HEPES (730 ml), penicillin/streptomycin solution 100 × (10 ml).

Adjust volume to 800 ml, filter-sterilize, and add heat-inactivated fetal bovine serum (100 ml) and Serum Plus Medium Supplement (100 ml) (from JRH Biosciences). Store at 4°C for up to 6 months.

Other Solutions

1. *Cytomix* — 120 mM KCl, 0.15 mM $CaCl_2$, 10 mM K_2HPO_4, 25 mM HEPES, 2 mM EDTA, 5 mM $MgCl_2$.
2. *Freezing medium* — 50% glycerol in cytomix, filter sterilized.
3. *Antibiotics* — filter-sterilize (with a 0.22-µM filter) all antibiotics prior to use. Stock solutions (1000 ×) are made in cytomix. Working concentrations are as follows (PF/BSF): G418 (15 /1.5 µg/ml), hygromycin (50/5 µg/ml), and phleomycin (2.5/2.5 µg/ml).

ABBREVIATIONS

BSF bloodstream form trypanosome
dsRNA double-stranded RNA
FACS fluorescence-activated cell sorting
FBS fetal bovine serum
GPI glycosylphosphatidyl inositol
LB Luria-Bertani broth
PF procyclic form trypanosome
RNAi RNA interference
RT room temperature
SP Serum Plus
tet tetracycline

ACKNOWLEDGMENTS

The authors wish to thank Drs. Elizabeth Wirtz and George Cross (Rockefeller University) for providing the transgenic trypanosomes PF 29-13 and BSF 90-13, without which this project would not have been possible. We also thank current and past members of the Englund lab for their patience as we developed this system. The helpful input and suggestions of Drs. Michele Klingbeil, Tina Saxowsky, and Kim Paul were greatly appreciated. Also, we thank Theresa Westhead for her excellent technical assistance. Last, we all are deeply indebted to our wives, who have shown incredible patience as we pursued this work. The majority of the research described in this manuscript was performed in the lab of P.T.E. and was supported by Grants AI21334 and GM27608 from the National Institutes of Health. J.C.M. is supported by the Charles E. Culpeper Biomedical Pilot Initiative.

REFERENCES

1. Vickerman, K., Developmental cycles and biology of pathogenic trypanosomes. *Br. Med. Bull.*, 41, 105, 1985.
2. Ferguson, M.A.J. et al., Glycosyl-phosphatidylinositol moiety that anchors *Trypanosoma brucei* variant surface glycoprotein to the membrane. *Science*, 239, 753, 1988.
3. Masterson, W.J. et al., A novel pathway for glycan assembly: biosynthesis of the glycosyl-phosphatidylinositol anchor of the trypanosome variant surface glycoprotein. *Cell*, 56, 793, 1989.
4. Simpson, L., S. Sbicego, and R. Aphasizhev, Uridine insertion/deletion RNA editing in trypanosome mitochondria: a complex business. *RNA*, 9, 265, 2003.
5. Boothroyd, J.C. and G.A. Cross, Transcripts coding for variant surface glycoproteins of *Trypanosoma brucei* have a short, identical exon at their 5' end. *Gene*, 20, 281, 1982.
6. Mair, G. et al., A new twist in trypanosome RNA metabolism: *cis*-splicing of pre-mRNA. *RNA*, 6, 163, 2000.

7. Wirtz, E. et al., A tightly regulated inducible expression system for conditional gene knock-outs and dominant-negative genetics in *Trypanosoma brucei. Mol. Biochem. Parasitol.*, 99, 89, 1999.
8. Seebeck, T. et al., Tubulin genes of *Trypanosoma brucei*: a tightly clustered family of alternating genes. *Proc. Natl. Acad. Sci. USA*, 80, 4634, 1983.
9. Thomashow, L.S. et al., Tubulin genes are tandemly linked and clustered in the genome of *Trypanosoma brucei. Cell*, 32, 35, 1983.
10. Bringaud, F. and T. Baltz, Differential regulation of two distinct families of glucose transporter genes in *Trypanosoma brucei. Mol. Cell Biol.*, 13, 1146, 1993.
11. Sanchez, M.A. et al., Six related nucleoside/nucleobase transporters from *Trypanosoma brucei* exhibit distinct biochemical functions. *J. Biol. Chem.*, 277, 21499, 2002.
12. Ngo, H., C. Tschudi, K. Gull, and E. Ullu, Double-stranded RNA induces mRNA degradation in *Trypanosoma brucei. Proc. Natl. Acad. Sci. USA*, 95, 14687, 1998.
13. Bastin, P., T. Sherwin, and K. Gull, Paraflagellar rod is vital for trypanosome motility. *Nature*, 391, 548, 1998.
14. LaCount, D.J., B. Barrett, and J.E. Donelson, *Trypanosoma brucei* FLA1 is required for flagellum attachment and cytokinesis. *J. Biol. Chem.*, 277, 17580, 2002.
15. Chen, Y. et al., Development of RNA interference revertants in *Trypanosoma brucei* cell lines generated with a double stranded RNA expression construct driven by two opposing promoters. *Mol. Biochem. Parasitol.*, 126, 275, 2003.
16. Wang, Z. et al., Inhibition of *Trypanosoma brucei* gene expression by RNA interference using an integratable vector with opposing T7 promoters. *J. Biol. Chem.*, 275, 40174, 2000.
17. El-Sayed, N.M. et al., The African trypanosome genome. *Int. J. Parasitol.*, 30, 329, 2000.
18. Morris, J.C. et al., Glycolysis modulates trypanosome glycoprotein expression as revealed by an RNAi library. *EMBO J.*, 21, 4429, 2002..
19. Drew, M.E. et al., The adenosine analog tubercidin inhibits glycolysis in *Trypanosoma brucei* as revealed by an RNAi library. *J. Biol. Chem.*, 278, 465ab 2003.
20. Morris, J.C. et al., Inhibition of bloodstream form *Trypanosoma brucei* gene expression by RNA interference using the pZJM dual T7 vector. *Mol. Biochem. Parasitol.*, 117, 111, 2001.
21. Wirtz, E. and C. Clayton, Inducible gene expression in trypanosomes mediated by a prokaryotic repressor. *Science*, 268, 1179, 1995.
22. van den Hoff, M.J., A.F. Moorman, and W.H. Lamers, Electroporation in "intracellular" buffer increases cell survival. *Nucleic Acids Res.*, 20, 2902, 1992.
23. Hirumi, H. and K. Hirumi, Continuous cultivation of *Trypanosoma brucei* blood stream forms in a medium containing a low concentration of serum protein without feeder cell layers. *J. Parasitol.*, 75, 985, 1989.
24. Brun, R. and M. Shonenberger, Cultivation and *in vitro* cloning of procyclic culture forms of *Trypanosoma brucei* in a semi-defined medium. *Acta Tropica*, 36, 289, 1979.
25. Shi, H. et al., Genetic interference in *Trypanosoma brucei* by heritable and inducible double-stranded RNA. *RNA*, 6, 1069, 2000.
26. Timmons, L. and A. Fire, Specific interference by ingested dsRNA. *Nature*, 395, 854, 1998.
27. Tzfati, Y. et al., A single-stranded DNA-binding protein from *Crithidia fasciculata* recognizes the nucleotide sequence at the origin of replication of kinetoplast DNA minicircles. *Proc. Natl. Acad. Sci. USA*, 89, 6891, 1992.

17 RNAi in the Malaria Parasite *Plasmodium*

Anke Brüning-Richardson and Glenn A. McConkey

CONTENTS

17.1 INTRODUCTION

RNAi has emerged recently as one of the most promising tools in studying gene function (summarized in Chapter 1). It was initially believed to be based on an evolutionary defense mechanism of silencing RNA in eukaryotes to protect their genomes from exogenous (viral) and endogenous (transposon) elements. Related mechanisms with small RNAs are involved in cell homeostasis by regulating translation (*Caenorhabditis elegans*) and structure of heterochromatic domains in the genome. In many organisms, including *C. elegans*, *Drosophila*, zebra fish, and *Xenopus,* RNAi is activated by the presence of foreign dsRNA in the cell which triggers sequence-specific mRNA degradation.[1] The dsRNA is enzymatically cleaved with the help of Dicer RNase III to give rise to small interfering RNA duplexes (siRNA). These siRNAs, usually 21 to 25 nucleotides in length and characterized by a two-nucleotide 3'-overhang, form a complex with a nuclease complex, the RNA-induced silencing complex (RISC), where they become unwound and aid in the homology-dependent degradation of the target RNA. Similar processes occur in plants as posttranscriptional gene silencing and cosuppression and quelling in fungi and algae.[2] *Plasmodium* was the first intracellular organism in which RNAi was successfully demonstrated.[3] Subsequent studies have shown that small RNAs may be used and that siRNAs are generated, but the mechanism of RNA silencing in *Plasmodium* has yet to be elucidated.[4,5]

0–8493–2141–7/05/$0.00 + $.50

Members of the genus *Plasmodium,* a single-celled protozoan, are the etiologic agents of malaria. There are more than 250 million cases of malaria annually. *Plasmodium falciparum* is the species responsible for the highest mortality, which claims the lives of 1 to 2 million people each year.[6] Resistance to antimalarial drugs is increasing, and multiple drug resistance has been observed. Hence, there is a great need for new antimalarial drugs and the development of an effective vaccine. The complete genome sequence of *P. falciparum* was released in October 2002.[7] This is a landmark achievement as the first protozoan genome completely sequenced and it presented a particular challenge with its A/T rich genome (80.6% A/T). The genome of the mosquito vector for human malaria, *Anopheles gambiae*, has also been sequenced.[8] With the human genome sequence completed, the three genomes in the triad of malaria are now known. Additionally, a rodent malaria species has been sequenced and incomplete genome sequences for four other *Plasmodium* species are publicly available, permitting comparative genomic studies. Combined with microarray and proteomic data on *P. falciparum*, there is a plethora of information for design of RNAi studies.[9–11] There are numerous possible avenues for the application of RNAi technology for the study of *P. falciparum*. It can be applied to the study of gene function, related genes, duplicated genes, or gene families. On the large scale, RNAi may aid the development of vaccines or novel drugs in combating disease or in gene therapy through application on a genome-wide scale. Reagents for RNAi are easily obtained and are fairly inexpensive. RNAi procedures are usually not time consuming nor labor intensive compared with other gene-disruption techniques in *Plasmodium*. Still, it is a relatively new technique and future developments will increase efficiency and optimize procedures.

17.2 GENERAL CONSIDERATION FOR RNAi IN *PLASMODIUM*

The efficiency of RNAi in the study of *P. falciparum* seems to be dependent on several factors. One must consider the gene targeted and its expression as well as the sequences to target. Importantly, as this parasite is principally intracellular, the method for delivery of dsRNA to the parasite cytosol must be considered. In organisms such as trypanosomes, the dsRNA has been expressed from a transfected plasmid with expression under the control of an inducible promoter. In *Plasmodium*, inducible expression systems have not been described. The methods developed have been targeting predicted essential genes and, therefore, are based on direct application of dsRNA.

17.2.1 The Target Gene and Parasite Stage

The efficiency of RNAi is dependent on the choice of gene to be studied and the stage in the life cycle of the parasite. In other systems, extremely abundant or stable proteins have been reported to be difficult to silence. This may prove to also be a difficulty for abundant or stable proteins in *P. falciparum,* although a thorough study has yet to be performed. It is thus recommended to choose a gene for interference with an obvious phenotypic effect. This was effective with the FAT phenotype of disrupting alpha-tubulin in trypanosomes and vacuole enlargement by disrupting falcipain in *P. falciparum*.[4,12] Other genes have less obvious phenotypes but are required for growth or development.

The life cycle of the parasite includes stages in the mosquito vector and in the mammalian host (Figure 17.1). The asexually reproducing stages of *Plasmodium* in erythrocytes, responsible for the pathology of malaria, are most readily accessible for manipulation. These stages of *P. falciparum* (and *P. berghei*) can be cultivated *in vitro*. Most studies have involved these stages. The only extracellular stage during cycling in the blood stages is the short-lived merozoite, released by lysis of the host erythrocyte that invades fresh erythrocytes. In the trophozoite stages, the parasite is an intracellular parasite in the erythrocyte surrounded by essentially three membranes, the erythrocyte membrane, the parasitophorous vacuole membrane, and the parasite cell membrane, although there are parasite-induced mechanisms for interaction with the extracellular environment.[13] The trophozoites are the stage that have been transfected with plasmid DNA.[14] Transfection of *Plasmodium* appears to require electroporation. During asexual reproduction in erythrocytes, a low percentage of the parasites differentiate into the sexual forms, gametocytes. Gametocytes of the avian malaria parasite *P. gallinaceum* were the first stages of malaria parasites successfully transfected with DNA.[15] The gametocyte stages mature into gametes when taken up by the mosquito. In the mosquito, fusion and reduction division precede replication on the midgut wall (oocyst stage). The thousands of parasites produced in an oocyst are released as sporozoites that migrate through the haemocoel to the salivary gland. Sporozoites can be purified from dissected salivary glands, although downstream experiments to assess

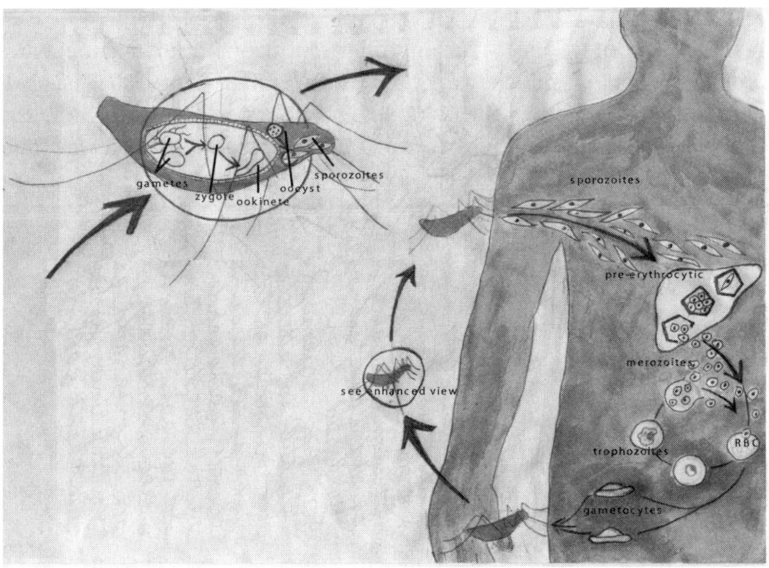

FIGURE 17.1 Life cycle of *Plasmodium*.

their viability are difficult. Technical obstacles preclude manipulations of the stages of the parasite in the liver. The methods described below focus on the trophozoite stages in the asexual life cycle.

The species and strain of *Plasmodium* are a consideration. RNAi of *P. falciparum,* the only human malaria that can be cultivated *in vitro*, has been performed by three groups. One of these groups has recently published the first application of RNAi to the rodent malaria *P. berghei*. In the studies with *P. falciparum,* the clonal strain 3D7 was used. There are potential differences in efficiency of RNAi in different strains.

17.2.2 Sequence Selection

The choice of sequences for the production of dsRNA for silencing seems crucial, as not all sequences have the desired "interfering" effect (Table 17.1 and reference [16]). dsRNAs of 21 bp to 1.7 kb have been efficient in silencing genes expressed in *P. falciparum*. This is similar to other invertebrate systems and embryonic mammalian cells.[2] Longer dsRNAs cover larger areas of the gene and increase the chances of interference. For selecting short (< 30 bp) dsRNAs, some publications suggest choosing and testing up to 20 sequences along the target gene.[17] This can be time consuming and laborious. dsRNA can be commercially obtained or produced in the laboratory. Recently, recombinant human Dicer enzyme has been used to digest longer dsRNAs into 21- to 25-bp dsRNAs in mammalian systems, although the effect on efficiency in *P. falciparum* has yet to be shown.

17.2.3 Delivery System

The route of delivery of the dsRNA is important. Electroporation and incubation have been applied to the RNAi studies in *P. falciparum.*[3–5] Other approaches, such as transfection with plasmids encoding the dsRNA, have not been used due to the lack of an inducible promoter. In the methods described herein, electroporation was found to be

TABLE 17.1
Summary of RNAi Approaches in Malaria Research

Organism	Application	Gene target	dsRNA size	Amount	Delivery	Reference
P. falciparum	*in vitro*	Pf DHODH	1.2 kb	1.2–2.5 μg/ml	electroporation	[3]
	in vitro	PfCS	0.9 kb	1.2–2.5 μg/ml	electroporation	
P. falciparum	*in vitro*	PfPPI	21 bp	3.75 μg/ml	electroporation	[5]
P. falciparum	*in vitro*	falcipain-1	1.7 kb	25–50 μg/ml	incubation	[4]
	in vitro	falcipain- 2	1.4 kb	25–50 μg/ml	incubation	
P. berghei	*in vivo*	berghepain 1	21 bp	25 μg	injection	[18]
	in vivo	berghepain 2	21 bp	25 μg	injection	

required to deliver the dsRNA past the three membranes to the parasite cytosol to observe the RNAi effect (unpublished results). This is analogous to transfection with plasmid DNA. Other groups have observed RNAi by simply soaking parasites in dsRNA. Apparent differences may be due to the amount of dsRNA used, the sensitivity of the gene product to RNAi, or the size of the dsRNA, although Kumar et al.[5] used electroporation to deliver 21-bp dsRNAs for interference of protein serine/threonine phosphatase (PfPP1).

17.2.4 Assessment of RNAi

Various techniques have been applied to monitor the effect of the RNAi on the parasite, including observing morphological changes or abnormalities in stained parasites, measuring mRNA levels by quantitative RT-PCR, monitoring growth by ^{3}Hypoxanthine incorporation studies, and assessing protein expression with antisera. Appropriate controls are essential to standardize experiments and assess specificity, particularly with partial degradation. Two negative controls developed in our studies are the use of a heterologous gene, *Toxoplasma gondii* dihydrofolate reductase-thymidylate synthase (DHFR-TS), and a gene expressed at a different stage in the life cycle, *P. falciparum* circumsporozoite protein (CSP).

17.3 DELIVERY TECHNIQUES

Ours laboratory, and others, have been successful in establishing RNAi in the erythrocytic stage of the parasite (Table 17.1). In our study, dsRNA of a segment of the gene encoding dihydroorotate dehydrogenase (PfDHODH) was introduced into synchronized *P. falciparum* cultures (3D7 strain) by electroporation. PfDHODH is an enzyme involved in pyrimidine biosynthesis.[3] Disruption of this enzyme was predicted to interrupt parasite growth. After electroporation with the dsRNA, the parasitized red blood cells were cultured and parasite growth was monitored (Figure 17.2). Both parasite number and incorporation of radio-labelled hypoxanthine reflect effects of RNAi. The specific effect of the dsRNA was demonstrated by measuring the level of endogenous PfDHODH mRNA by RT-PCR (Figure 17.3). Controls included treatment with single-stranded RNA, dsRNA encoding the circumsporozoite protein, and dsRNA encoding a gene from *T. gondii*. A decrease in parasite growth (60%), correlating with a decrease in levels of PfDHODH mRNA, was demonstrated. Also tested was dsRNA encoding the enzyme chorismate synthetase (PfCS), which is believed to be involved in folate synthesis. Again, a decrease in growth was detected, though it was less marked than by the dsRNA encoding PfDHODH. Kumar et al. investigated PP1 serine/threonine protein phosphatase (PfPP1) with RNAi technology, except that they utilized dsRNA oligonucleotides.[5] They observed a 70% inhibition of parasite growth and loss of immunostaining on Western blots. RNAi has also been applied to study two cysteine protease genes (*falcipain-1* and *falcipain-2*) involved in hemoglobin degradation using

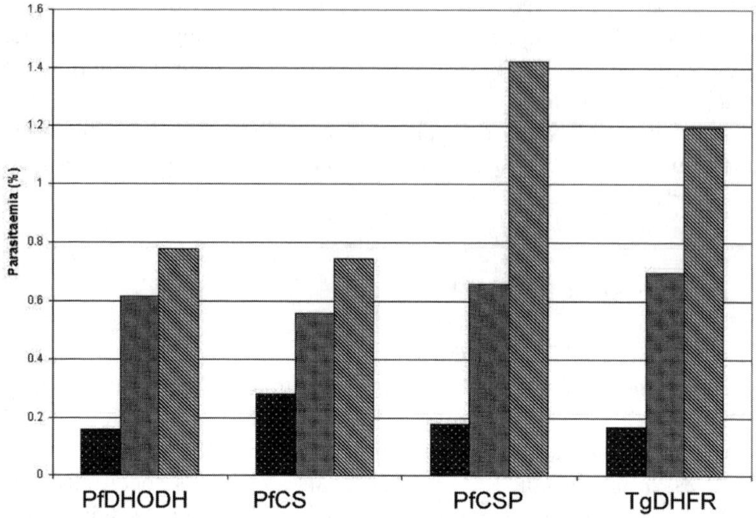

FIGURE 17.2 Time course of inhibition of growth by dsRNA. The effect of exposing synchronized ring-stage trophozoites to dsRNA at 24, 48, and 72 h (dark, medium, and light stipple, respectively) plotted as percent parasitemia. The dsRNAs used were encoding PfDHODH, PfCS, PfCSP, and TgDHFR.

a soaking method[4] (also see Chapter 18). In this study, soaking of synchronized parasite cultures in a solution of dsRNA was sufficient for RNAi. Morphological changes and decrease in growth were observed with a coincident decrease in endogenous falcipain mRNAs. Interestingly, the inhibition was more pronounced when both genes were targeted simultaneously.[4]

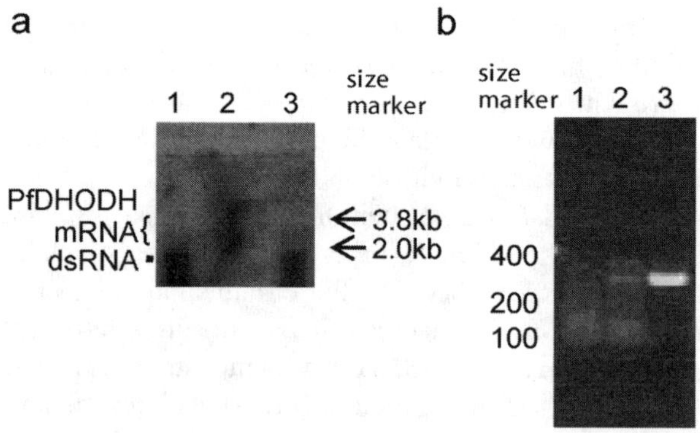

FIGURE 17.3 PfDHODH mRNA levels in dsRNA-treated parasites. A Northern blot (a) and quantitative RT-PCR (b) of *P. falciparum* treated with dsRNA encoding a fragment of PfDHODH vs. control cultures. Total RNA from dsRNA-treated parasites was blotted from a denaturing polyacrylamide gel and probed with radio-labelled PfDHODH (a, lanes 1 and 3) or subjected to PCR (b, lane 2). Water-treated parasites served as controls (a, lane 2 and b, lane 2) Size markers are numbered. Primers for PCR of PfDHODH map outside the region of the gene targeted by the dsRNA, with genomic DNA serving as a control (b, lane 3). From McRobert, L., Ph.D. thesis, 2001, School of Biology, University of Leeds, Leeds, UK.

17.4 OTHER PLASMODIA

The application of RNAi to the rodent malaria parasite *P. berghei* was recently described.[18] The siRNAs were delivered by intravenous injection into *P. berghei*–infected mice. Two cysteine protease genes, analogous to the falcipains investigated by RNAi in *P. falciparum*, were targeted. Morphological changes in the parasites obtained by blood smears from treated animals and a reduction in the level of the corresponding mRNA were found with treatment. However, under *in vivo* conditions, only a 40 to 50% reduction in the berghepain mRNA levels was observed in the treated animals. Unfortunately, there was no significant reduction in parasitemia as compared to the controls. The authors suggested that these results might have been due to the sequences chosen for the siRNA or the dsRNA delivery system. They also suggested that there are three classes of enzymes involved in hemoglobin degradation. Targeting enzymes from all three classes may lead to a significant reduction in parasitemia. This is an exciting area for potential *in vivo* therapy that needs development.

17.5 FUTURE DIRECTIONS AND APPLICATIONS

The mechanism of RNAi in *Plasmodium* is unclear. The observation that the dsRNAs were broken into short interfering RNAs of approximately 25 nucleotides in size indicates that a Dicer-like enzyme exists in the malarial parasite (see Chapter 18). Searches of the annotated *P. falciparum* genome have not identified a Dicer homologue, although there are several homologues of the section of Dicer associated with RNA helicase activity (unpublished observations). These are annotated as RNA helicases and lack the RNase III domain found in Dicer.[19] Possibly *Plasmodium* Dicer is too distantly related to be identified by homology. A potential Dicer homologue has been found in the *T. gondii* genome sequence.[20]

RNAi technology in the study of malaria is in its infancy. With more commercial products available and a vast wealth of sequence data, RNAi technology is an indispensable tool for understanding gene function in the malarial parasite. Comparative genomic approaches between the several species of *Plasmodium* and related Apicomplexa, with genome-sequencing projects underway, are being applied to predict gene function and identify putative missing enzymes in pathways.[21] RNAi provides a practical approach to testing *in silico* predictions and validating potential targets for antimalarial chemotherapy. The power of comparative genomic predictions with RNAi and expression data from microarrays and proteomic surveys will greatly increase our understanding of this deadly parasite and may guide development of antimalarial drugs and vaccines. RNAi has also been applied to the vector *Anopheles*.[22] It should now be feasible to study gene function in the vector stages of the parasite and even in the exoerythrocytic stages in the mammalian host.

PROTOCOLS

See Chapter 18 for additional protocols.

Protocol 17.1. Preparation of Double-Stranded RNA

Materials

- Plasmid pPCR-Script SK (Stratagene, Inc.)
- Appropriate restriction endonucleases (Stratagene, Inc., Promega, New England Biolabs)
- *In vitro* transcription system (Promega)
- Stop solution (1 M NH_4Ac, 10 mM Na_2EDTA, 0.2% SDS) plus 14 μg/ml glycogen
- Phenol/chloroform
- Ethanol

Experimental Procedures

Plasmid containing the gene of interest is used as a template for RNA synthesis. In our studies, we used a 1.2-kb fragment of the *P. falciparum* dihydroorotate dehydrogenase (*PFDHODH*) gene and a 915-bp fragment of the chorismate synthetase (aroc) gene.

1. Clone the gene or gene fragment of interest into a plasmid containing a bacteriophage promoter (T7 or T3 RNA polymerase) using standard cloning procedures. We cloned into the multiple cloning site (MCS) of plasmid pPCR-Script SK (Stratagene, Inc.) using *Bam* H1 and *Eco* R1 restriction endonuclease sites. This vector contains T7 and T3 RNA polymerase promoters adjacent to the MCS. As controls, we used *T. gondii* dihydrofolate reductase thymidylate synthetase (DHFR-TS) and *P. falciparum* circumsporozoite protein CSP genes inserted in pBluescript II SK (+/–). The coding sequence of *T. gondii* DHFR-TS is divergent from *P. falciparum* DHFR-TS, and the CSP gene has an A/T content similar to other *P. falciparum* genes but is not required during erythrocyte stages.
2. Linearize plasmid DNA (5 μg) with either of two restriction enzymes that cleave at an end of the inserted sequence, providing templates for transcription with T7 and T3 polymerases.
3. Purify linear plasmid DNA using a DNA purification kit (e.g., Qiaquick DNA Cleanup System, Qiagen).
4. Synthesize RNA from the linear templates.[23] Set up the following reaction:

Linear template DNA	1 μg
Transcription buffer	1 ×
rNTPs	0.5 mM
RNasin	40 U
DTT	10 mM
T3 /T7 RNA polymerase	(17 U)/(15 U)
Final volume	20 μl

5. Incubate the reactions for 90 min (T3 RNA polymerase at 25°C and T7 RNA polymerase at 37°C).
6. Mix reactions to combine sense and antisense RNA transcripts.
7. Add 380 μl of stop solution plus 14 μg/ml glycogen to stop the transcription reaction.

8. Purify RNA by standard phenol/chloroform extractions, followed by incubation at 68°C for 10 min and 37°C for 30 min to anneal the strands of RNA.
9. Precipitate with two volumes absolute ethanol for several hours at –20°C.
10. Pellet precipitate by centrifugation (10 min, full speed in microcentrifuge), rinse the precipitate with ice-cold 70% (v/v) ethanol, lyophilize, and store dry at –70°C.
11. Rehydrate in diethyl pyrocarbonate (DEPC) treated water (10 μl) prior to experimentation. Check the quality of the dsRNA product immediately before the RNAi experiment by agarose gel electrophoresis.

Technical note. Alternatively, short RNA oligonucleotides may be used. Complementary 21-nucleotide oligonucleotides can be commercially obtained or synthesized as described.[24]

Protocol 17.2 Parasite Cultivation and Preparation

P. falciparum asexual parasites (3D7 strain) are maintained *in vitro* following a modified protocol for continuous culture.[25]

Materials

- Complete RPMI medium (RPMI 1640 (InVitrogen), 20 to 40 mM HEPES, pH 7.6, 0.2% (w/v) NaHCO3, 0.5% (v/v) Albumax (InVitrogen))
- Human O^+ blood collected in citrate-phosphate dextrose solution (store at 4°C)
- Giemsa stain/methanol (Sigma)
- Sorbitol
- Cytomix (120 mM KCL, 0.15 mM CaCl, 2 mM EGTA, 5 mM $MgCl_2$, 10 mM K_2HPO_2, 10 mM KH_2PO_4, 25 mM HEPES, pH 7.6)

Experimental Procedures

1. Prepare erythrocytes by serially washing human blood several times with RPMI 1640, with removal of leukocyte-enriched buffy coat after each wash.
2. Culture parasites in complete medium containing 5% hematocrit of human O^+ erythrocytes and incubate in low oxygen tension (1% (v/v) O_2, 3% (v/v) CO_2, and 96% (v/v) N_2) at 37°C.
3. Replace the medium daily and maintain parasitemia (percent of erythrocytes infected) between 0.5 to 5% as monitored by microscopy of thin smears (fixed with methanol and Giemsa stained).

Synchronization of parasites with the sorbitol method[26]

4. Synchronize cultured parasites by centrifuging infected erythrocytes (3000 rpm (500 × g), 5 min in a tabletop centrifuge) and replacing the supernatant with ½-culture volume of 5% (w/v) sorbitol.
5. Incubate at room temperature for five minutes and dilute the infected erythrocytes with complete RPMI medium at least two-fold to stop the lysis.
6. Harvest by centrifugation and resuspend the parasite pellet in the original volume of complete RPMI medium. Only early trophozoite stage ("ring-stage") parasites survive this procedure.

7. Cultivate the parasites under normal conditions for 40 to 42 h. Repeat the synchronization protocol to recover a highly synchronous (> 95%) culture of ring-stage parasites with parasitemia of 5 to 8%.
8. After a brief recovery period in standard culture conditions, collect the infected erythrocytes by centrifugation (3000 rpm (500 × g); 5 min in a tabletop centrifuge).
9. Resuspend the parasites in five packed cell volume (PCV) filter-sterilized incomplete cytomix (120 mM KCL, 0.15 mM CaCl, 2 mM EGTA, 5 mM $MgCl_2$, 10 mM K_2HPO_2, 10 mM KH_2PO_4, 25 mM HEPES pH 7.6).

Protocol 17.3 RNAi Assay

Materials

- 0.2-cm electroporation cuvette (BioRad)
- Electroporator (BioRad) Gene Pulser II with capacitance extender
- Rehydrated dsRNA

Experimental Procedures

1. Place an aliquot of the prepared parasites (0.4 ml) in a 0.2-cm electroporation cuvette and add 10 μl of rehydrated dsRNA or DEPC-treated water (as a control); keep on ice.
2. Electroporate using 1.25 kV 200W 25 μF setting in a BioRad electroporator.
3. Dilute the contents of each cuvette with 6 ml complete RPMI medium containing washed erythrocytes at 5% hematocrit.
4. To measure the effect on growth, aliquot cultures into 96-well plates (200 μl/well), testing each sample in triplicate. Replace the growth medium daily. Measure effects on growth by adding [^{3}H]-Hypoxanthine (1 μCi/well) to samples at 24-h intervals.
5. Harvest microplates 16 to 18 h later and measure incorporation (see Protocol 18.6).

Technical notes. For measuring effects on RNA, samples are maintained in 25-cm^2 tissue culture flasks following electroporation to be processed for RNA at 24-h intervals. Total RNA is isolated by extraction with guanidinium isothiocyanate as in the RNagents kit (Promega, Inc.). For measuring relative levels of mRNA, RT-PCR may be used with limited cycles of PCR. The primers should map outside the fragment of the gene used for dsRNA treatment.

REFERENCES

1. Cogoni, C. and Macino, G., Post-transcriptional gene silencing across kingdoms, *Curr. Opin. Genet. Dev.,* 10, 638, 2000.
2. Cottrell, T.R. and Doering, T.L., Silence of the strands: RNA interference in eukaryotic pathogens, *Trends Microbiol.,* 11, 37, 2003.
3. McRobert, L. and McConkey, G.A., RNA interference (RNAi) inhibits growth of *Plasmodium falciparum, Mol. Biochem. Parasitol.,* 119, 273, 2002.
4. Malhotra, P. et al., Double-stranded RNA-mediated gene silencing of cysteine proteases (falcipain-1 and -2) of *Plasmodium falciparum, Mol. Microbiol.,* 45, 1245, 2002.

5. Kumar, R. et al., Characterisation and expression of a PP1 serine/threonine protein phosphatase (PfPP1) from the malaria parasite, *Plasmodium falciparum*: demonstration of its essential role using RNA interference, *Malar. J.,* 1, 5, 2002.
6. World Health Organization Report, WHO, Geneva, Switzerland, 2002.
7. Gardner, M.J. et al., Genome sequence of the human malaria parasite Plasmodium falciparum, *Nature,* 419, 498, 2002.
8. Holt, R.A. et al., The genome sequence of the malaria mosquito *Anopheles gambiae*, *Science,* 298, 129, 2002.
9. Ben Mamoun, C. et al., Co-ordinated programme of gene expression during asexual intraerythrocytic development of the human malaria parasite *Plasmodium falciparum* revealed by microarray analysis, *Mol. Microbiol.,* 39, 26, 2001.
10. Bozdech, Z. et al., Expression profiling of the schizont and trophozoite stages of *Plasmodium falciparum* with a long-oligonucleotide microarray, *Genome Biol.,* 4, R9, 2003.
11. Florens, L. et al., A proteomic view of the *Plasmodium falciparum* life cycle, *Nature,* 419, 520, 2002.
12. Ngo, H. et al., Double-stranded RNA induces mRNA degradation in *Trypanosoma brucei*, *Proc. Natl. Acad. Sci. U.S.A.,* 95, 14687, 1998.
13. Lauer, S.A. et al., A membrane network for nutrient import in red cells infected with the malaria parasite, *Science,* 276, 1122, 1997.
14. Wu, Y. et al., Transfection of *Plasmodium falciparum* within human red blood cells, *Proc. Natl. Acad. Sci. U.S.A.,* 92, 973, 1995.
15. Goonewardene, R. et al., Transfection of the malaria parasite and expression of firefly luciferase, *Proc. Natl. Acad. Sci. U.S.A.,* 90, 5234, 1993.
16. Tuschl, T. and Borkhardt, A., Small Interfering RNAs: A Revolutionary Tool for the Analysis of Gene Function and Gene Therapy, *Molecular Interventions,* 2, 158, 2002.
17. Tuschl, T., http://www.rockefeller.edu/labheads/tuschl/sirna.html, revised May 6, 2004.
18. Mohmmed, A. et al., *In vivo* gene silencing in Plasmodium berghei — a mouse malaria model, *Biochem. Biophys. Res. Commun.,* 309, 506, 2003.
19. Bahl, A. et al., PlasmoDB: the *Plasmodium* genome resource. A database integrating experimental and computational data, *Nucleic Acids Res.,* 31, 212, 2003.
20. Kissinger, J.C. et al., ToxoDB: accessing the *Toxoplasma gondii* genome, *Nucleic Acids Res.,* 31, 234, 2003.
21. McConkey, G.A. et al., Annotating the *Plasmodium* genome and the enigma of the shikimate pathway, *Trends Parasitol.,*20, 60, 2004.
22. Brown, A.E. et al., Stable and heritable gene silencing in the malaria vector *Anopheles stephensi*, *Nucleic Acids Res.,* 31, e85, 2003.
23. Fire, A. et al., Potent and specific genetic interference by double-stranded RNA in *Caenorhabditis elegans*, *Nature,* 391, 806, 1998.
24. Mellor, H., http://www.bch.bris.ac.uk/staff/Mellor, 2003.
25. Trager, W. and Jensen, J.B., Human malaria parasites in continuous culture, *Science,* 193, 673, 1976.
26. Lambros, C. and Vanderberg, J.P., Synchronization of *Plasmodium falciparum* erythrocytic stages in culture, *J. Parasitol.,* 65, 418, 1979.

18 Silencing the Expression of Cysteine Proteases (Falcipains) in *Plasmodium Falciparum* using RNA Interference Approaches

Palakodeti V.N. Dasaradhi, Asif Mohmmed, Manzar J. Hossain, Virander S. Chauhan, and Pawan Malhotra

CONTENTS

18.1 INTRODUCTION

The key approaches to elucidate the function of a gene in an organism are either to look for the dominant mutant of the gene that ablates the activity of that gene or to experimentally ablate the genetic message *in vivo*. Several loss-of-function (LOF) approaches, such as gene knockout by homologous recombination, antisense approach, and ribozyme technology, have been applied in the past. Of these arsenals, gene targeting by homologous recombination is the most commonly used technology to determine gene function; however, it is expensive and time consuming and many organisms are not amenable to it. RNA interference (RNAi) is a new tool of reverse genetics that has coalesced during the last decade. RNA silencing is a posttranscriptional regulatory

0-8493-2141-7/05/$0.00 + $.50

mechanism induced by double-stranded RNA (dsRNA), which inhibit the gene expression through specific degradation of homologous mRNA.[1] Even though RNAi was first discovered in plants, its first elegant demonstration was first shown in an animal, the nematode worm *Caenorhabiditis elegans.* Fire, Mello, and colleagues, while attempting to use antisense RNA as an approach to inhibit the gene expression, found that the mixture of sense and antisense RNA (dsRNA) was at least tenfold or more potent than either sense or antisense RNAs alone. They also showed that the inhibition by dsRNA treatment closely resembled the loss-of-function or gene knockout phenotypes.[2] RNAi was shown to be similar to PTGS triggered by a transgenic DNA, called co-suppression in plants,[3] and to a related phenomenon termed quelling, observed in filamentous fungus *Neurospora crassa.*[4] In the beginning, utility of RNAi was limited to mammalian cell lines due to the innate immune and interferon-related responses triggered by dsRNA.[5] However, the discovery by Tuschl and colleagues that chemically synthesized short interfering RNAs (siRNAs) of 21 to 25 nt can bring about effective silencing of gene expression without inducing the interferon response opened the gate for large-scale application of RNAi in a wide range of organisms.[6] RNA interference by double-stranded RNA has now been demonstrated in diverse groups of organisms such as protozoa, animals (invertebrates and vertebrates), plants, and fungi.[7]

A two-step mechanism has been proposed for RNAi: an initiation step in which the large dsRNA is digested into ~21- to 25-nt-long RNA species referred to as small interfering RNAs (siRNAs) or guide RNAs, and an effectors step that involves binding of siRNA to a nuclease complex, referred to as RNA-induced silencing complex (RISC), and subsequent cleavage of homologous mRNA by RISC complex.[8–12] An amplification step, which involves either copying of input dsRNA or replication of siRNAs themselves, has also been proposed within the RNAi pathway.[13,14] Importantly, RNAi appears to be an evolutionary conserved mechanism, and its natural function is to protect the genome against invasion by mobile genetic elements such as transposons and viruses.[14,15] RNAi machinery also affects gene expression through additional mechanisms. In plants, exposure to dsRNA induces genetic methylation of sequences homologous to the silencing trigger.[16] Recently, RNAi machinery has been shown to be involved in a number of cellular regulatory systems such as developmental regulation and regulation at the level of chromatin structure.[17] MicroRNAs/short temporal RNAs (miRNA/stRNA), another class of tiny RNAs, bring about these regulations. These microRNAs are endogenously encoded small noncoding RNAs, derived by processing of short RNA hairpins, and are partially complementary to target mRNA. Mostly, they inhibit the translation of mRNAs without inducing mRNA cleavage.[14,18]

18.2 RNAi IN PARASITES AND LOWER ORGANISMS

Because of its efficacy, efficiency and cost effectiveness, RNAi has emerged as a powerful tool for knocking down the gene expression, with a potential utility for investigating the gene function for high-throughput, function-based genetic screens and for therapeutic usage.[19] Although RNAi has not been able to substitute gene knockout technology, it nevertheless produces hypomorphic mutants that are extremely useful for understanding the gene function(s). RNAi has tremendous impact in carrying out functional genomic studies in organisms that are not amenable to other classical genetic approaches. Among these genetically intractable organisms are the majority of parasites — both protozoa and helminthes. Genetic crosses in these organisms are often difficult or virtually impossible, either because of the complexity of the parasite lifecycle or the absence of a known mechanism of genetic exchange.[20] Among these protozoa and helminthes, RNAi has been demonstrated in *Trypanosomas* (see chapter 16), *Plasmodium* (also see chapter 17), *Toxoplasma gondii, Paramecium, Ceanorhabiditis elegans* (see chapters 11 and 12), *C. bergiae, Burgio malayi, Schistosoma mansoni,* and *Planaria* and *Hydra*.[7,14] Even though RNAi was tried in *Leishmania*, it did not work efficiently in this parasitic organism. It has been proposed that *Leishmania* may be naturally deficient in RNAi machinery.[21] Among these lower eukaryotes, RNAi was first reported in *Trypanosoma brucei* in transient fashion by electroporating the *in vitro* synthesized dsRNAs.[22] Later studies used inducible, stably maintained constructs that express stem-loop transcripts. These constructs were heritable and provided prolonged expression of dsRNAs/siRNAs, allowing interference of expression of genes of highly stable proteins. RNAi has been successfully used to study the different aspects of *Trypanosoma* biology such as flagellum ontogeny, flagellum attachment, studies of mitochondrial enzymes, and glycosome enzymes.[23] Studies on the mechanism of RNAi in *T. brucei* have revealed that siRNAs produced in this organism are a few nucleotides longer (~25 nt long) than the siRNAs produced in *Drosphila,* which are 22 nt long. It has been speculated that in *Trypanosoma,* RNAi evolved as a central mechanism for mRNA surveillance that functions to down-regulate expression of deleterious gene products.[24]

18.3 RNAi IN *PLASMODIUM*

RNAi has also been successfully demonstrated in *Plasmodium.* However, in comparison to extensive RNAi studies carried out in *Trypanosoma*, there are very few reports of RNAi in *Plasmodium.* The parasite *Plasmodium falciparum* is one of the four human malaria parasites and alone is responsible for 100 million cases of malaria and the death of more than 1 million African children annually. *P. falciparum* has a 23-megabase genome consisting of 14 chromosomes and encodes about 5300 proteins. Of these predicted proteins of *P. falciparum*, about 60% are of unknown functions.[25] The main bottleneck in the functional genomic studies of *P. falciparum* and other malaria species is the low success rate in gene targeting by homologous recombination. Although many examples of successful transfection experiments exist in *Plasmodia*, currently it takes 2 to 3 months to obtain a single transfectant. Moreover, only single crossover insertions have been observed in most cases in *P. falciparum.*[20] RNAi thus represents an exciting breakthrough in the quest to inhibit gene expression on the level of individual gene, gene families, and entire genome in *P. falciparum.* So far, there are only three reports of successful application of RNAi, in *P. falciparum*[26–28] and one in *P. berghei.*[29] Since the malaria parasite depends upon *de novo* synthesis of pyrimidine for its growth, McRobert and McConkey first applied RNAi to silence the expression of dihydroorotate dehydrogenase (DHODH), an enzyme involved in pyrimidine biosynthesis. This group showed significant reduction in parasite growth (~63%) upon introduction of PfDHODH dsRNA by electroporation into the synchronized *P. falciparum* ring-stage-infected erythrocytes.[26] An independent study has reported the abrogation of *P. falciparum* serine/threonine phosphatase activity (pp) by synthetic short interfering RNAs.[27] These siRNAs were introduced into *P. falciparum*–infected erythrocytes by the electroporation procedure developed by Wellem's group.[30]

The malaria group at ICGEB carried out in-depth studies to establish RNAi in *P. falciparum* and *P. berghei*.[28,29] We used dsRNAs of two cysteine protease genes of *P. falciparum* (*falcipains* 1 and 2) to evaluate their roles in the erythrocytic stages of the parasite. These genes were selected because the effects of different cysteine protease inhibitors on the morphology and biochemistry of the parasite have been well established. Moreover, recombinant *falcipain* proteins have been shown to degrade hemoglobin *in vitro*. Addition of each of the two *falcipain* dsRNAs to the *P. falciparum* culture resulted in enlargement of the food vacuoles and inhibition in the parasite growth. An additive effect was seen when the two dsRNAs were added together. In addition, each of the *falcipain* dsRNAs produced significant block in the degradation of hemoglobin in the treated parasites (Figure 18.1). All of these effects were similar to those seen upon addition of the well-known cysteine protease inhibitor E-64 to the parasite culture. A nonspecific dsRNA from an aminopeptidase-N gene of insect origin did not produce any of these effects. The inhibitory effects produced by *falcipain* dsRNAs on the parasite culture coincided with the marked reduction in the levels of endogenous mRNAs homologous to the dsRNAs in the treated parasite (Figure 18.2). We also tried to define the mechanism of RNAi in *P. falciparum*. We wondered whether the general mechanism of RNAi holds true in *P. falciparum*. Small RNA species were isolated from *P. falciparum* culture treated with labeled *falcipain* dsRNAs, and these RNA species were resolved on 15% (w/v) denaturing polyacrylamide gel. As shown

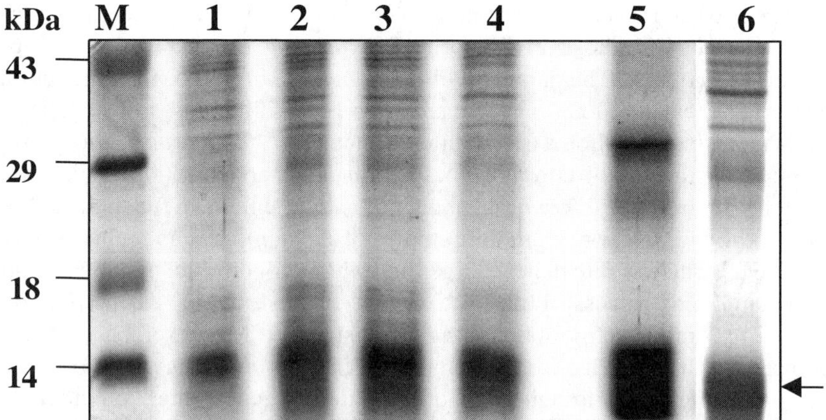

FIGURE 18.1 Inhibition of Hb degradation by *falcipain* dsRNA. Lane 1: dsRNA-untreated parasite; lane 2: Fal-1 and 2 dsRNA-treated parasite; lane 3: Fal-2 dsRNA-treated parasite; lane 4: Fal-1 dsRNA-treated parasite; lane 5: Hb control; and lane 6: E64-treated parasites.

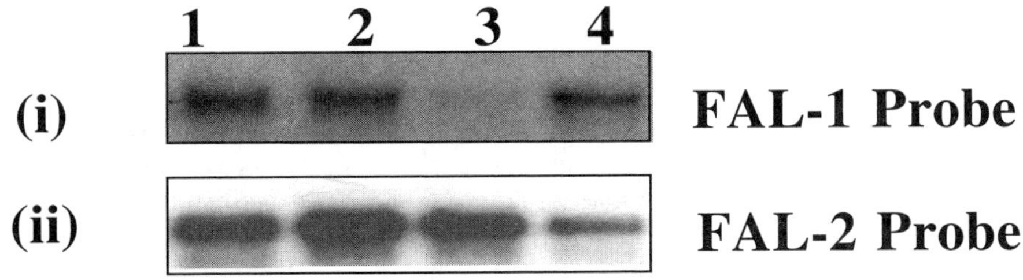

FIGURE 18.2 Northern blot showing gene-specific degradation of endogenous mRNA. Lane 1: untreated parasites; lane 2: aminopeptidase-N treated parasites; lane 3: fal-1 dsRNA-treated parasites; and lane 4: fal-2 dsRNA-treated parasites.

in Figure 18.3, the *falcipain* dsRNA–treated parasites showed the generation of ~25-nt RNA species, similar to what was observed in *Trypanosoma*. Further, the *falcipain* dsRNA–treated parasite extract was able to cleave the labeled mRNA, thereby suggesting that the mechanism of RNAi is conserved in *P. falciparum*. These experiments suggested that RNAi could be a method that can be used to silence a gene(s), at least during the blood stages of *P. falciparum*.

Here we present detailed protocols to produce and evaluate dsRNAs/siRNAs-induced protein knockdown in *P. falciparum*. We believe that these protocols will serve as incisive tools to gain useful insight into the biology of *P. falciparum* through functional genomic studies.

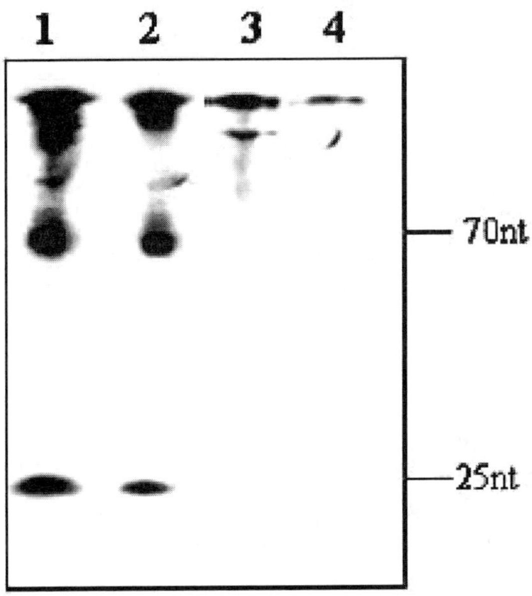

FIGURE 18.3 Generation of ~25-nt RNA upon incubation with *falcipain* dsRNA in parasite culture. Lane 1: *falcipain* 1 dsRNA-treated parasites; lane 2: *falcipain* 2 dsRNA-treated parasites; lane 3: control-labeled dsRNA from *falcipain*; and lane 4: control-labeled dsRNA from *falcipain*.

18.4 SELECTION OF dsRNA/siRNA SEQUENCES

For the selection as well as design of dsRNA/siRNA sequences, we usually follow the guidelines provided by Tuschl's group.[31] The target region should be selected from the open reading frame of a given cDNA sequence, preferably 50 to 100 nt downstream of the start codon. It has been shown that the length of dsRNA can affect the potency of RNAi; however, the precise minimum length of dsRNA to achieve strong RNA interference is not yet worked out. Tuschl's group analyzed the length requirement of dsRNA and showed that short dsRNA (< 150 bp) is less effective than large dsRNA in degrading target mRNA.[32] We recommend use of dsRNAs greater than 500 bp.

SiRNAs are selected by analyzing the target gene sequences using software available at http.//www.ambion.com, or guidelines provided by T. Tuschl (http://www.mbibpc.gwdg.de/abteilungen/100/105). A general rule is that the sequence should be AA(19N)TT (where N is any nucleotide) with a GC content of 30 to 70%. In our laboratory, we routinely use siRNA sequences with GC content of 40 to 50%. The selected siRNA sequences are screened against EST libraries, or to mRNA sequences of the respective organism as well as human to ensure that they only target the specific gene (http://www.ncbi.nlm.nih.gov/BLAST). Although the available rules for the selection of siRNAs are reliable, they do not guarantee that each of the selected siRNA will produce the desired effects. Moreover, it has been demonstrated that different siRNAs designed across the sequence of the target gene vary in their silencing efficiency.[33] We strongly recommend using two or three different siRNAs across the target gene sequence to carry out RNAi experiments.

18.5 PREPARATION OF siRNA

We generally use chemically synthesized siRNA from the supplier, Dharmacon. The supplier provides the three options for the purchase of siRNAs. We use option C, which provides siRNA duplexes that are ready to use. The other two options include single-stranded siRNAs, which should be annealed according to the manufacturers' instructions.

One can also synthesize siRNAs according to the procedure developed by Donze and Picard[34] (also see Chapter 4). This method is less expensive, and large amounts of siRNAs can be obtained. We have also used this procedure to synthesize siRNAs corresponding to GFP.

18.6 CONCLUDING REMARKS

RNAi has emerged as a powerful tool to inhibit gene expression on the level of single gene, gene families, and entire genome. RNAi also holds promise for *in vivo* genetic applications. In *P. falciparum,* use of RNAi to dissect gene function has several advantages over other methods, such as gene targeting using homologous recombination. Compared to transfection experiments, RNAi technology is technically simple and quick, and the result of gene silencing can be obtained in 2 to 3 days. For RNA interference studies, one does not require complete knowledge of the entire sequence of a gene and its flanking sequences. RNAi in *P. falciparum* also provides a tool for studying the functions of a number of genes together; e.g., all of the genes involved in a hemoglobin degradation pathway can be studied individually as well as together to compare their role. In comparison to gene-targeting studies, RNAi can be successfully applied to study the functions of redundant proteins. With the availability of a large amount of sequence information for *P. falciparum*, RNAi can be a technique of choice to investigate quickly the biological functions of novel genes. Application of RNAi to *P. falciparum* may thus become useful to advance our knowledge of its basic biology and possibly for antimalarial therapies.

PROTOCOLS

Protocol 18.1 *P. falciparum* Culture Protocols

Materials

- RPMI media powder (Invitrogen)
- Human serum (Blood Bank) (heat-denature at 56°C for 1 h)
- O^+ Blood (Blood Bank)
- Triple distilled water
- 3.6% (w/v) sodium bicarbonate
- Gentamicin (10 mg/ml) (Gibco BRL)
- Thawing solution (3.5% (w/v) NaCl sterilize by autoclaving)
- 5% (w/v) sorbitol in water (filter-sterilized)
- 0.15% (w/v) saponin in RMPI (Saponin from Sigma)
- Phosphate buffered saline, pH 7.4 (8.0 g NaCl, 0.2 g KCl, 1.14 g anhydrous Na_2HPO_4, 0.2 g NaH_2PO_4 per liter, pH to 7.4 and autoclave before use)

Experimental Procedures

Preparation of complete medium

1. Dissolve RPMI powder (10.43 g) in 1 liter of distilled water. To make complete medium, mix 100 ml of RPMI solution, 5.8 ml of 3.6% (w/v) sodium bicarbonate solution, 200 µl of Gentamicin stock, and 10 ml of human sera. Filter-sterilize. To make 4% (v/v) hematocrit, add 400 µl of RBCs in 10 ml of complete medium.

Thawing frozen parasite

2. Thaw the frozen stock by vigorous agitation at 37°C and transfer the content to a 15-ml tube.
3. Add equal volume of prewarmed (37°C) thawing solution drop by drop.
4. Spin at 1200 rpm for 2 min at 20°C and discard the supernatant by aspiration (some lysis may occur at this step).
5. Resuspend the pellet in 1 ml of thawing solution; spin and discard the supernatant by aspiration.
6. Repeat the washing steps until there is no more detectable lysis.
7. Resuspend in fresh complete medium containing 2% (v/v) RBCs and transfer into a 5-ml culture dish.

8. Keep the culture plate in an airtight gas chamber containing gas mixture (90% (v/v) N_2, 5% (v/v) CO_2, and 5% (v/v) O_2) and incubate in a 37°C incubator.
9. Once the parasite is revived, scale up the culture to 10 ml and transfer it to a culture plate. Incubate the culture as in step 9.

Maintenance of parasite culture

10. Parasite is maintained by regularly changing the old medium with fresh complete medium. Monitor the culture regularly by examining thin smears stained with Geimsa stain.
11. Subculture when the parasitemia reaches considerably high levels (5 to 10% parasitemia).

Subculture/dilution of parasite culture

12. Mix the culture (at 5% parasitemia) thoroughly and gently so that no RBCs are attached to the substratum.
13. To dilute the parasitemia from 5% to 1%, add 2 ml of the culture to 8 ml of fresh complete medium containing 4% (v/v) hematocrit (dilution resulting in 1% parasitemia).
14. Incubate at 37°C in a gas chamber filled with gas mixture as in step 9.

Synchronization of parasites by sorbitol treatment

15. Harvest the culture at about 10% parasitemia (with majority of the parasites at ring stage) by centrifuging at 1200 rpm for 5 min at RT.
16. Add 5 volumes of sorbitol solution, prewarmed at 37°C, to the cell pellet and mix gently.
17. Incubate the mix for 5 min at 37°C.
18. Centrifuge the mix at 1500 rpm for 5 min at RT.
19. Remove the supernatant completely by aspiration and add fresh, complete medium containing 2% (v/v) hematocrit. Transfer to a fresh culture dish.
20. Incubate the culture dish at 37°C as in step 9.
21. Repeat the sorbitol treatment once after one cycle of growth (~48 h) to obtain completely synchronized culture.

Isolation of parasite from infected blood by saponin lysis

22. Harvest the culture by centrifuging at 1200 rpm for 5 min at RT.
23. Add 1.5 volumes of 0.15% (w/v) saponin solution to the cell pellet, mix properly, and incubate for 10 min in ice.
24. Centrifuge the mix at 2000 rpm for 15 min at RT.
25. Collect the parasite pellet and wash with cold 1 × PBS.
26. Aliquot the washed parasite in freezing vials and store at –20°C till further use.

Protocol 18.2 Genomic DNA Isolation from *P. falciparum*

Materials

- *P. falciparum* erythrocyte culture in the trophozoitre-schizont stage
- Phosphate buffered saline, pH 7.4

- 0.15% (w/v) saponin in RPMI (Saponin from Sigma)
- Lysis buffer (10 mM Tris, 1 mM EDTA, pH 8.0, 0.5% (w/v) SDS, 100 mM NaCl, 200 μg/ml proteinase K (Sigma))
- RNase A (Ambion)
- Phenol:chloroform:isoamyl alcohol (25:24:1)
- Chloroform:isoamyl alcohol(24:1)
- 3 M Sodium acetate, pH 5.2
- Absolute ethanol

Experimental Procedure

1. Centrifuge *P. falciparum*–infected erythrocytes at 1200 rpm for 5 min at 2 to 8°C. Wash cells once in ice-cold 1 × PBS.
2. Lyse the RBCs by saponin as described previously (Protocol 18.1).
3. After saponin lysis, wash the pellet with cold 1 × PBS to remove RBC membrane debris.
4. Add 100 μl of lysis buffer for 1×10^6 pelleted cells. Incubate at 37°C for 2 h to overnight.
5. Add RNase A (100 μg/ml) and incubate at 37°C for 1 h.
6. Add equal volume of Phenol:chloroform:isoamyl alcohol, mix by inverting, and spin at 10,000 rpm for 5 min.
7. Collect the upper aqueous layer very carefully, add equal volumes of chloroform:isoamyl alcohol, mix, and spin at 10,000 rpm for 5 min. Collect the upper aqueous layer.
8. Add 0.1 volume of sodium acetate (pH 5.2) and 2.5 volumes of absolute ethanol. Incubate for 1 h to overnight at –20°C.
9. Centrifuge at 12,000 rpm for 10 min to precipitate the genomic DNA.
10. Aspirate the supernatant and wash the pellet once with 70% (v/v) ethanol.
11. Centrifuge at 12,000 rpm for 10 min, aspirate the supernatant, and air-dry the pellet.
12. Gently resuspend the pellet in 25 to 100 μl sterile distilled water.
13. Analyze the DNA on a 0.7% (w/v) agarose gel.

Protocol 18.3 Preparation of Double-Stranded RNA

Materials

- *P. falciparum* genomic DNA
- Taq DNA polymerase (Perkin Elmer Cetus)
- Pfu DNA polymerase (Stratagene)
- Plasmid pGEM-T (Promega)
- HiScribe RNAi transcription kit (New England Biolabs)
- DNase1 and T7/SP6 RNA polymerase(Promega)
- RNasin (MBI) and RNaseT1 (Invitrogen)
- 3 M sodium acetate, pH 512
- Ethanol

Experimental Procedures

1. Amplify the gene fragment of interest by PCR using *P. falciparum* genomic DNA, and clone it into pGEM-T or Litmus 28i vectors (included in HiScribe Kit) using standard cloning procedures.
2. Linearize the plasmid containing the gene fragment in two sets using the restriction sites present at the 5' and 3' ends to the cloned fragment.
3. Generate sense and antisense RNA transcripts using these two linearized plasmid DNAs with appropriate RNA polymerases following the reaction conditions given below:

DNA	X µl (5 µg)
5 × transcription buffer	20 µl
rNTP mix (25 mM each)	8 µl
T7/SP6 Polymerase	3 µl (60 U)
RNAsin	1 µl (40 U)
DEPC water	X µl (make up to 100 µl)

4. Incubate the reaction at 37°C for 2 h.
5. Freeze both the reactions at –20°C for 15 min to inactivate the enzymes.
6. Mix sense and antisense reaction mixtures, spin, and incubate at 65°C for 40 min.
7. Slowly cool to room temperature to anneal the two ssRNAs.
8. Incubate the reaction with 10 U of DNase1 for 15 min at 37ºC.
9. Extract the dsRNA with phenol:chloroform:isoamyl alcohol (25:24:1) and then with chloroform:isoamyl alcohol (24:1).
10. Precipitate the dsRNA by adding 1/10 volume of 3 M sodium acetate and 2.5 volume of ethanol at –20°C overnight.
11. Centrifuge at 13,000 rpm and wash the pellet with 70% (v/v) ethanol.
12. Dry the pellet by leaving the tube open in laminar flow hood.
13. After complete drying, resuspend the dsRNA pellet in sterile DEPC water.
14. Analyze dsRNA on agarose gel. If there is any contamination of residual ssRNAs, treat the samples with RNAseT1; otherwise, proceed with RNAi reaction.

Protocol 18.4 Parasite Culture for RNAi

A widely used method to introduce DNA/RNA into the *P. falciparum* is electroporation, and this has been used by McConkey's and Barik's groups. However, one of the methods to introduce dsRNA in *C. elegans* is simple soaking of the worms in dsRNA containing solution,[35] and we, therefore, intended to develop an equally simple procedure for *P. falciparum,* which is described below.

Materials

- Material required for cell culture and synchronization (see Protocol 18.1)
- dsRNA (prepared as explained in Protocol 18.3) or siRNA (obtained from a commercial supplier such as Dharmacon or Qiagen)

Experimental Procedures

1. Synchronize *P. falciparum* culture as explained in Protocol 18.1.
2. Adjust the synchronized culture (late ring stage) to 5% hematocrit with 1% infected red blood cells and dispense 1-ml aliquots of this culture into 24-well culture plate. For small-scale study (200 μl), one can also use 96-well culture plates.
3. Treat the culture with 50 to 200 μg/ml of specific or control dsRNA/siRNA in serum-free medium (complete medium except human sera) for 30 min with intermittent shaking, and then add serum to a final concentration of 10% (v/v).
4. Maintain the parasite further for 24 h, 48 h, or 56 h, depending upon the type of experiment.

Protocol 18.5 Examination of Morphological Effects on Parasite by Gene-Silencing Activity

Materials

dsRNA/siRNA treated parasite cultures
Giemsa stain (Sigma)
Inverted microscope

Experimental Procedures

1. Smear aliquots from the treated as well as control parasite cultures and stain with Giemsa stain every 24 h.
2. Count number of infected and uninfected RBCs in different fields of the microscope (minimum of 2000 erythrocytes in all) and calculate the % parasitemia for each culture.
3. Look for the morphological abnormalities, if any, in the treated parasites. We observe food vacuole enlargement in the *falcipain* dsRNA-treated parasites.
4. After 48 h, one should also compare the % ring parasitemia for each culture of the treated samples with that of untreated controls. To do this, make a thin smear of the culture and stain with Giemsa stain. Count the number of parasites at ring stage and uninfected RBCs in different fields of the microscope (minimum of 2000 erythrocytes in all), and calculate the % ring parasitemia.

Protocol 18.6 Parasite Viability Assay — [^{3}H] Hypoxanthine Uptake Assay

[^{3}H] hypoxanthine incorporation assay provides a quantitative means for the measurement of inhibitors or drug effects on parasite growth.[36] Our group and other workers[26,28] have used this assay to assess the effect of different dsRNAs/siRNAs on parasite growth. The following protocol has been optimized for the [^{3}H] hypoxanthine uptake assay.

Materials

- Synchronized culture
- [^{3}H] hypoxanthine (Amersham Biosciences)

- Scintillation counter (Beckman)
- Glass-fiber filter (Beckman)

Experimental Procedures

1. Dispense 100 μl of synchronized cultures (0.8 to 1% parasitemia) at late ring stage in 96-well culture plate (in triplicates) and treat with required amounts of dsRNA/siRNA (Protocol 18.3).
2. Maintain the cultures for 24 h and then add [^{3}H] hypoxanthine (1.2 μCi per well).
3. Maintain the culture for additional 24 h, and then freeze the plate to lyse the infected RBCs.
4. Harvest the lysed culture on a glass-fiber filter, wash the filter with water and ethanol, and count it on scintillation counter.
5. Compare the uptake of [^{3}H] hypoxanthine incorporation in control with dsRNA/siRNA treated parasites.

Protocol 18.7 Northern Blotting for the Analysis of Endogenous RNA Transcripts

To ascertain that the effects seen in dsRNA/siRNA–treated parasite cultures are due to RNAi, it is necessary to measure the mRNA levels of the cognate gene against which dsRNA/siRNA has been used. This is done usually by RT-PCR or by northern blot analysis. If desired, one can also analyze the generation of siRNAs in the dsRNA/siRNA treated parasites. We assess the RNAi effects by northern blot analysis as well as by analyzing the generation of 25-nt RNA species using the following protocols.

Materials

- dsRNA/siRNA
- Parasite culture
- RNA extraction kit (Qiagen Ltd.)
- Agarose gel electrophoresis buffers and apparatus for RNA gels
- Hybond N^+ (Amersham Biosciences)
- Random primer labeling kit (Invitrogen)

Experimental Procedures

1. Treat 3 to 4 ml of parasite culture with dsRNA/siRNA (Protocol 18.3).
2. After treatment with dsRNA/siRNA for 48 h, extract total RNA from the parasite pellet. We use Qiagen RNA isolation kit to extract total RNA.
3. For northern blot analysis, we follow the protocol described by Kyes et al.[37] Briefly, separate the total RNA (2 to 5 μg) from the treated as well as untreated parasite (control) on 1.2% (w/v) agarose gel, transfer it onto nylon membrane (Hybond N^+) by standard procedures, and hybridize with the random prime-radiolabelled DNA probe.
4. Expose the blot to x-ray film and develop the autoradiogram. We usually develop it after 48 to 96 h, depending upon the signal intensity.

Protocol 18.8 Evaluation of Generation of Small Interfering RNAs

For the evaluation of small interfering RNAs, one can use either labeled or unlabeled dsRNAs. When using unlabeled dsRNAs, northern blot analysis should be done to detect specific siRNAs. We have used both of these procedures to detect siRNA species and usually follow the protocol described by Hamilton and Baulcombe, 1999.[8]

Materials

- Total parasite RNA
- PEG/NaCl solution [5% (w/v) polyethylene glycol 8000 in 0.5 M NaCl)
- Ethanol
- Standard denaturing polyacrylamide gel apparatus and solutions

Experimental Procedures

1. Precipitate high-molecular-weight RNA species from total RNA with PEG/NaCl solution.
2. Transfer the supernatant into a fresh tube containing low-molecular-weight RNA. The pellet containing high-molecular-weight RNA can be used for northern blot analysis to check for the reduction in specific mRNA.
3. Ethanol precipitate low-molecular-weight RNA from the supernatant of PEG/NaCl at –20°C for at least 2 hours and pellet down at 20,000 × g for 15 min. This pellet mainly comprises tRNA, rRNA, and other small RNA species.
4. Dissolve the pellet in formamide and separate low-molecular-weight RNAs on 15% (w/v) polyacrylamide gel made with 7 M urea and 0.5X Tris-Borate-EDTA buffer.
5. If labeled dsRNA was used for the experiment, gel could be directly dried and exposed to x-ray film.
6. To detect unlabeled siRNA species, northern blot analysis should be done.
7. For the northern blot analysis of siRNA species, ^{32}P-labeled RNA probe transcribed *in vitro* from the plasmid templates is preferred.

Protocol 18.9 Detection of Reduction in Target Protein by Western Blotting

Materials

- SDS-PAGE Mini-PROTEAN III system (BioRad)
- 10 × PBS, pH 7.4
- Hybond nitrocellulose membrane (Amersham)
- Blot transfer buffer (25 mM Tris, 192 mM glycine, and 20% (v/v) methanol)
- Ponceau S stain (Sigma)
- Saponin (Sigma)
- PBST (1 × PBS + 0.2 % (v/v) Tween-20)
- Fat-free milk powder
- Reducing SDS-PAGE sample buffer (BioRad)
- ECL detection system (Amersham Biosciences)

Experimental Procedures

1. Collect the infected RBCs from dsRNA/siRNA treated, as well as from untreated, culture by centrifugation at 1200 rpm.
2. Lyse RBCs with 0.15% (w/v) saponin for 10 min on ice.
3. Wash the pellet in PBS twice and dissolve it in 25 to 50 μl reducing SDS-PAGE sample buffer.
4. Separate the parasite extracts in 10% SDS-PAGE and transfer the proteins onto nitrocellulose membrane using transfer buffer. Verify the protein transfer by staining the membrane with 0.5% (v/v) Ponceau S in water and then wash with 1 × PBS.
5. Incubate the membrane in blocking solution 5% (w/v) milk powder in PBST for at least 2 hours at 37°C or overnight at 4°C.
6. Wash the blot 2 × 5 min in PBST.
7. Incubate the blot with primary antibody diluted appropriately in 1% (w/v) milk powder in PBST for 1 h at 37°C with gentle shaking.
8. Wash the blot 3 × 10 min in PBST.
9. Incubate the blot with secondary antibody conjugated to HRP and diluted appropriately in 1% (v/v) milk powder/PBS/Tween-20 with gentle shaking.
10. Repeat step 8.
11. Perform ECL detection according to the protocol described by the supplier.

Protocol 18.10 Immunofluorescence Detection of Protein Knockdowns

Materials

- Standard microscope glass slides
- Acetone and methanol
- 10 × PBS, pH 7.4
- 10% (v/v) fetal calf serum in PBS (Invitrogen)
- DAPI stain (sigma)
- Antifade reagent (BioRad)

Experimental Procedures

1. Smear the parasitized RBCs from treated as well as untreated samples onto glass slides and allow them to dry.
2. Fix the cells in acetone for 5 min at RT and then in methanol for 10 sec at 4°C.
3. Wash the fixed slides in PBS and block them in 10% fetal calf serum in PBS at 37°C for 30 min in a humid chamber.
4. Incubate the slides in primary antibodies diluted appropriately in blocking solution for 1 h at 37°C.
5. Wash slides in 1 × PBS for 1 h at RT with intermittent shaking.
6. Incubate slides with appropriately diluted, fluorescently labeled secondary antibodies for 45 min at 37°C.
7. Repeat step 5.

8. Stain the cell nuclei with DAPI stain for 15 min at RT.
9. Repeat the washing step 5.
10. Mount the slides using antifade reagent and examine the immunofluorescence.

Technical notes. One can also perform different biochemical assays to monitor the protein knockdown. In the study in which we knocked down the expression of cysteine protease genes involved in hemoglobin degradation, we compared the hemoglobin accumulation in the treated and untreated parasites by running the parasite lysates on SDS-PAGE.

ACKNOWLEDGMENTS

The authors thank Amit Kumar, Dr. Raj K. Bhatnagar, and Dr. Sunil K. Mukherjee for their valuable comments on the preparation of this chapter. P.V.N.D. and M.J.H. were supported by a CSIR research fellowship.

REFERENCES

1. Dykxhoorn, D.M., Novina, C.D., and Sharp, P.A., Killing the messenger: Short RNAs that silence gene expression, *Nat. Rev. Cell Biol.*, 4, 457, 2003.
2. Fire, A. et al., Potent and specific genetic interference by double-stranded RNA in *Caenorhabiditis elegans*, *Nature*, 391, 806, 1998.
3. Jorgensen, R., Altered gene expression in plants due to *trans* interactions between homologous genes, *Trends Biotechnol.*, 8, 340, 1990.
4. Romano, N. and Macino, G., Quelling: transient inactivation of gene expression in *Neurospora crassa* by transformation with homologous sequences, *Mol. Microbiol.*, 6, 3343, 1992.
5. Manche, L. et al., Interactions between double-stranded RNA regulators and the protein kinase DAI, *Mol. Cell. Biol.*, 12, 5238, 1992.
6. Elbashir, S.M., Lendeckel, W., and Tuschl, T., Duplexes of 21- and 22-nucleotide RNAs mediate RNA interference in cultured mammalian cells, *Nature*, 411, 494, 2001.
7. Fjose, A. et al., RNA interference: mechanism and applications, *Biotech. Ann. Rev.*, 7, 31, 2001.
8. Hamilton, A.J. and Baulcombe, D.C., A species of small antisense RNA in post-transcriptional gene silencing in plants, *Science*, 286, 950, 1999.
9. Bernstein, E. et al., Role for a bidentate ribonuclease in the initiation step of RNA interference, *Nature*, 409, 263, 2001.
10. Hammond, S.M. et al., An RNA-directed nuclease mediates post-transcriptional gene silencing in *Drosophila* cells, *Nature*, 404, 293, 2000.
11. Zamore, P. et al., RNAi-double-stranded RNA directs the ATP-dependent cleavage of mRNA at 21–23 nucleotide intervals, *Cell*, 101, 25, 2000.
12. Hutvagner, G. and Zamore, P.D., RNAi: nature abhors a double-strand, *Curr. Opin. Genet Dev.*, 12, 225, 2002.

13. Cogoni, C. and Macino, G., Gene silencing in *Neurospora crassa* requires a protein homologous to RNA-dependent RNA polymerase, *Nature*, 399, 166, 1999.
14. Agrawal, N. et al., RNA interference: biology, features, mechanism and application, *Mol. Micro. Rev.*, 67, 657–685, 2003.
15. Tuschl, T., RNA interference and small interfering RNAs, *Chembiochem,* 2, 239, 2001.
16. Pal-Bhadra, M., Bhadra, U., and Birchler, J.A., RNAi related mechanisms effect both transcriptional and posttranscriptional transgene silencing in *Drosophila*, *Mol. Cell*, 9, 315, 2002.
17. Fischle, W., Wang, Y., and Allis, C.D., Histone and chromatin cross-talk, *Curr. Opin. Cell Biol.*, 15, 172, 2003.
18. McManus, M.T., and Sharp, P.A., Gene silencing in mammals by small interfering RNAs, *Nat. Rev. Genet.*, 3, 737, 2002.
19. Paddison, P.J. and Hannon, G.J., RNA interference: the new somatic cell genetics? *Cancer Cell*, 2, 17, 2002.
20. Ullu, E., and Tschudi, C., RNAi-applications in parasitology, *New Tech. Life. Sci.*, 6, 43, 2000.
21. Robinson, K.A., and Beverley, S.M., Improvements in transfection efficiency and tests for RNA interference (RNAi) approaches in the protozoan parasite *Leishmania, Mol. Biochem. Parasitol.*, 128, 217, 2003.
22. Ngo, H. et al., Double-stranded RNA induces mRNA degradation in *Trypanosoma brucei, Proc. Natl. Acad. Sci. U.S.A.*, 95, 14687, 1998.
23. Cottrell, T.R. and Doering, TL. Silence of the strands: RNA interference in eukaryotic pathogens, *Trends Microbiol.*, 11, 37, 2003.
24. Djikeng, A. et al., RNA interfernce in *Trypanosoma brucei:* Cloning of small interfering RNAs provides evidence for retroposon-derived 24–26 nucleotide RNAs, *RNA*, 7, 1522, 2001.
25. Gardner, M.J. et al., Genome sequence of the human malaria parasite *Plasmodium falciparum*, *Nature*, 419, 498, 2002.
26. McRobert, L. and McConkey, G.A., RNA interference (RNAi) inhibits growth of *Plasmodium falciparum, Mol. Biochem. Parasitol.*, 119, 273, 2002.
27. Kumar, R. et al., Characterization and expression of a PP1 serine/threonine protein phosphatase (PfPP1) from the malarial parasite, *Plasmodium falciparum:* demonstration of its essential role using RNA Interference, *Malaria J.*, 1, 5, 2002.
28. Malhotra, P. et al., Double-stranded RNA-mediated gene silencing of cysteine proteases (falcipain-1 &-2) of *Plasmodium falciparum, Mol. Microbiol.*, 45, 1245, 2002.
29. Mohmmed, A. et al., *In vivo* gene silencing in *Plasmodium berghei* — a mouse malaria model, *Biochem. Biophys. Res. Com.*, 309, 506, 2003.
30. Wu, Y. et al., Transfection of *Plasmodium falciparum* strain within human blood cells, *Proc. Natl. Acad. Sci. U.S.A.*, 92, 973, 1995.
31. Tuschl, T. et al., The siRNA user guide, 12 July 2002 (http:/www.mpibpc.gwdg.de/abteilungen/100/105/sirna.html)
32. Elbashir, S.M., Lendeckel, W., and Tuschl,T., RNA interference is mediated by 21–22-nucleotide RNAs, *Genes Dev.*, 15, 188, 2001.

33. Holen, T. et al., Positional effects of short interfering RNAs targeting the human coagulation trigger tissue factor, *Nucleic Acids Res.*, 30, 1757, 2002.
34. Donze, O. and Picard, D., RNA interference in mammalian cells using RNA synthesized with T7RNA polymerase, *Nucleic Acids Res.*, 30, e46, 2002.
35. Tabara, H., Grishok, C., and Mello, C.C., RNAi in *C. elegans*: soaking in the genome sequence, *Science*, 282, 430, 1998.
36. Rosenthal, P.J. et al., Antimalarial effects of vinyl sulfone cysteine protease inhibitor, *Antimicrob. Agents Chemother.*, 40, 1600, 1996.
37. Kyes, S., Pinches, R., and Newbold, C., A simple RNA analysis method shows var and rif multigen family expression patterns in Plasmodium falciparum, *Mol. Biochem. Parasitol.*, 105, 311, 2000.

19 RNAi in Transgenic Animal Models

Hidetoshi Hasuwa and Masaru Okabe

CONTENTS

19.1 INTRODUCTION

With completion of the "genome project" in human and mouse, we have entered the postgenome era, centered around the analysis of 30,000 to 40,000 gene functions. At present, ES cells and homologous recombination serve as powerful tools for many researchers aiming to elucidate the function of target genes in live animals. However, such methods are slow and cumbersome and are limited to mouse because of the technical difficulty in establishing ES cells in other species. Any improved technique should meet the following criteria: (a) simplicity of method to facilitate analysis of many genes in a short period of time, and (b) applicability to species other than mouse. RNAi could thus serve as an alternative, or supplemental, method to homologous recombination.

The first reports of successful RNAi in mammals were published by Svoboda et al.[1] and Wianny and Zernicka-Goetz.[2] They used microinjection into mouse oocytes of double-stranded RNA (dsRNA) homologous to the target genes and demonstrated target-specific inhibition of protein translation. However, many researchers assumed that these successes represented exceptions for RNAi in mammals since dsRNA evoked an interferon response in somatic cells through activation of PKR (RNA-dependent protein kinase) and 2', 5'-oligoadenylate.[3] To our surprise, the introduction of small duplexes of synthetic 21- to 23-nt RNAs (siRNA) into mammalian cells was found to have a gene-specific silencing function without interferon-related side effects.[4] This represented a major step forward in performing RNAi in mammals. However, the synthetic siRNAs were effective for only a few days in mammalian cells posttransfection. To overcome this limitation, a number of plasmid vectors were developed for long-term delivery of siRNAs in cells.[5–11] Thus, the application of this system, in combination with the transgenic animal techniques, would be expected to result in a continuous supply of siRNAs. The integrated transgene is likely to have the RNAi activity not only through the lifetime of the transgenic animal but also into the subsequent generations.

0-8493-2141-7/05/$0.00 + $.50

We chose green fluorescent protein (EGFP) as a model target in our studies to show RNAi activity in animal models and developed the "green mice" and "green rats" characterized by EGFP expression throughout the body.[12,13] Here we describe successful silencing of the ubiquitously expressed EGFP *in vivo* by injecting vectors that consisted of RNA polymerase III (pol III) promoter (H1 or tRNA), designed to express 21-nt EGFP-specific siRNAs.[14]

Our success of the transgene-based RNAi system suggests that the method may be used as an alternative to gene knockout. Moreover, gene silencing in rats[14] indicates that the method may be applicable to animals for which homologous recombination is not possible due to the unavailability of ES cell lines. As shown in methods below, it was possible to control the translation of certain proteins through a transgenically introduced RNAi system in live animals. This implies that the transgenic RNAi system could be used to modify gene function in the body, or only in targeted cells or organs, by choosing vectors (to control expression) or gene delivery systems to limit the target cells or organs. In any case, successful control of a specific protein translation by RNAi would open the door not only in basic research in studying gene function but also for a clinical gene therapy.

PROTOCOLS

PROTOCOL 19.1 CONSTRUCTION OF RNAI VECTOR

Materials

- pH1 vector[14] (Figure 19.1a)
- ptRNA vector (The tRNA^{val} promoter was amplified from piGENE tRNA Pur vector (Clontech) and cloned into the pBluescript II SK+ (Stratagene) (Figure 19.1b)
- DNA oligos (C-18 cartridge purified grade, Figure 19.2)

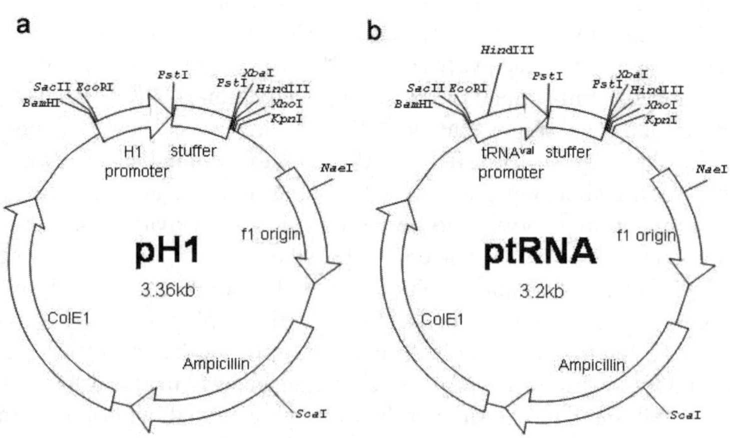

FIGURE 19.1 Map of pH1 and ptRNA vector. The annealed oligos are inserted into the *Pst* I site and *Xba* I , *Xho* I, or *Kpn* I site.

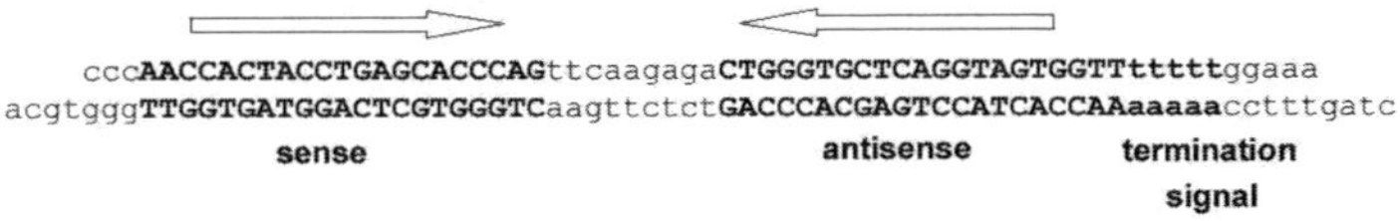

FIGURE 19.2 Sequence of oligos encoding the siRNA and termination signal. Sense and antisense sequence is shown in bold upper case, and termination signal is shown in bold lower case.

- pRed[14]
- *E. coli* strain JM109, SURE2 (Stratagene)

Experimental Procedures

Construction of pH1/siRNA$_{EGFP}$ *and ptRNA/siRNA*$_{EGFP}$

1. Digest 5 μg of pH1 or ptRNA vector with *Xba* I and *Pst* I.
2. Electrophorese the digested vector and extract from the gel.
3. Anneal the synthesized siRNA oligos. Mix 10 nM sense and antisense oligos, heat at 95°C for 5 min, and let stand at room temperature overnight.
4. Ligate the annealed oligos and digested pH1 or ptRNA vector.
5. Transform into the *E. coli* strain JM109 or SURE2 following the supplier's protocol.
6. Prepare the plasmid from the transformed colonies and sequence to confirm the inserted oligos.

Construction of pGtoR

To detect the RNAi transgene at early embryo stages in living mice, a maker gene (Red fluorescence gene: *HcRed1*) was ligated to RNAi plasmid.

7. Digest the pH1/siRNAEGFP and pRed with *Bam*H I and *Hin*d III.
8. Electrophorese both digested vectors and gel-extract about 300 bp RNAi fragment and 5 kb HcRed1 fragment.
9. Ligate the RNAi and HcRed1 gene fragments.
10. Transform into the *E. coli* strain JM109 or SURE2 according to the supplier's protocol.
11. Prepare the plasmid and analyze.

Technical notes: The double-stranded sequences are not stable in *E. coli*. They tend to be eliminated during the cultivation. Care should be taken regarding overgrowth. In some cases, it may not be possible to obtain clear sequence data. Sequencing from both ends of inserted oligos was sometimes effective in solving this problem. In some cases, the usage of CUGA Sequencing Kit (NIPPON GENETECH) was effective.

Protocol 19.2 Purification of Transgene

Materials

- SeaKem Gold Agarose (TaKaRa 50150)
- QIAEX II Gel Extraction kit (Qiagen 20021)

- Phenol/chloroform/isoamyl alcohol; prepare at 25:24:1.
- SUPREC-01 (TaKaRa 9040)

Experimental Procedures

1. Digest transgene with restriction enzymes (pGtoR: *Bam*H I and *Sal* I, ptRNA/siRNA$_{EGFP}$: *Sca* I).
2. Apply 30 to 50 µg of the DNA on agarose gel.
3. Recover the band and purify with QIAEX II kit.
4. Dissolve in 400 µl of 10 mM Tris-HCI, pH 8.5.
5. Extract with phenol-chloroform and ethanol precipitate.
6. Dissolve in TE buffer (10 µM Tris-HCl, 0.1 mM EDTA (pH 8.0)).
7. Filter with SUPREC-01.
8. Measure DNA concentration.
9. Adjust concentration to 0.54 µg × MW (kb) of DNA. (If the fragment is over 5 kb, use 2.7 µg/ml as an injection concentration.)

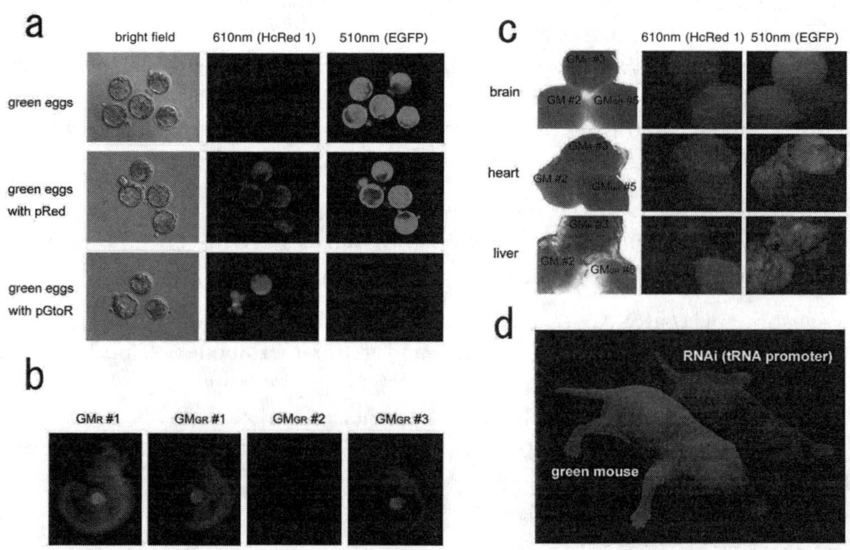

COLOR FIGURE 19.3* **a.** Green eggs obtained from wild-type females mated with homozygous "green" males were injected with pGtoR, cultured in kSOM for 3 days, and observed for HcRed1 (middle column) and EGFP (right column) expression by fluorescence microscopy. "Green" blastocysts with no additional transgene (top row), "green" blastocysts with the pRed transgene (middle row), and "silenced" blastocysts with the pGtoR transgene (bottom row). **b.** Day 10.5 embryos were recovered and their fluorescence observed (from the left, "green" embryo with pRed GM_R #1, with pGtoR GM_{GR} #1, 2, and 3, respectively). **c.** Various organs were removed from newborn founder mice (GM #2, GM_R #3, GM_{GR} #5) containing the genetic backgrounds indicated above and photographed under normal lighting (left), 510 nm (right), and 610 nm (center) band pass filter. The exposure time for each panel was adjusted to obtain optimal images. **d.** A newborn "green" mouse and a "silenced green" mouse with the ptRNA/siRNA$_{EGFP}$ transgene.

---

* Color figures follow page 204.

Technical Notes: The agar used for DNA purification should be a highly purified grade and should be used at a low concentration as much as possible to minimize the contamination of agar. If the recovery of the DNA was less than 2 μg in step 8, we redid the purification step from scratch.

Protocol 19.3 Preparation of Eggs

Materials

- Mouse (female: BDF1, male: green mouse: C57BL/6TgN(act-EGFP)OsbC14-Y01-FM131[12])
- Rat (female: SD, male: green rat: SD TgN(act-EGFP)Osb4[13])
- Pregnant mare serum gonadotropin (PMSG, Sigma G-4877) and human chorionic gonadotropin (HCG, Sigma, C-1063) for superovulation
- Hyaluronidase type IV-S (Sigma H-4272) for removal of cumullus cells
- Tweezers
- Egg-handling pipette (finely drawn capillary tube (FUNAKOSHI 1-40-7500) with mouth pipette (Sigma A5177))
- Dissection microscope (OLMPUS SZX12)
- CO_2 incubator (37°C, 5%CO_2, Asahi 4020)
- Modified kSOM medium[15] (for culture of mouse eggs)
- FHM medium[16] (for collection of the mouse eggs); mix the following per 500 ml Modified HER medium[17] (for rat eggs); mix the following per 500 ml:

Reagent	Source	Modified kSOM	FHM
NaCl	Sigma S-5886	2.775 g	2.775 g
KCl	Sigma P-5405	0.093 g	0.093 g
KH_2PO_4	Sigma P-5655	0.0238 g	0.0238 g
$MgSO_4$	Sigma M-2643	0.012 g	0.012 g
Sodium lactate (60%, w/v)	Sigma L-7900	1.18 ml	1.18 ml
Sodium pyruvate	Sigma P-4562	0.011 g	0.011 g
Glucose	Sigma G-6152	0.018 g	0.018 g
Glutamine (200 mM stock)	Sigma G-7513	2.5 ml	2.5 ml
BSA	Sigma A-3311	0.5 g	0.5 g
EDTA.4Na (10 mM stock)	Sigma E-6511	0.5 ml	0.5 ml
$NaHCO_3$	Sigma S-5761	1.05 g	0.168 g
HEPES	Sigma H-6147	—	2.38 g
$CaCl_2/2H_2O$	Sigma C-7902	0.126 g	0.126 g
Essential AA	GibcoBRL 11130-051	10 ml	10 ml
Nonessential AA	GibcoBRL 11140-050	5 ml	5 ml
Penicillin-Streptomycin	GibcoBRL 15140-148	2.5 ml	2.5 ml
Phenol Red 0.5%	Dai-ichi 1.3 ml × 10	1.0 ml	1.0 ml
1.0N NaOH	Nacalai tesque 315-11	—	Adjust pH 7.4
Water	GibcoBRL 15230-162	Adjust 500 ml	Adjust 500 ml

i. Solution A, HAM-F12 (NISSUI Pharmaceutical Co., Ltd., 05910), 3.18 g/300 ml
ii. Solution B, RPMI1640 (NISSUI Pharmaceutical Co., Ltd., 05911), 1.04 g/100 ml

iii. Solution C, mix the following components per 100 ml:

HANKS (NISSUI Pharmaceutical Co., Ltd., 05905)	0.98 g
EAGLE MEM (NISSUI Pharmaceutical Co., Ltd., 05904)	0.088 g
$NaHCO_3$ (Sigma S-5761)	0.78 g
Penicillin G (MEIJI SEIKA KAISHA, Ltd.)	0.0375 g
Streptomycin (MEIJI SEIKA KAISHA, Ltd.)	0.025 g

Inject mixed gas (5% CO_2, 5% O_2, 90% N_2) into solutions A, B, and C for 5 min, respectively. Mix solutions A, B, and C (total 500 ml). Add 125 ml FCS (final 20% (v/v) FCS).

These medium were filtrated with 0.2 μm filter, divided into 10 ml and stored in –30°C freezer.

(*N.B.* These "green" animals are available to the scientific community. (http://kumikae01.gen-info.osaka-u.ac.jp/tg/greenmouse.cfm)

Experimental Procedures

Mouse

1. Inject 5 IU of PMSG and 5 IU of HCG (i.p.) at 48-h intervals to 8-week-old female BDF1 mice and mate with green male mice.
2. Sacrifice the mice at 22 to 24 h after HCG injection.
3. Collect eggs from ampullar portion of the oviduct into FHM medium.
4. Treat eggs with hyaluronidase for 5 min, wash, and cultivate in modified kSOM medium.

Rat

5. Inject 25 IU of PMSG and 25 IU of HCG (i.p.) at 48-h intervals to 4-week-old female SD rats and mate with green male rats.
6. Sacrifice the rats at 22 to 24 h after HCG injection.
7. Collect eggs from ampullar portion of the oviduct into modified HER medium.
8. Treat eggs with hyaluronidase for 5 min, wash, and cultivate in modified HER medium.

Protocol 19.4 Microinjection of RNAi Transgene

Materials

- Inverted microscope (OLYMPUS IX70) with micromanipulator (Narishige)
- Injector (Eppendorf, FEMT-JET 5247)
- Holding and injection needle: Capillary puller (Sutter P-97/IVF) with capillaries (Sutter B100-75-10 and BF100-78-10)
- Microforge (Technical Products International, MF1)

Experimental Procedures

1. Place 20 eggs in 20 μl of FHM medium drops prepared on a 60-mm dish and cover with paraffin oil (use lid of the dish for easier manipulation).
2. Set under a micromanipulator.

3. Hold the eggs with holding pipette by suctioning.
4. Focus on the rim of male pronuclei and inject about 1 to 2 pl transgene into pronuclei.
5. Transfer eggs to kSOM or HER medium after injection and cultivate until transplantation into oviducts of pseudopregnant mothers.

Technical notes. The size of the pipette tip is critical for survival of injected eggs. We break the tip by making contact with the holding pipette. The size can be estimated by the injection pressure. We use a pipette that works well with pressures from 700 to 1000 psi.

(*N.B.* A protocol for embryo transfer is beyond the scope of this chapter; see reference [18], pages 173–178.)

Protocol 19.5 Detection of siRNAs

Materials

- TRIzol Reagent (Invitrogen, #15596-026)
- Mixer Mill MM300 (Qiagen, #85100)
- Zirconium balls, ϕ 5mm (AS ONE, #YTZ-5)
- DEPC-treated water
- XCell SureLock™ Mini-Cell with XCell II™ Blot Module (Invitrogen, #EI0002)
- 15% TBE-Urea precast gel, 1.0 mm 12 well (Invitrogen, #68852)
- TBE-Urea Sample Buffer 2 × (Invitrogen, #LC6876)
- TBE buffer (Invitrogen, #LC6675)
- Hybond-N+ (Amersham Biosciences, #RPN303B)
- DNA oligonucleotides (sense and antisense sequence of the siRNA)
- ^{32}P rATP (Amersham Bioscience, #AA0018)
- MEGALABEL™ (TAKARA, #6070)
- MicroSpin G-25 columns (Amersham Bioscience, #27-5325-01)
- PerfectHyb Hybridization Solution (TOYOBO HYB-101)
- 2 × SSC (0.3 M NaCl, 30 mM sodium citrate) with 0.1% (w/v) SDS
- BAS 1500 system (FUJI FILM) (or an alternative)

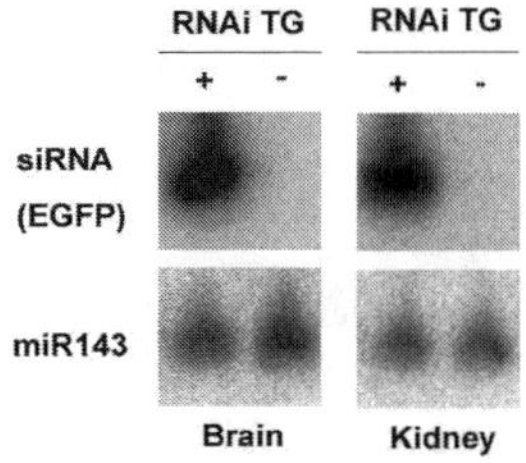

FIGURE 19.4 Expression of siRNA. Brain and kidney were harvested from ptRNA/siRNA$_{EGFP}$ mouse and control (nonTG mouse). The ^{32}P-labelled EGFP siRNA sense probe and miRNA143 antisense probe were used for northern blotting.

Experimental Procedures

RNA extraction

1. Collect the tissue in a 2-ml Eppendorf tube.
2. Add 1 ml TRIzol and a zirconium ball.
3. Set the tube into Mixer Mill MM300 and shake for 2 min.
4. Add 200 μl chloroform and mix by vortex.
5. Centrifuge at 15,000 rpm for 15 min.
6. Recover the aqueous fractions (~400 μl) and add the 400 μl isopropanol.
7. Centrifuge at 15,000 rpm for 15 min.
8. Rinse and air-dry.
9. Dissolve in an appropriate volume of DEPC-treated water.

Electrophoresis and blotting

10. Mix 30 μg total RNA (~10 μl) with equal volume of TBE-Urea Sample Buffer (2 ×).
11. Heat at 95°C for 2 min and immediately place on ice for 5 min.
12. Load 30 μg total RNA sample (~20 μl) into the wells.
13. Resolve at ~180 V for 60 min.
14. Stain with ethidium bromide and capture image.
15. Electro-blot to Hybond-N+ membrane by XCell II™ Blot Module (~30 V, 90 min) following the supplier's information.

Oligonucleotide labeling

16. Mix the following reaction components.

DNA oligo (10 pmol/μl)	1 μl
10 × phosphorylation buffer	1 μl
[γ-^{32}P] ATP	5 μl
T4 polynucleotide kinase	1.5 μl
Water	2 μl
Total	10 μl

18. Incubate at 37°C for 30 min.
19. Desalt through a G25 column as directed by the supplier. Save the flow through ^{32}P-labeled probe.

Hybridization and wash

20. Prehybridize blotted membrane with 10 ml PerfectHyb Hybridization Solution at 50°C for 30 minutes.
21. Add ^{32}P-labeled probe onto prehybridized membrane.
22. Hybridize at 50°C 3h ~ overnight.
23. Wash the hybridized membrane with 2 × SSC 0.1% (w/v) SDS at 50°C for 30 min.
24. Expose the membrane to IP plate for 1 to 3 days.
25. Read the IP plate as directed by the manufacturer.

Technical note. In some cases, RNAi transgene was silenced for an unknown reason. It is best to check both the transgene (genomic PCR or southern hybridization) and siRNA (northern hybridization).

REFERENCES

1. Svoboda, P. et al., Selective reduction of dormant maternal mRNAs in mouse oocytes by RNA interference, *Development,* 127, 4147, 2000.
2. Wianny, F. and Zernicka-Goetz, M., Specific interference with gene function by double-stranded RNA in early mouse development, *Nat. Cell Biol.,* 2, 70, 2000.
3. George R. Stark. et al., How cells respond to interferons. *Annu. Rev. Biochem.,* 67, 227, 1998.
4. Elbashir, S. M. et al., Duplexes of 21-nucleotide RNAs mediate RNA interference in cultured mammalian cells, *Nature,* 411, 494, 2001.
5. Brummelkamp, T.R., Bernards, R., and Agami, R., A system for stable expression of short interfering RNAs in mammalian cells, *Science*, 296, 550, 2002.
6. Miyagishi, M. and Taira, K., U6 promoter-driven siRNAs with four uridine 3' overhangs efficiently suppress targeted gene expression in mammalian cells, *Nat. Biotechnol.*, 20, 497, 2002.
7. Paul, C.P. et al., Effective expression of small interfering RNA in human cells. *Nat. Biotechnol.*, 20, 505, 2002.
8. Lee, N.S. et al., Expression of small interfering RNAs targeted against HIV-1 rev transcripts in human cells, *Nat. Biotechnol.*, 20, 500, 2002.
9. Paddison, P.J. et al., Short hairpin RNAs (shRNAs) induce sequence-specific silencing in mammalian cells, *Genes Dev.*, 16, 948, 2002.
10. Yu, J.Y., DeRuiter, S.L., and Turner, D.L., RNA interference by expression of short-interfering RNAs and hairpin RNAs in mammalian cells, *Proc. Natl. Acad. Sci. U. S. A.*, 99, 6047, 2002.
11. Sui, G. et al., A DNA vector-based RNAi technology to suppress gene expression in mammalian cells, *Proc. Natl. Acad. Sci. U. S. A.*, 99, 5515, 2002.
12. Okabe, M. et al., Green mice as a source of ubiquitous green cells, *FEBS Lett.,* 407, 313, 1997.
13. Ito, T. et al., Application of bone marrow-derived stem cells in experimental nephrology, *Exp. Nephrol.,* 9, 444, 2001.
14. Hasuwa, H. et al., Small interfering RNA and gene silencing in transgenic mice and rats, *FEBS Lett.,* 532, 227, 2002.
15. Liu, Z. and Foote, R.H., Development of bovine embryos in KSOM with added superoxide dismutase and taurine and with five and twenty percent O_2, *Biol. Reprod.,* 53, 786, 1995.
16. Lawitts, J. A. and Biggers, J. D., Optimization of mouse embryo culture media using simplex methods, *J. Reprod. Fertil.,* 91, 543, 1991.
17. Ogawa, S., Sato, K., and Hashimoto, H., *In vitro* culture of rabbit ova from the single cell to the blastocyst stage, *Nature,* 233, 422, 1971.
18. Hogan, B. et al., *Manipulating the Mouse Embryo: A Laboratory Manual,* 2nd ed., Cold Spring Harbor Laboratory Press, New York, 1994.

20 RNA Interference in Mouse Models

Erwei Song, Patrick Stern, Deborah Palliser, Luk Van Parijs, and Judy Lieberman

CONTENTS

20.1 INTRODUCTION

Soon after Tuschl and colleagues demonstrated that RNA interference (RNAi) operates in mammalian cells via short RNA duplexes of 19 to 23 nucleotides in length,[1] tremendous efforts were taken to study the feasibility of this new technique to treat human disease. RNAi has been shown to mediate specific and effective gene silencing. The specificity and effectiveness, coupled with the fact that every animal gene can potentially be targeted via this mechanism, suggest that this technique may provide an exciting new therapeutic tool with a wide array of potential disease targets.[2] However, two recent reports suggest that siRNAs, particularly at high concentrations, may also silence genes with incomplete homology and induce an interferon (IFN) response with global inhibition of translation. These off-target effects may lead to toxicity and need to be considered in all *in vivo* experiments. Induction of an IFN response is likely to be more of an issue when RNAi is induced via viral vectors.[3,4]

To date, most studies investigating RNAi in mammalian cells have been carried out *in vitro*, often using transformed cell lines, which may have different properties than primary cells. However, this year, the first report to apply RNAi to an animal disease model was published.[5] In this study, hydrodynamically injected duplex siRNA directed against Fas effectively silenced endogenous gene expression for 10 days without diminution, and mice with silenced Fas expression were protected from hepatic destruction, and death in two models of autoimmune hepatitis. Thereafter, RNAi-based therapies have been implemented in a variety of animal disease models, including viral and autoimmune hepatitis,[6,7] septic shock,[8] cancer,[9,10] ocular neovascularization,[11]

0–8493–2141–7/05/$0.00 + $.50

and neurodegenerative disease.[12] So far, mice are the only mammalian models for which RNAi has been reported.

Animal models have rapidly become a key component in investigating RNAi. Specific gene function can be validated *in vivo* by using RNAi to silence gene expression in animals.[13] When *in vivo* delivery methods have been improved, injection of siRNAs to induce RNAi will provide a powerful tool to study *in vivo* gene function. RNAi transgenic mice with constitutive or regulated silenced gene expression can also be produced more rapidly than knockout mice using gene targeting via homologous recombination.[14,15] Methods of RNAi delivery and its pharmacokinetics *in vivo* can be studied for therapeutic use.[16,17] The feasibility of RNAi-based cell therapy can be evaluated by transplanting RNAi modified cells into animals.[18] The therapeutic potential of RNAi can be further tested by tackling disease models.[19] Nonspecific silencing or induction of an IFN response by RNAi or unanticipated toxicities can only be thoroughly studied *in vivo*.[3] In this chapter, we will discuss some of the protocols for using RNAi in animal models.

There are two basic ways of using RNAi *in vivo*, either by injecting small RNAs (siRNAs) as drugs or by expressing stem-loop RNAs from plasmids or viral vectors that are processed intracellularly into siRNAs. Each of these approaches has its advantages and drawbacks and suitability for different situations.

20.2 PROTOCOLS FOR DELIVERING siRNAs AS DRUGS

A major strategy to administer siRNAs as drugs is to use synthetic duplex siRNA molecules. These are small in size, can diffuse to some extent through biological membranes,[20] are relatively stable in biological fluids,[21,22] and apparently lack toxicity.[19,23] Duplex siRNA can be injected systemically into the vascular system of mice via the tail vein or introduced directly to target tissue by local infusion through catheterization and local injection. In addition, siRNAs can be chemically modified and complexed with other molecules to enhance stability and intracellular delivery.

Mice expressing reporter genes are a valuable research tool for investigating siRNA delivery to target tissues.[16,17] There are two possible approaches — one involves hydrodynamic injection (see below) of a plasmid encoding a luciferase reporter gene. The effectiveness of delivery of siRNAs targeting luciferase and their silencing can then be assessed using sensitive cameras imaging the whole mouse. The disadvantage of this approach is that luciferase is only expressed where the hydrodynamic delivery is effective. Another approach is to deliver siRNAs targeting EGFP to a mouse in which EGFP is ubiquitously expressed in all cells, such as the JAX® GEMM® GFP transgenic mouse from Jackson Laboratories.[24] With this mouse, using flow cytometry and/or fluorescence microscopy, the effectiveness of silencing can be dissected for all tissues in the body and can also identify how well silencing is working in particular cell types by co-staining for phenotypic markers.

PROTOCOLS

Protocol 20.1 Hydrodynamic Injection

Hydrodynamic injection was first used to transfer naked DNA genes into hepatocytes[25] and has been adapted to deliver siRNA.[16,17] This maneuver aims to deliver a large bolus of liquid into the vena cava of animals via their tail veins. The volume is too large for the heart to handle rapidly, so the fluid containing siRNA backs up into highly vascular organs such as the liver, spleen, heart, and lungs, resulting in siRNA delivery through unclear mechanisms.[20]

Materials

- BALB/c or C57BL/6 mice, 8 to 10 weeks old, weighing more than 22 g (Jackson Laboratory) (see Technical note 1)
- Infrared heat lamp (Fisher)
- 1-ml latex-free syringe with 28½-G needles (BD)
- Isofluorane (Fisher)
- Injection PBS (Fisher)

Experimental Procedures

1. Dissolve 50 μg of siRNA (i.e., 2 mg/kg) in 1–2 ml of PBS (see Technical note 2) and draw the solution into the syringe. Mount a new 28½-G needle onto the syringe for injection (see Technical note 3).
2. Anesthetize the mice by isofluorane inhalation.
3. Place the tail in a 50 to 60°C water bath for 30 sec until the tail vein is dilated, and then place the mouse on an operating board under infrared lamp to keep the tail vein dilated.
4. Puncture the tail vein (see Technical note 4) and push the liquid into the vein within 10 sec (5 to 7 sec is optimal) (see Technical note 2). If there is no resistance as you inject the fluid, you can be sure that the needle is in the vein. With a successful injection, the vein turns pale and blood returns with a flush after injection.

Technical notes. 1: Hydrodynamic injection is very tricky. For beginners trying to master hydrodynamic injection, white BALB/c mice are preferred to black C57BL/6 mice because the tail veins are better visualized in white mice. Larger male mice are the easiest to inject, as the penile vein is an alternative to the tail vein. Lack of resistance during injection is the key indication of successful delivery. 2: The delivery volume and speed are two major factors determining success. Several groups injected a volume corresponding to 10% of the body weight of the animal (2-ml injection), while we found that injection of a volume corresponding to 5% body weight (1-ml injection) results in efficient siRNA delivery and reduces immediate mortality due to cardiac failure. 3: An insulin syringe with a preattached needle is preferred, as this avoids excessive dead volume in the syringe. 4: The tail veins distribute on the lateral and dorsal sides of the tail, while the vessel on the ventral side is the tail artery.

PROTOCOL 20.2 INTRAPORTAL VEIN INFUSION

Vascularized organs, such as the liver and kidney, with relatively large perfusing or draining vessels can be chosen for local infusion to deliver siRNA. As compared to hydrodynamic injection, local perfusion may be more readily adapted for clinical use via therapeutic catheterization.[26] However, puncturing the portal veins or renal veins via abdominal surgery is traumatic and requires surgical skills. Here we present a protocol for local siRNA infusion to the liver via the portal veins.

Materials

- Stereo operating microscope (SMZ1500, Nikon)
- 1-ml latex-free syringe with 30½-G needles (BD)
- Sterile cotton (BD)
- One sterile disposable scalpel (Feather)
- Two pairs of full curved 4" iris tissue forceps without teeth (Fisher)
- One pair of 5" standard tissue forceps with teeth (Fisher)
- Two pairs of 4" ½ curved iris dissection scissors (Fisher)
- One pair of 5" ½ straight surgical scissors (Fisher)
- 7-0 silk surgical suture (Fisher)
- Pentobarbital sodium (Sigma)
- Injection PBS (Fisher)
- FloSeal Matrix Hemostatic Sealant (FloSeal)

Experimental Procedure

1. Dissolve 50 μg of siRNA in 100 μl of injection PBS and draw the solution into a 1-ml latex-free syringe with a 30½-G needle.
2. Anesthetize the mouse with an intraperitoneal injection of pentobarbital (400 mg/kg).
3. Make an upper midline abdominal incision for 2 cm. Lift up the duodenum and left lateral lobe of the liver to expose the portal vein that is situated to the left of the common bile duct under the caudal lobe of the liver.
4. Puncture the portal vein with the aid of a 10 × magnification stereo microscope (see technical note below). Slowly push the syringe piston to deliver the solution into the portal vein. Make sure that no leakage occurs during the injection.
5. Following needle removal from the portal vein, apply pressure with sterile cotton for about 10 sec to prevent bleeding. Alternatively, spray FloSeal Matrix hemostatic sealant to the puncture site for hemostasis.
6. Close the incision in the abdominal wall using 7-0 surgical sutures.

Technical note. Attention should be paid to ensure that the needle tip is in the lumen of the portal vein by checking for blood return and verifying that it has not punctured the vein, before injecting the siRNA.

PROTOCOL 20.3 LOCAL INJECTION INTO THE BRAIN

Organs with relatively loose interstitial tissue, such as brain, pancreas, bone marrow, and tumor, can be chosen for local injection. This method directly delivers high concentrations of siRNA into target tissues and minimizes the exposure time of siRNA to serum RNase. However, local injection of most visceral organs requires surgery that is traumatic to the animals.

Materials

- Bone wax (KMedic)
- See Protocol 20.2 for other materials

Experimental Procedures (adapted from published protocol;[27] see Technical note 1)

1. Dissolve 20 μg of siRNA in 20 μl of injection PBS and draw the solution into a 1-ml latex-free syringe with a 27½-G needle.
2. Anesthetize the mouse with an intraperitoneal injection of pentobarbital (400 mg/kg).
3. Make a midline scalp incision and expose the parietal bone of the skull (see Technical note 2).
4. Puncture the brain with the 27½-G needle at the incision through the parietal bone and inject the solution slowly.
5. Plug the puncture hole in the skull with bone wax and close the incision in the scalp with 7-0 surgical sutures.

Technical notes. 1: Alternative delivery methods using pumps to deliver siRNA into the CSF at various locations are available. 2: Puncturing through the scalp is not recommended for siRNA injection due to high level of RNases in the skin.

PROTOCOL 20.4 LOCAL INJECTION IN THE BONE MARROW

Materials

See Protocol 20.3.

Experimental Procedures

1. Dissolve 20 μg of siRNA in 20 μl of injection PBS and draw the solution into a 1-ml latex-free syringe with a 27½-G needle.
2. Anesthetize the animals by isofluorane inhalation.
3. Make a 1 cm incision on the skin of the hind limbs. Separate the posterior tibial muscle and the triceps muscle using blunt dissection to expose the tibia.
4. Insert the needle into the bone marrow cavity at the midpoint of the tibia.
5. When no resistance is felt, inject the solution into the bone marrow cavity.
6. Plug the puncture hole with bone wax and close the incision with 6-0 surgical sutures.

Protocol 20.5. Subretinal Injection

Materials

See Protocol 20.2 for materials in addition to the following:

- Stereo operating microscope (SMZ1500, Nikon)
- Ophthalmological surgery instrument set (Fisher)

Experimental Procedures (adapted from the protocol of reference [28])

1. Dissolve 20 pmol of siRNA in 1 µl of injection PBS.
2. Under an operating microscope, open the conjuctiva and make a shelving puncture of the sclera behind the limbus with a 30-G needle in the superior quadrant of the eye.
3. Insert a 33-G needle along the premade tunnel.
4. Observe the advancement of the needle until it reaches the subretinal space.
5. Slowly push the solution into the space. Observe the appearance of partial retinal detachment, using indirect ophthalmoscopy to confirm successful administration.

Protocol 20.6 Intratumoral Injection

Intratumoral injection of anticancer drugs may enhance pharmacological efficacy and minimize adverse effect in normal tissues. This technique can be adapted to efficiently deliver siRNAs into local nonmetastatic cancers *in vivo*. Subcutaneous implantation of human tumor cells into immunodeficient mice sets up a model for evaluating intratumoral injection of siRNAs and their antitumor effect.

Materials

- Human tumor cell lines (ATCC)
- NOD/LtSz-scid/scid (NOD/SCID) mice (Jackson Lab)
- Athymic nude mice (Taconic Farms)
- See Protocol 20.2 for other materials

Experimental Procedures

Subcutaneous implantation of tumor cells

1. Seed tumor cells in 15-cm tissue culture dishes and grow to approximately 75% confluency.
2. Harvest cells with cell scraper, centrifuge and wash with HBSS, and resuspend in HBSS at a density of 3 to 4 × 10^7 cells/ml for human tumor cells.
3. For NOD/SCID mice shave the flank to be injected 1 to 2 d before tumor cell injection. When shaving the SCID mice with an electric razor, detach and sterilize the razor blade by autoclaving prior to use.
4. Weigh the animals prior to injection and maintain a body-weight record over the course of the experiment.
5. Inject 0.1-ml aliquots of tumor cells subcutaneously using a 25-G needle.

6. Check the injection site of the animals the day after tumor cell inoculation for any bleeding, bruising, or infection caused by the injection. Examine the mice serially daily to check if a tumor mass is palpable under the skin. Record tumor size using calipers to record diameters in three dimensions.

Intratumoral injection

7. When the subcutaneous tumor areas reach approximately $3 \times 3 \times 3$ mm^3 in size, the mice are ready for intratumoral siRNA injection.
8. Dissolve 50 µg of siRNA in 100 µl of injection PBS and draw the solution into a 1-ml latex-free syringe with a 27½-G needle.
9. Anesthetize the mice with an intraperitoneal injection of pentobarbital (400 mg/kg).
10. Place the mouse on a dissecting board. Use one hand to hold the tumor and one hand to hold the syringe.
11. Push the needle directly into the tumor and let it go as deeply as possible without passing through the tumor.
12. Smoothly push the solution into the tumor. After injection, leave the needle in the tumor for about 20 sec and then slowly pull out.
13. If multi-injections are required, after the first injection, slowly pull the needle back but not out of the tumor, change the direction of the needle inside the tumor and then push in for additional injections.
14. Follow the mice by measuring the size of tumor and body weights of the mice daily until the siRNA-treated tumors regress or the subcutaneous tumor areas in control (irrelevant siRNA-treated) mice reach over 1000 mm^2 or until the mice die or exhibit signs of morbidity.
15. At the end of the experiment, determine tumor-growth delay, defined as the difference between the median time in days required for the therapeutic siRNA-treated tumors and for the control siRNA-treated tumors to reach a predetermined size.[29]
16. Sacrifice the mice by cervical dislocation and harvest the subcutaneous tumors and visceral organs including liver, lungs, and lymph nodes according to Protocol 20.9 for histological and biochemical examination to evaluate tumor growth and metastasis.

Protocol 20.7 Enhancing siRNA Delivery via Cationic Liposomes (adapted from published protocol[8])

siRNAs must gain access to the cytoplasm of target cells where RNA interference takes place. SiRNAs are not efficiently taken up into cells *in vitro* without transfection reagents. To enhance siRNA delivery, the oligonucleotides may be complexed with peptides bearing positive charges,[30] incorporated into liposomes, or complexed with lipids before injection.[8] Chemical modification of siRNAs enhances *in vivo* stability to endogenous RNases by modifying the phosphate backbone to phosphorothioates, modifying the 2'OH group of the ribose, or capping the 5' ends. This is a rapidly evolving field. Although not yet commercially available at the time of this writing, modified siRNAs are soon likely to be marketed for use in animal experiments.

Materials

- DOTAP (N-[1-(2,3-Dioleoyloxy)propyl]-N,N,N-trimethylammonium methylsulfate; 400 μg in 400 μl) (Roche)

Experimental Procedure

1. Dilute 10 μg of siRNA into 10 ml of PBS in a sterile reaction tube. In a separate sterile reaction tube, mix 60 μl (60 μg) of DOTAP (1 μg/μl) in PBS to a final volume of 90 μl.
2. Transfer the siRNA to the reaction tube already containing DOTAP in PBS and carefully mix the reagents by gently pipetting the mixture several times. Do not vortex or centrifuge.
3. Incubate the mixture for 15 min at room temperature.
4. Prepare the mice for local injection or local perfusion according to Protocols 20.2 to 20.4 and slowly inject the DOTAP/siRNA mixture into target organs.

Technical note. DOTAP/siRNA mixture is not suitable for hydrodynamic injection, as lipid thrombosis may occur. Other lipids and lipid complexes may also be tried after demonstrating *in vitro* efficacy.

Protocol 20.8 Enhancing siRNA Delivery via Complexing with MPG Peptides

Basic peptides can be used to deliver negatively charged proteins or nucleic acids into the cytosol. One paper reported using an MPG peptide (GALFLGFLGAAG STMGAWSQPKKKRKV) to deliver siRNA into mammalian cells *in vitro*.[30] However, it is not yet clear whether it may be adapted for *in vivo* delivery.

Materials

- MPG peptide synthesized by a commercial source

Experimental Procedure (adapted from published protocol[30])

1. Mix 50 μg of siRNA with MPG at a molar ratio of 1:10 (MPG:siRNA) and dissolve the mixture in 100 or 200 μl of injection PBS.
2. Incubate the mixture for 30 min at 37°C.
3. The mixture is suitable for local injection or local infusion according to Protocols 20.2 and 20.3.

Protocol 20.9 Organ Harvesting, Preservation, and Fixation for Detecting siRNAs and Measuring Gene Silencing

After siRNA delivery, efficient detection methods are needed to validate *in vivo* distribution of RNAi and its silencing effect. Two different approaches have been used to measure the efficiency of *in vivo* RNAi delivery: direct detection of siRNA using modified northern blot analysis or fluorescence-labeled siRNA molecules and evaluation of specific gene silencing of mRNA and protein levels in target tissues. To facilitate monitoring siRNA delivery *in vivo*, mice expressing GFP can be used. Following injection of siRNAs, target tissues with down-regulated GFP expression will indicate silencing (ES, DP, and JL, unpublished data).

To measure the delivery efficiency and silencing effect of RNAi in animal models, target organs should be harvested at appropriate times. Synthetic siRNAs can be detected in various organs as early as 24 h after injection, and their silencing effect may initiate at 48 h, depending on factors such as injection rate, delivery vehicle, and the half-lives of target proteins.[5,7] Peak silencing is usually at about 3 days. In quiescent cell types not actively proliferating, gene silencing can be maintained for as long as 10 to 14 days after injection.[5] However, while fluorescently labeled siRNA may not be detected for as long as the siRNAs persist due to cleavage of the fluorophore from the siRNA. RNAi-expressing vectors such as retroviruses potentially provide a sustainable source of siRNA *in vivo*.[13] However, complications from insertional mutagenesis may preclude their use for *in vivo* therapy.[31] Since siRNA may be susceptible to RNase degradation, it is important to harvest fresh animal organs with RNase-free manipulation.

Materials

- Scissors, forceps, long-handled scissors, scalpels (Fisher)
- O.C.T. (10.24% polyvinyl alcohol, 4.26% polyethylene glycol, 85.5% nonreactive ingredients) (Tissue Tek)
- Isofluorane (Fisher)
- Ethanol
- 37% (v/v) paraformaldehyde (Sigma)
- Xylene (Fisher)
- Paraffin (Fisher)
- 2% (w/v) trypan blue in PBS (Sigma)
- Endotoxin-free collagenase (Boehringer-Mannheim)
- DNase I (Boehringer-Mannheim)
- Complete Dulbecco's Modified Eagle Medium (DMEM) (Life Technologies) supplemented with 10% (v/v) heat-inactivated (56°C for 30 min) fetal calf serum, 3.7 g/L of sodium bicarbonate, 100 U/ml of penicillin, and 100 μg/ml streptomycin (Sigma)
- Petri dish (Corning)
- Vinyl specimen mold (Tissue Tek)
- Cryovials (Nalgene)
- 50 μl centrifuge tubes (Falcon)
- 70 μ-Cell strainer (Falcon)
- Hemacytometer (Sigma)

Experimental Procedures

1 Anesthetize the mouse by isofluorane inhalation.
2. Clean the abdomen with 75% (v/v) ethanol. Lift the skin and dissect with sharp scissors without penetrating the peritoneum. Remove the skin flap to expose the

subcutaneous area of the chest and abdomen. Clean the exposed area with 75% (v/v) ethanol. Open the peritoneal cavity and expose the aorta and inferior vena cava. Cut the inferior vena cava or aorta to kill the animals (see Technical note 1).

3. Remove the target organ and place onto a clean 15-cm^2 Petri dish (see Technical note 2). Divide the organ into three large pieces (for sectioning and cell isolation) and two small pieces (for RNA and protein isolation).
4. For cryosectioning: immediately place one large piece of tissue in liquid nitrogen. After freezing, place it in a vinyl specimen mold mounted on dry ice. Add several drops of O.C.T. solution to cover the tissue. Once the tissue lump is completely frozen, it is ready for sectioning. Fix the samples in 100% acetone following cryosectioning (see Technical note 3).
5. For RNA and protein isolation: immediately place two small pieces of tissue into two separately labeled cryovials and submerge them in liquid nitrogen. After freezing, the vials can be stored at –80°C until analyzed.
6. For paraffin sectioning: place one large piece of tissue into 4% (v/v) paraformaldehyde overnight for fixation. Following fixation, place it serially in 75% (v/v), 80% (v/v), 95% (v/v), and 100% ethanol for 5 min each and finally in xylene for 10 min. After dehydration, transfer the tissue to paraffin that has been premelted and maintained at 56°C for 14 h. Then, ramp down the temperature to room temperature. The paraffin block is now ready for sectioning.

Cell isolation from target tissue or tumor

7. Once the tissue for cell isolation is excised, work in a sterile environment using a laminar flow hood.
8. Wash the tissue twice in 20 ml of complete DMEM in 50-ml conical tubes.
9. Place the tissue in 10 ml of fresh complete DMEM in 50-ml conical tubes and mince it into small pieces using sterile long-handled scissors or a scalpel.
10. Spin the pieces of tissue at 200 g for 3 min at 4°C. Remove the supernatant.
11. Resuspend the pellet in 8 ml of complete DMEM. Add 1 ml of endotoxin-free collagenase solution and 1 ml of DNase I solution.
12. Incubate the tissue pieces for 15 min at 37 °C while shaking at high speed using a horizontal waterbath shaker.
13. Spin at 200 g for 5 min at 4°C. Remove the supernatant and add 10 ml of complete DMEM to the pellet.
14. Pass the cell suspension through a sieve and collect the cells in a 50-ml conical tube. Wash the cell strainer and original tube containing the disaggregated tissue with 5 ml of complete DMEM, and add the wash to harvested cells.
15. Spin the cells at 200 g for 3 min at 4°C.
16. Count the viable cells using trypan blue and a hemacytometer. The cells are ready for intracellular staining and flow cytometry analysis.

Technical notes. 1: To remove blood in harvested organs, perfusion with 1 to 2 ml PBS through the open aorta may be performed prior to dissection. 2: To harvest organs in the thoracic cavity, the diaphragm and rib cage should be opened. 3: Alternative fixation may be required for staining of some antigens.

PROTOCOL 20.10 DETECTION OF FLUORESCENCE-LABELED SIRNA

To track the distribution of the siRNA administered *in vivo*, siRNAs can be labeled with various fluorophores or radioactive isotopes. The 5' and 3' ends of the sense strand of siRNA molecules are usually chosen for the labeling, as such modifications do not affect silencing.[32] These can be purchased from Dharmacon Research, Inc. Commonly used fluorophores for siRNA labeling are FITC, Cy3, and Cy5, and detection of the labeled siRNA is by fluorescence microscopy or flow cytometry. Here, we outline methods to detect fluorescence-labeled siRNA. Additionally, siRNA molecules taken up by target tissues can also be detected by northern blot (see Protocol 20.10 below) modified to detect RNA molecules of fewer than 100 nucleotides in length.

Materials

- Fluorescence microscope (Leica)
- Flow cytometer such as FACScalibur (BD)
- Acetone (99.9%, HPLC Grade, Sigma)
- Vectorshield mounting solution (Vector Lab)
- FACS buffer (PBS supplemented with 2% (v/v) FCS)
- Fixation buffer (PBS supplemented with 1% (v/v) formalin, 0.1% (w/v) NaN_3, and 5% (v/v) FCS) (Fisher)

Experimental Procedures

1. Harvest target organs or tumor tissue as described in Protocol 20.8.
2. For fluorescence microscopy: perform consecutive cryosectioning and prepare sections at approximately 5 to 10 microns of thickness. Hematoxylin and eosin (H&E) staining and immunohistochemistry staining for specific cellular markers can be used to indicate target tissues and cell types.
3. Fix the sections in acetone for 5 min at room temperature (see Technical notes).
4. Air-dry the sections and mount the slides with cover slips using Vectorshield mounting solution.
5. Evaluate fluoresence intensity and distribution under fluorescence microscope.
6. For FACS analysis: isolate target tissue cells according to Protocol 20.9.
7. Wash cells with FACS buffer.
8. Pellet the cells by centrifuging at 200 × g at 4°C for 5 min.
9. Resuspend cell pellets in 100 μl of fixation buffer for FACS analysis.
10. Stain with cell surface or intracellular markers after permeabilization according to standard protocol as required to indicate uptake of siRNA into specific cell types.

Technical notes. Fixation of fluorescence sections should be minimized. Alternatively, without acetone fixation, wash slides with PBS after sectioning and evaluate fluorescence intensity immediately under fluorescence microscopy. It is important to ascertain that fixation and permeabilization do not reduce signals of fluorescently tagged siRNA.

Protocol 20.11 Modified Northern Hybridization Procedure for siRNA Detection

Materials

- Tissue homogenizer (Fisher)
- UV cross-linker (Fisher)
- Model 785 Vacuum Blotter (Bio-Rad)
- phosphorImager (Molecular Dynamics)
- Sephadex G-25 MicroSpin Column (Ambion)
- Nytran Plus membrane (Schleicher & Schuell)
- Trizol reagent (Molecular Research Center)
- Isoamyl alcohol (IAA, Sigma)
- 75% (v/v) ethanol
- Denaturing acrylamide gel (15%, w/v) (BioRad)
- [γ^{32}P] ATP (DuPont)
- Sodium dodecyl sulfate (SDS) (Sigma)
- Low-stringency hybridization buffer (UltraHyb buffer) (0.36 M Na_2HPO_4, 0.14 M NaH_2PO_4, 1 mM EDTA, and 7% (v/v) SDS) (Ambion)
- 2 × SSC and 0.1 × SSC (1 × SSC: 0.15 M NaCl, 0.15 M sodium citrate)
- 10 × TBE buffer (Tris-borate/EDTA electrophoresis buffer; National Diagnostic)

Experimental Procedures

1. Homogenize tissue and extract total RNA using a modified Trizol protocol as follows:
2. Add 1 ml of Trizol reagent to the tissue homogenate and incubate for 5 min at room temperature, followed by adding 100 µl ammonium acetate (10 M) and 200 µl chloroform:IAA (49:1). Vortex and incubate on ice for 15 min.
3. Centrifuge at 1200 × g at 4°C for 15 min and carefully transfer the clear supernatant into a clean microcentrifuge tube.
4. Add 500 µl Ethanol (100%) and mix thoroughly by vortexing. Incubate the mixture at –70°C for 20 min and pellet the RNA by centrifuging at 1200 × g at 4°C for 20 min.
5. Wash the pellets with 70% (v/v) ethanol and air-dry.
6. Resuspend the pellets in nuclease-free water and quantify RNA by UV absorption.
7. Load 5 µg of RNA onto a 15% (w/v) denaturing acrylamide (tris-borate-EDTA urea) gel and electrophorese at 10 W in 1 × TBE. Prepare radio-labeled oligonucleotide as a probe by end-labeling with [^{32}P] as described in *Molecular Cloning, a Laboratory Manual,*[33] and use 1 and 10 fmol of known siRNA that will hybridize with the probe as positive control and for quantification.
8. After electrophoresis, electrotransfer the gel to Nytran Plus membrane for 2.5 h at 350 mA in 0.5 TBE using a Vacuum Blotter.
9. UV cross-link the RNA to the membrane at 1200 µF.
10. Prehybridize the blot for 30 min at 40°C in low-stringency hybridization buffer (UltraHyb buffer) in a shaking water bath.

11. Prepare the probe by 5'-end labeling 100 pmol of the sense strand of the target siRNA with 150 μCi of [γ^{32}P] ATP (6000 Ci/mol). Purify the labeled probe on a G-25 MicroSpin Column.
12. After heating to 65°C for 5 min, add the labeled probe (6000 Ci/mol) to the prehybridization buffer and perform the hybridization overnight at 37°C in a shaking waterbath.
13. Wash the blot twice with 2 × SSC/0.2% (w/v) SDS for 20 min and three times with 0.1% SSC/0.1% SDS for 30 min, followed by film exposure and analysis by phosphorImager.

Technical note. The Trizol protocol has been modified to precipitate the RNA pellets with high-salt buffer after extraction. This significantly enriches small RNA species of less than 100 nt.

Protocol 20.12 Evaluation of Gene-Silencing Effects by RNase Protection Assay

Measurement of mRNA and protein expression are used to measure the efficacy of RNAi-mediated silencing in target tissue. Target gene at mRNA can be quantified using real-time RT-PCR, northern hybridization, or RNase protection assay of tissue homogenates, or by *in situ* hybridization of slides. Real-time RT-PCR is the easiest of the mRNA assays. Protein expression of silenced genes can be quantified by immunoblot or detected *in situ* using immunohistochemistry. Furthermore, protein expression in suspension cells such as bone marrow cells, peripheral blood cells, and splenocytes can be studied on a cell-by-cell basis using flow cytometry. Because of the possibility of off-target effects, measurement of the effects on other genes besides the targeted genes is critical for all of these assays. Microarray may also provide a means to scan for unanticipated silencing.[3] Here, we present protocols for RNase protection assay and immunohistochemistry.

Materials

- [α^{32}P] UTP (Dupont)
- RPA *in vitro* transfection kit (BD)
- Multiple mouse RPA template set (BD)
- Denaturing polyacrylamide gel (12%, w/v) (BioRad)
- Phenol:chloroform (50:1; Gibco)
- Proteinase K (Sigma)

Experimental Procedures (adapted from the manufacturer's protocol)

1. Extract total RNA from tissue sample using Trizol Reagent. Wash the pellets with 75% (v/v) ethanol and vacuum-dry.
2. Synthesize [α^{32}P] UTP-labeled RNA probe by *in vitro* transcription. In a 20-μl *in vitro* transcription reaction, mix 10 μl of [α^{32}P] UTP (20 μCi) with 1μl RNasin, 1 μl nucleotide pool, 2 μl DTT, 4 μl 5 × transcription buffer, 1 μl RPA template set, and 1 μl T7 RNA polymerase, and incubate at 37°C for 1 h.

3. After terminating the reaction with 2 µl DNase, extract the labeled RNA probe in Tris-saturated phenol and chloroform:isoamyl alcohol (50:1). Precipitate the RNA probe in 4 M ammonium acetate and 100% ice-cold ethanol (1:5) and wash with 90% ethanol.
4. Mix 10 to 20 µg of sample RNA with 15 µl of hybridization buffer from the kit and 2 µl [α^{32}P] UTP-labeled probe RNA (100 cpm).
5. Place the samples in a heat block prewarmed to 90°C and immediately turn down the temperature to 56°C for incubation overnight.
6. Add 100 µl of high-salt buffer containing 0.2 µl of RNase A (10 mg/ml) and 0.01 µl of RNase T1 (80 U/µl) from the kit, and incubate the mixture for 45 min at 30°C.
7. Add 15 µl of 10% (w/v) SDS and 5 µl of 10 mg/ml proteinase K, and incubate the mixture for 15 min at 37°C.
8. Extract the hybridized duplex RNA with Tris-saturated phenol and chloroform:isoamyl alcohol (50:1) and precipitate in 4 M ammonium acetate and 100% ice-cold ethanol (1:5), followed by washing with 90% (v/v) ethanol.
9. Load the RNA onto a 12% (w/v) denaturing polyacrylamide gel and run the gel at 55 W constant power for 2 h. After electrophoresis, vacuum-dry the gel, expose to film, and analyze with a phosphorImager to detect the RNA signals.

Protocol 20.13 Evaluation of Gene-Silencing Effects by Immunohistochemistry (Labeled Streptavidin-Biotin Method)

Materials

- Microwave oven (Sharp)
- Plastic staining set including 13 staining chambers, a rack, and two slide holders (Tissue Tek)
- Hydrophobic slide marker (Super^{HT} Pap pen; Kiyota)
- 30% (v/v) hydrogen peroxidase (Sigma)
- Citrate buffer (420 mg citric acid in 200 ml deionized water, pH 6.0)
- Primary antibody
- Immunohistochemistry staining kit including biotinylated secondary antibody, streptavidin biotin complex, and normal serum for blocking (Vector Lab)
- Antibody diluent (PBS-based, BD)
- Diaminobenzidine substrate (DAB substrate kit, Vector Lab)
- Hematoxylin solution (Sigma)
- Xylene (Fisher)
- Cytoseal mounting medium (Vector Lab)

Experimental Procedures

1. For paraffin sections, de-wax and rehydrate the slides as follows before staining. Incubate sections twice in xylene for 5 to 10 min at room temperature and sequentially incubate in 100% ethanol twice, in 95% (v/v), 85% (v/v), and 70% (v/v) ethanol once, and finally in deionized water twice for 5 min each at room temperature.

2. Immerse the slides in citrate buffer for heating in a microwave oven. First, heat the samples at high power for 3 to 5 min to boiling and then incubate them at medium power for another 15 to 30 min (depending on specific antibodies as described in the manufacturer's datasheet). After incubation, remove the chambers with sample slides from the microwave oven to cool down at RT for at least 30 min (see Technical note 1).
3. Wash the slides with deionized water twice and then incubate them in 3% (v/v) H_2O_2 for 30 min at RT.
4. Wash the slides with deionized water twice and circle staining areas with a hydrophobic Pap pen. Add normal serum into the circled area for blocking and incubate the slides in a moisture chamber at RT for 30 min (see Technical note 2).
5. Blot excessive serum from sections. Add one drop of primary antibody dissolved in antibody diluent at appropriate concentration (refer to antibody datasheet for recommended range, but titration may be needed to limit background) to the circled area, and incubate the slides in a moisture chamber at room temperature for 1 h (see Technical note 3).
6. Wash slides 3 times for 15 min in PBS. Add one drop of ready-to-use biotin-conjugated secondary antibody from the immunohistochemistry staining kit to the circled area and incubate the slides in a moisture chamber at room temperature for 30 min.
7. Wash slides 3 times for 15 min in PBS. Add one drop of ready-to-use streptavidin from the immunohistochemistry staining kit to the circled area and incubate the slides in a moisture chamber at room temperature for 30 min.
8. Wash slides 3 times for 15 min in PBS, and meanwhile prepare the DAB solution (1 drop of buffer stock, 2 drops of DAB, and 17 μl of H_2O_2 from the DAB substrate kit dissolved in 1.5 ml deionized water).
9. Add DAB solution to the circled area to incubate for 2 to 5 min, and stop the reaction by immersing in deionized water.
10. Counterstain the sections by immersing the slides in hematoxylin solution for 10 sec, and wash with hot tap water for 10 min.
11. Dehydrate the samples sequentially in 70% (v/v), 85% (v/v), 95% (v/v), and 100% ethanol to xylene for 5 min each, and mount them with cover slips using a xylene-based Cytoseal mounting medium.

Technical notes. 1: For immunohistochemistry staining on cryosections, skip steps 1 and 2. If the frozen tissues were fixed with 4% (v/v) paraformaldyhyde rather than acetone after cryosectioning, which is required for some specific antigens, step 2 is needed to expose antigens. 2: Blocking serum is chosen according to the animal species of secondary antibody. 3: Isotype control antibody should be added to separate section of the tissue for negative control.

PROTOCOL 20.14 RNAI-KNOCKDOWN MICE: PRODUCTION OF LENTIVIRUS AND TITERING

In the postgenomic era with the advent of large-scale genome sequencing, thousands of novel genes have been identified.[34] Reverse genetics has become the most effective way to assess the function of a gene. Knockout mice, using homologous recombination to delete a segment of the target gene segment, was once the only *in vivo* way to perform reverse genetics in mammals. This technique only provides models with "all or none" phenotypes, producing homozygous knockout mice is time consuming and costly.[35] By contrast, RNAi technology can be used to rapidly create transgenic mice with a "knock down" of target genes.[35] Plasmids, or lentiviruses engineered to express short hairpin RNAs that are processed into siRNAs by Dicer within cells, can be used to create a stably silenced phenotype by transfection or infection of embryonic stem (ES) cells or early embryos.[15,35–38] When RNAi sequences with different potencies of gene silencing are used, a spectrum of distinct "knock down" phenotypes can be generated in mice.[15]

The following protocols to generate knockdown mice are adapted from the Web site of Dr. Parijs's laboratory at MIT http://web.mit.edu/ccrhq/vanparijs)

Materials

- Lentilox vector (Luk Van Parijs Lab: http://www.mit.edu/ccrhq/vanparijs)
- VSVG and Delta VPR plasmids (InVitrogen)
- 293 T cells (ATCC), NIH 3T3 fibroblasts (ATCC)
- 0.45-μm low-protein binding filter (Millipore)
- Calcium phospate transfection kit (GIBCO)
- 100 × stock of polybrene (hexadimethrine bromide, Sigma)
- Petri dish (15-cm, Corning)

Experimental Procedures

Lentivirus production

1. Construct lentivirus vector (lentilox) carrying U6 promoter for stem-loop siRNA[39] expression and CMV promoter for reporter GFP gene expression according to the protocol described by Dr. Luk Van Parijs on his Web site.
2. Plate the packaging 293 T cells in 15 cm tissue culture dish at a density of 1×10^7 cells/dish in 20 ml DMEM supplemented with 10% (v/v) FCS 1 day prior to transfection.
3. Introduce the lentivirus vector DNA, VSVG, and delta VPR plasmid DNAs (50 μg in total, ratio 2:1:1) into 293 T cells by calcium phosphate transfection using the standard protocol published in *Short Protocols in Molecular Biology*.[40] Remove medium 4 h later and add warm DMEM medium supplemented with 10% (v/v) FCS.
4. Collect the supernatant 48 h posttransfection and filter through 0.45-μm low-protein binding filter. For embryo infections, the virus must be concentrated by centrifugation. Spin at 4°C in an SW-28 rotor of a Beckman ultracentrifuge at 25,000 rpm for 90 min. Decant the supernatant and add cold PBS (5 to 15 ml) for storage at 4°C overnight. Resuspend by gentle pipetting over the pellet 20 to 50 times without touching the pellet. After resuspension, the pellet, which does not contain virus, can be discarded.
5. Flash-freeze aliquots of lentivirus in liquid nitrogen and store at –80°C for future use.

Determination of virus titer by FACS analysis

6. Plate 1 × 10^5 NIH-3T3 cells per well of a 6-well plate 1 day prior to the assay.
7. Thaw a frozen aliquot of lentivirus.
8. Prepare serial dilutions of viral supernatant (total volume: 500 µl):

Neat:	500 µl supernatant	+	0 µl medium
1/3:	166 µl supernatant	+	334 µl medium
1/10:	50 µl supernatant	+	450 µl medium
1/33:	16 µl supernatant	+	486 µl medium

9. Remove medium from the cells and replace by dilutions of viral supernatant.
10. Add polybrene to each well at a final concentration of 4 µg/ml.
11. Add 2 to 3 ml prewarmed medium to each well 4 h postinfection.
12. Incubate for 2 days in 5% CO_2 at 37°C for maximal GFP expression in infected NIH-3T3 cells.
13. Trypsinize the cells and wash once with PBS 2% (v/v) FCS, and determine GFP-positive percentage by FACS.
14. Calculate titer as follows:
    % GFP^+ cells × number of cells/well (on day 0) × 2 × 1/dilution = virus particles/ml.

Protocol 20.15 RNAi-Knockdown Mice: Obtaining Single-Cell Embryos

Materials

- Female C57BL/6 mice (Jackson Lab)
- Fertile (B6´DBA) F1 hybrid male mice (Jackson Lab)
- Pregnant mare serum gonadotrophin (PMSG) (Roche)
- hCG (Teikoku Hormone Manufacturing Collected)
- Whitten's M16 medium (Gibco)
- See Protocol 20.9 for other materials

Experimental Procedures

1. Intraperitoneally inject 7.5 IU pregnant mare serum gonadotrophin (PMSG) into female mice and intraperitoneally inject 7.5 IU hCG 40 to 48 h later.
2. Immediately after the hCG injection, cage the females with male mice.
3. Sacrifice the females 20 h after the hCG injection by cervical dislocation.
4. Remove the oviducts and place them in a 60-mm tissue culture dish containing PBS supplemented with 10% (v/v) FCS, penicillin 100 U/ml, and streptomycin 100 µg/ml.
5. Flush the oviducts with PBS supplemented with 10% (v/v) FCS, penicillin 100 U/ml, and streptomycin 100 mg/ml, and recover the single-cell embryos. Culture the single-cell embryos in Whitten's M16 medium (WM) overnight in 5% CO_2 at 37°C.

Protocol 20.16 RNAi-Knockdown Mice: Lentivirus Infection of Single-Cell Embryos

Materials

- Infectious lentivirus (generated from Protocol 20.13)
- Micromanipulator system attached to a Nikon inverted microscope (Beckman Instrument)

Experimental Procedures

1. Thaw one vial of lentivirus (virus titer: $5 \times 10^8 - 5 \times 10^9$ IU)on ice and dilute it in injection medium.
2. Transfer a group of embryos (from 10 to 20) to a microinjection chamber containing HEPES-buffered injection medium supplemented with 10% (v/v) FBS.
3. Focus on the zona pellucida at the highest power, use a fine glass needle to cut a small slit in the zona of the recipient embryo, and inject high-titer RNAi lentivirus (4 to 10×10^8 IU μl^{-1}) into the perivitelline space in an adequate volume to see an obvious swelling of the space.
4. After about 30 cells have been injected, place the cells into the 35-mm dish with warm Whitten's M16 medium (WM) supplemented with 10% (v/v) FCS. Culture the injected embryos overnight in 5% CO_2 at 37°C.
5. The infected single-cell embryos are now ready for oviduct implantation.

Protocol 20.17 RNAi-Knockdown Mice: Oviduct Transfer of Infected Embryos

Materials

See Protocol 20.9 and Protocol 20.16.

Experimental Procedures

1. Pipette 50 μl of WM media on the bottom of a 33-mm Petri dish.
2. Cover the media with embryo-tested mineral oil and leave at room temperature.
3. Anesthetize a female mouse with an intraperitoneal injection of pentobarbital (400 mg/kg).
4. Shave the caudal dorsal area of the anesthetized female and sterilize the area with 70% (v/v) ethanol.
5. Place the prepared pseudopregnant female under the dissecting microscope with ventral side down.
6. Make a 5-mm skin incision with a pair of scissors and watchmaker forceps on the abdominal wall, and identify the ovaries surrounded by white fat pad. Carefully pull the ovaries out of the incision by holding the fat pad with a pair of forceps without teeth. Position the ovary for easy access to the oviduct.
7. Pipette two to three embryos for one recipient into the microdrop of WM media in the Petri dish.

8. Gently tear open the bursa surrounding the oviduct with two pairs of forceps, allowing access to the infundibulum of the oviduct.
9. Load a transfer pipette (the tip diameter should be only slightly (see Technical note) larger than the embryos) with oil from the transfer dish by using mouth suction. Then draw a 2- to 5-mm air bubble into the pipette, followed by medium with the embryos.
10. Slide the tip of the loaded pipette in and blow into the pipette until the air bubble is visible within the oviduct.
11. Check the pipette tip underneath the microscope to make sure that the embryos have been transferred.
12. Gently replace the ovary/fat pad into the abdominal cavity.
13. Close the abdominal wall with 5-0 silk surgical sutures. The pups are expected on day 20 following surgery. The resulting neonates can be screened for lentiviral integration by Southern blotting using standard protocols.[33]

Technical note: Mouth pipetting is necessary to maintain viable embryos. For biosafety reasons use a mouth pipette with barrier protection.

Protocol 20.18 Isolation of Hematopoietic Stem Cells

Cell therapy involves injection of autologous or allogeneic cells that have been genetically modified *ex vivo*.[41] These cells can then be adoptively transferred to replace abnormal hematopoietic cells in conditioned recipients or transplanted to repopulate tissues. RNAi can be introduced into hematopoietic stem cells by lentivirus infection, with viruses encoding shRNAs that are processed within cells into siRNAs. Hematopoietic stem cells either of human or murine origin are now widely used for cell therapy, and transduction of these cells with RNAi targeting HIV virus genome or cell receptors important for viral entry can generate HIV-resistant hematopoietic cells. Here we present, as an example, methods using RNAi-transduced hematopoietic stem cells to repopulate immunodeficient SCID mice with human blood cells following the experimental design of Banerjea et al.[42]

Materials

- Human cord blood mononuclear cells (StemCell Tech)
- Selection medium (PBS supplemented with 2% (v/v) FCS)
- Expansion medium (IMDM supplemented with 2 mM L-glutamine, 20 ng/mL human recombinant IL-3, 20 ng/ml human recombinant GM-CSF, 20 ng/ml of human recombinant cf., and 20% (v/v) FCS)
- Human recombinant GM-CSF (Stem Cell Tech)
- Human recombinant SCF (Stem Cell Tech)
- Falcon 5-ml polystyrene round-bottom tubes (Becton Dickinson)
- EasySep positive selection cocktail, EasySep Magnetic Nanoparticles, and selection magnet (StemCell Tech)
- CD34 antibody (BD)

Experimental Procedures

1. Thaw 2 × 10^7 cord blood mononuclear cells and wash with selection media (PBS containing 2% (v/v) FCS).
2. Resuspend the cells in 100 μl of selection media in a 12- × 75-mm Falcon 5-ml polystyrene round-bottom tube.
3. Add 10 μl of EasySep Positive Selection Cocktail to the cells-, and incubate at RT for 15 min.
4. Add 10 μl of Mix EasySep Magnetic Nanoparticles to the cells and incubate at room temperature for 10 min.
5. Add 2.5 ml of selection media to the cells and place the tube into the magnet to incubate for 5 min at room temperature. After incubation, discard the supernatant.
6. Repeat step 5 another five times.
7. Resuspend the enriched cells in expansion medium, count the cells, and stain for CD34 to confirm the enrichment efficiency.
8. Resuspend the enriched cells in expansion medium at low density (< 10^3 cells/ml). The cells are now ready for lentivirus transduction.

Protocol 20.19 Transduction with Lentivirus and Validation of Silencing

Materials

See Protocol 20.18.

Experimental Procedures

1. Generate of lentivirus vector carrying RNAi and GFP reporter gene (lentilox) (see Protocol 20.16).
2. The day after CD34 enrichment, resuspend the cells in expansion medium at a density of 2000 cells/100 μl.
3. Add the lentivirus lentilox vector to the cells at an MOI of either 1:1 (low-titer virus) or 5:1 (high-titer virus).
4. Perform spin-transduction at 600 × g without braking at room temperature for 2 h.
5. Add 1 volume of expansion medium to the cells and incubate in 5% CO_2 at 37°C for 2 h.
6. Wash the cells twice and resuspend in expansion medium at low density (< 10^3 cells/mL).
7. Transduction efficiency can be determined by flow cytometry analysis of GFP expression or modified Northern blot for siRNA, as previously described.
8. Repeat the transduction the next day to increase the efficiency.

Protocol 20.20 Adoptive Transfer

Materials

- NOD/SCID mice (Takonic Farm)
- IMDM (StemCell Tech)
- Ciprofloxin (Glaxo)

- Human G-CSF (StemCell Tech)
- Antihuman CD45
- CD19 and CD66b/CD15/CD33 antibodies (BD)

Experimental Procedures

1. Seven days after expansion, the transduced cells are ready for adoptive transfer into NOD/SCID mice.
2. Irradiate normal human bone marrow cells with 1500 cGy for use as carrier cells.
3. Prepare cell mixtures in IMDM supplemented with 2% (v/v) FCS, so that 0.25 ml contains $1–1.5 \times 10^5$ transduced CD34 cells and 10^6 carrier cells.
4. Within 24 h before adoptive transfer, irradiate the NOD/SCID mice by exposure to 350 cGy of total body γ-irradiation administered in a single dose at < 250 cGy/min.
5. Inject 0.25 ml of the cell mixture intravenously into the tail vein of the irradiated mice.
6. After injection, give the mice sterile food and prophylactic ciprofloxin (100 mg/ml) in the drinking water until analysis. Inject 1 µg human G-CSF intraperitoneally every other day for a total of four injections.
7. To harvest bone marrow, 9 to 10 weeks after transplantation, aspirate bone marrow cells from both femora and tibiae using a sterile 21-G needle and a 3-ml syringe, and distribute into 2 ml of cold DMEM supplemented with 2% (v/v) FBS. Count viable nucleated bone marrow cells.
8. Stain cells with antihuman CD45, CD19, and CD66b/CD15/CD33 for flow cytometry analysis to determine the repopulation of human progenitors.
9. Perform flow cytometry analysis for GFP expression in repopulated human progenitors, and measure siRNA production in these cells according to Protocol 20.11 described above.

ACKNOWLEDGMENT

We thank Sang-Kyung Lee for useful discussions and critical reading of the manuscript.

REFERENCES

1. Elbashir, S. M. et al., Duplexes of 21-nucleotide RNAs mediate RNA interference in cultured mammalian cells, *Nature,* 411, 494, 2001.
2. Lieberman, J. et al., Interfering with disease: opportunities and roadblocks to harnessing RNA interference, *Trends Mol. Med.,* 9, 397, 2003.
3. Jackson, A. L. et al., Expression profiling reveals off-target gene regulation by RNAi, *Nat. Biotechnol.,* 21, 635, 2003.
4. Sledz, C. A. et al., Activation of the interferon system by short-interfering RNAs, *Nat. Cell Biol.,* 5, 834, 2003.
5. Song, E. et al., RNA interference targeting Fas protects mice from fulminant hepatitis, *Nat. Med.,* 9, 347, 2003.

6. McCaffrey, A. P. et al., Inhibition of hepatitis B virus in mice by RNA interference, *Nat. Biotechnol.,* 21, 639, 2003.
7. Zender, L. et al., Caspase 8 small interfering RNA prevents acute liver failure in mice, *Proc. Natl. Acad. Sci. U.S.A.,* 100, 7797, 2003.
8. Sorensen, D. R., Leirdal, M., and Sioud, M., Gene silencing by systemic delivery of synthetic siRNAs in adult mice, *J. Mol. Biol.,* 327, 761, 2003.
9. Brummelkamp, T. R., Bernards, R., and Agami, R., Stable suppression of tumorigenicity by virus-mediated RNA interference, *Cancer Cell,* 2, 243, 2002.
10. Verma, U. N. et al., Small interfering RNAs directed against beta-catenin inhibit the in vitro and in vivo growth of colon cancer cells, *Clin. Cancer Res.,* 9, 1291, 2003.
11. Reich, S. J. et al., Small interfering RNA (siRNA) targeting VEGF effectively inhibits ocular neovascularization in a mouse model, *Mol. Vis.,* 9, 210, 2003.
12. Xia, H. et al., siRNA-mediated gene silencing *in vitro* and *in vivo*, *Nat. Biotechnol.,* 20, 1006, 2002.
13. Shi, Y., Mammalian RNAi for the masses, *Trends Genet.,* 19, 9, 2003.
14. Rubinson, D. A. et al., A lentivirus-based system to functionally silence genes in primary mammalian cells, stem cells and transgenic mice by RNA interference, *Nat. Genet.,* 33, 401, 2003.
15. Hemann, M. T. et al., An epi-allelic series of p53 hypomorphs created by stable RNAi produces distinct tumor phenotypes *in vivo*, *Nat. Genet.,* 33, 396, 2003.
16. McCaffrey, A. P. et al., RNA interference in adult mice, *Nature,* 418, 38, 2002.
17. Lewis, D. L. et al., Efficient delivery of siRNA for inhibition of gene expression in postnatal mice, *Nat. Genet.,* 32, 107, 2002.
18. Banerjea, A. et al., Inhibition of HIV-1 by lentiviral vector-transduced siRNAs in T lymphocytes differentiated in SCID-hu mice and CD34(+) progenitor cell-derived macrophages, *Mol. Ther.,* 8, 62, 2003.
19. Caplen, N. J., RNAi as a gene therapy approach, *Expert Opin. Biol. Ther.,* 3, 575, 2003.
20. Herweijer, H. and Wolff, J. A., Progress and prospects: naked DNA gene transfer and therapy, *Gene Ther.,* 10, 453, 2003.
21. Miyagishi, M., Hayashi, M., and Taira, K., Comparison of the suppressive effects of antisense oligonucleotides and siRNAs directed against the same targets in mammalian cells, *Antisense Nucleic Acid Drug Dev.,* 13, 1, 2003.
22. Bertrand, J. R. et al., Comparison of antisense oligonucleotides and siRNAs in cell culture and in vivo, *Biochem. Biophys. Res. Commun.,* 296, 1000, 2002.
23. Bi, F., Liu, N., and Fan, D., Small interfering RNA: a new tool for gene therapy, *Curr. Gene Ther.,* 3, 411, 2003.
24. Okabe, M. et al., "Green mice" as a source of ubiquitous green cells, *FEBS Lett.,* 407, 313, 1997.
25. Zhang, G., Budker, V., and Wolff, J. A., High levels of foreign gene expression in hepatocytes after tail vein injections of naked plasmid DNA, *Hum. Gene Ther.,* 10, 1735, 1999.
26. August, D. A., Theoretic and experimental aspects of regional liver infusion, *Surg. Oncol. Clin. N. Am.,* 5, 399, 1996.

27. Terashima, T. et al., Retrograde and anterograde labeling of cerebellar afferent projection by the injection of recombinant adenoviral vectors into the mouse cerebellar cortex, *Anat. Embryol. (Berl),* 196, 363, 1997.
28. Bennett, J. et al., Adenovirus vector-mediated *in vivo* gene transfer into adult murine retina, *Invest. Ophthalmol. Vis. Sci.,* 35, 2535, 1994.
29. Corbett, T. et al., *In vivo* methods for screening and preclinical testing-use of rodent solid tumors for drug discovery, in *Anticancer Drug Development Guide-Preclinical Screening Clinical Trials and Approval,* Teicher, B.A., Ed., Humana, Totowa, NJ, 1997, chap. 2.
30. Simeoni, F. et al., Insight into the mechanism of the peptide-based gene delivery system MPG: implications for delivery of siRNA into mammalian cells, *Nucleic. Acids Res.,* 31, 2717, 2003.
31. Hacein-Bey-Abina, S. et al., LMO2-associated clonal T cell proliferation in two patients after gene therapy for SCID-X1, *Science,* 302, 415, 2003.
32. Amarzguioui, M. et al., Tolerance for mutations and chemical modifications in a siRNA, *Nucleic. Acids Res.,* 31, 589, 2003.
33. Sambrook, J. and Russell, D.W., *Molecular Cloning: A Laboratory Manual,* 3rd ed., Cold Spring Harbor Laboratory Press, New York, 2001.
34. Barstead, R., Genome-wide RNAi, *Curr. Opin. Chem. Biol.,* 5, 63, 2001.
35. Rubinson, D. A. et al., A lentivirus-based system to functionally silence genes in primary mammalian cells, stem cells and transgenic mice by RNA interference, *Nat. Genet.,* 33, 401, 2003.
36. Grabarek, J. B. et al., RNA interference by production of short hairpin dsRNA in ES cells, their differentiated derivatives, and in somatic cell lines, *Biotechniques,* 34, 734, 2003.
37. Kunath, T. et al., Transgenic RNA interference in ES cell-derived embryos recapitulates a genetic null phenotype, *Nat. Biotechnol.,* 21, 559, 2003.
38. Lois, C. et al., Germline transmission and tissue-specific expression ot transgenes delivered by lentiviral vectors, *Science,* 295, 868, 2002.
38. Sui, G. et al., A DNA vector-based RNAi technology to suppress gene expression in mammalian cells, *Proc. Natl. Acad. Sci. U.S.A.,* 99, 5515, 2002.
40. Ausubel, F. et al., *Short protocols in molecular biology,* 3rd ed., John Wiley & Sons, New York, 1995.
41. Grompe, M., Therapeutic liver repopulation for the treatment of metabolic liver diseases, *Hum. Cell,* 12, 171, 1999.
42. Banerjea, A. et al., Inhibition of HIV-1 by lentiviral vector-transduced siRNAs in T lymphocytes differentiated in SCID-hu mice and CD34+ progenitor cell-derived macrophages, *Mol. Ther.,* 8, 52, 2003.

21 Virus-Induced RNA Silencing in Plants

M. Barciszewska-Pacak, B. Grabowska, J. Wojciechowicz, A. Pacak, and M. Figlerowicz

CONTENTS

21.1 INTRODUCTION

According to the central dogma of molecular biology, the flow of genetic information proceeds from DNA through RNA to proteins. In the case of viruses, this principle concerns only those whose genome is composed of DNA. In viruses using RNA as the only carrier of genetic information, the above scheme is subjected to a simplification: RNA → proteins. As a result, during the replication of viral genomes, not double-stranded (ds) DNA, but dsRNA, is formed. Recently, it has been postulated that plants took advantage of this unique feature of RNA-based viruses and developed a specific antiviral system exploiting the RNA silencing phenomenon.[1–3] It is thought that a plant immune system against viruses is based on the same general mechanism that is involved in other processes of sequence-specific RNA degradation, e.g., posttranscriptional gene silencing (PTGS), and gene cosuppression observed in plants, gene quelling in fungi, or RNA interference (RNAi) in animals.[4,5]

This chapter provides basic information concerning the mechanism and potential application of virus-induced RNA silencing. It also contains several protocols that we currently use in our laboratory to study this phenomenon.

0-8493-2141-7/05/$0.00 + $.50

21.2 RNA SILENCING AS A SPECIFIC PLANT IMMUNE SYSTEM

Viruses are obligatory parasites, which means that they cannot function without a host organism. To establish an infection, a virus has to overcome physical barriers, enter a plant cell, and start to multiply genomic molecules. In RNA viruses the latter process is catalyzed by an RNA-dependent RNA polymerase (the enzyme synthesizes RNA using RNA as a template). Consequently, the dsRNA replication intermediate, which triggers the RNA silencing pathway, is generated in the virus-infected cells.[6,7] Most of the collected data indicate that the specific degradation of a virus genome occurs according to a similar scenario as RNAi or PTGS in animals and plants, respectively.[4,5,8] The viral replication intermediate becomes a substrate for the RNase III-like enzyme (plant analog of Dicer), which digests dsRNA into small interfering RNAs (siRNA). Then siRNA becomes part of an RNA-induced silencing complex (RISC) and targets it against viral genomic molecules (either against (–) or (+) RNA strands). This way the plant can restrict, or even eliminate, viral infection.[1–3] It was also demonstrated that the RNA-silencing signal is able to spread from the site of infection to more distant tissues. If it moves faster than the virus, RNA-based defense can also protect uninfected cells; in the other case systemic infection is established. Interestingly, as far as specific siRNAs are present in cells, the plant is resistant to reinfection or infection by a closely related virus. The above mechanism explains two phenomena frequently observed by plant virologists, that is, plant recovery from viral infections and cross-protection.[9]

This does not, however, mean that RNA-based viruses are not dangerous for plants. In the course of evolution, viruses have developed different strategies that help them to evade the RNA-based plant immune system.[10–14] During the last few years, several virus-encoded proteins functioning as RNA silencing suppressors were identified. They are highly diverse in size and structure and target different steps of the RNA-silencing pathway. Taking into account the biological activity of RNAi suppressors, they were divided into three groups:[14] (1) inhibitors of the Dicer or Dicer-like enzymes, which prevent siRNA formation and/or accumulation, e.g., HC-Pro protein encoded by potato virus Y); (2) inhibitors of a host-derived silencing signal, which most likely slow down dsRNA production by a cellular RNA-dependent RNA polymerase e.g., p25 protein encoded by potato virus X); and (3) inhibitors of systemic silencing, which probably interfere with the mobile silencing signal, e.g., 2b protein encoded by cucumber mosaic virus).

The recently proposed model of brome mosaic virus (BMV) replication is another interesting example showing how viruses can protect their genomes from dsRNA-mediated host defense.[15] BMV is a model (+) RNA virus of plants.[16] It was demonstrated that BMV genomic molecules are synthesized in 50- to 70-nm spherules budding into the endoplasmic reticulum membranes. They are formed by BMV replicase proteins: 1a (helicase and methyltransferase) and 2a (polymerase). The inner surface of these spherules is covered by the 1a protein anchoring to the cytoplasmic face of membranes. Each spherule contains hundreds of 1a and only two or three 2a proteins and forms a ribonuclease-resistant shell in which RNA replication occurs. It seems that only one RNA template and one nascent strand are located inside the spherule.

Initially the genomic (+) RNA strand serves as a template. It is copied and then released through a narrow neck connecting the spherule with cytoplasm. The (–) RNA strand remains inside and is used as a template for (+) RNA strands synthesis. This way the exposure of the dsRNA forming during replication to the RNases circulating in cytoplasm is reduced.

Studies involving genetically modified viruses carrying host-derived exon sequences in their genomes (either the entire coding region or its fragment) revealed that RNA-based defense is capable of targeting both viral genomic molecules and host mRNA. If such a modified virus enters a plant cell posttranscriptional silencing of a host gene is induced.[17] Plant gene expression was also inhibited during infection with a virus bearing a promoter sequence in its genome.[18–20] More detailed studies demonstrated that, in this case, the host gene is silenced at the transcriptional level due to its promoter's methylation (RNA-directed DNA methylation). In general, these observations showed that modified viruses (RNA silencing-inducing viral vectors) could be used as effective tools to suppress host gene expression. Thus, a new technique called virus-induced gene silencing (VIGS) was developed.[17,21]

21.3 VIGS

At present, one cannot say that VIGS is a widely used tool in plant functional genomics. There are, however, several examples of successful VIGS application that have proven its effectiveness in studies on plant genomes (Table 21.1). Most of these examples concern gene silencing in *Nicotiana benthamiana*, commonly used by plant virologists as a model organism since it can be infected with an unusually broad spectrum of viruses. The usefulness of VIGS was also demonstrated with other plants including *Arabidopsis thaliana*. At the moment, the major hindrance to more extensive exploiting of VIGS in functional genomics is the lack of a universal viral vector that would be capable of inducing RNA silencing in any plant. The construction of such a vector is most probably impossible since virus host range, although sometimes very wide, is always limited. It seems that the best solution to this problem is the production of numerous vectors and then the selection of those most effective within each plant group or family.

There are many advantages of VIGS that contribute to its high effectiveness. First, it exploits the natural capacity of plants to degrade RNA, which has been identified as a foreign genetic material. Second, it can operate at both transcriptional and posttranscriptional levels, Third, the preparation of the VIGS-mediating construct is not a complicated process. The target gene–derived sequence is simply inserted into a virus vector in sense or antisense orientation. The construction of sophisticated vectors carrying long inverted repeats is not only unnecessary but also unadvisable. One hundred to 150 nt inverted repeats can be easily removed from single-stranded RNA genomes by recombination.[33] However, there are examples indicating that in some cases, relatively short (40 to 60 nt) direct inverted repeats could be very effective VIGS triggers.[23] Fourth, VIGS can be triggered at practically any moment of plant development, and, therefore, it can be used to investigate genes essential for cell growth

TABLE 21.1
Examples of Successful VIGS Application in Functional Genomics of Plants

Virus	Type of virus	Vector description	Viral vector/virus delivery	Plant/silenced gene	Literature
BSMV (barley stripe mosaic virus)	(+)RNA with tripartite genome: αRNA, βRNA, γRNA	γRNA with VIGS-inducing sequence located downstream of γb coding sequence in sense, antisense orientation or as a direct inverted repeat	Mechanical inoculation with *in vitro* transcribed RNA	*Hordeum vulgare/PDS*[1]	*Holzberg et al. 2002* [22]; *Lacomme et al. 2003* [23]
BWYV (beet western yellows virus)	(+)RNA with monopartite genome	BWYV	Aphid inoculation with BWYV virus	*N. benthamiana* transgenic line expressing P74[2] protein/*P74* transgene	*Brault et al. 2002* [24]
PVX (potato virus X)	(+)RNA with monopartite genome	Genomic RNA with VIGS-inducing sequence located upstream of *CP*[3] gene in sense or antisense orientation DNA plasmid containing viral genome cDNA with VIGS-inducing sequence located upstream of *CP* gene in sense orientation	Mechanical inoculation with *in vitro* transcribed RNA 2-step inoculation:1st - DNA plasmid was introduced into *Nicotiana clevelandii*, 2nd - isolated virus was used to inoculate transgenic plants	*N. benthamiana/PDS* *N. benthamiana* transgenic line expressing GFP[4]/*GFP*	*Ruiz et al. 1998* [17]
PVX	(+)RNA with monopartite genome	Genomic RNA with VIGS-inducing sequence located upstream of *CP* gene in sense orientation	Mechanical inoculation with *in vitro* transcribed RNA	*N. benthamiana*/putative cellulose synthase gene	*Burton et al. 2000* [25] *Voinnet 2001* [1]
PVX	(+)RNA with monopartite genome	Ti[5] plasmid containing viral genome cDNA with VIGS inducing sequence located upstream or downstream of *CP* gene in sense orientation	*Agroinoculation*	*N. benthamiana* transgenic line 16c/*GFP* *N. benthamiana/PDS, rbcS*[6], *FLO:LFY* (NFL)[7]	*Ratcliff et al. 2001* [26]

TMV (tobacco mosaic virus)	(+)RNA with monopartite genome	Genomic RNA with VIGS inducing sequence located between movement protein and *CP* genes in sense, antisense orientation or as a direct inverted-repeat or hairpin-loop	Mechanical inoculation with *in vitro* transcribed RNA	*N. benthamiana* expressing constitutively *GFP* transgene/*GFP* *N. benthamiana*/*PDS*	*Lacomme et al. 2003* [23]
TRV (tobacco rattle virus)	(+)RNA with bipartite genome: RNA1, RNA2	Ti[5] plasmid containing cDNA of RNA2 with VIGS inducing sequence located downstream of *CP* gene in sense orientation	*Agroinoculation*	*N. benthamiana* transgenic line 16c/*GFP* *N. benthamiana*/*PDS, rbcS, FLO:LFY*(NFL)	*Ratcliff et al. 2001* [26]
TRV	(+)RNA with bipartite genome: RNA1, RNA2	Plasmids containing cDNA of RNA2 with VIGS induced sequences located downstream of *CP* gene in sense orientation	*Agroinoculation*, cultures with RNA1 were mixed with cultures containing RNA2 in a 1:1 ratio prior to infiltration	*N. benthamiana*/*SGT1*[8] *N. benthamiana* transgenic lines containing different *R*[9] genes/transgenic lines containing different *R* genes/*R* genes	*Peart et al. 2002* [27] *SGTN. benthamiana*
TRV	(+)RNA with bipartite genome: RNA1, RNA2	Ti plasmids containing cDNA of RNA2 with VIGS inducing sequences located downstream of *CP* gene in sense orientation	*Agroinoculation*, cultures with RNA1 and RNA2 were mixed in a 1:1 ratio and infiltrated or sprayed	*Lycopersicon esculentum*/*PDS*, *CTR1*[10], *CTR2*	*Liu et al. 2002* [28]
TRV	(+)RNA with bipartite genome: RNA1, RNA2	Transcribable cDNA of genomic RNA2 with VIGS inducing sequence, located downstream of *CP* gene in sense orientation	2-step inoculation: 1st inoculation with transcripts (*N. benthamiana*) 2nd inoculation with TRV extracted from *N. benthamiana* leaves	*A. thaliana*/*PDS*	*Yu et al. 2003* [29]
CbLCV (cabbage leaf curl virus)	ssDNA[11] with bipartite genome: DNA A, DNA B	DNA plasmid containing DNA A with fragments of different sequences cloned into deleted *AR1*[12] gene	Biolistic delivery of two plasmids containing DNA A and DNA B	*A. thaliana/CH42*[13], *PDSA. thaliana* transgenic line expressing GFP/*GFP*	*Turnage et al. 2002* [30]

(*Continued*)

TABLE 21.1
Continued

Virus	Type of virus	Vector description	Viral vector/virus delivery	Plant/silenced gene	Literature
TGMV (tomato golden mosaic virus)	ssDNA with bipartite genome: DNA A, DNA B	DNA plasmid containing DNA A with fragments of sense or antisense sequences cloned into deleted *AR1* gene	Biolistic delivery of two plasmids containing DNA A and DNA B or only one plasmid with DNA A	*N. benthamiana/luc*[14] transgene, *su*[15]	*Kjemtrup et al. 1998* [31] *Voinnet 2001* [1]
TYDV (tobacco yellow dwarf virus)	ssDNA with monopartite genome	Ti[5] plasmid containing fragment of TYDV cDNA and *Chs*A[16] gene in sense orientation located down-stream of LIR-V1[17] of TYDV	*Agrobacterium tumefaciens* transformation	*Petunia hybrida* transgenic line producing viral episomes/*Chs*A gene and transgene	*Atkinson et al. 1998* [32] *Voinnet 2001* [1]

Appendix:
[1]phytoene desaturase; [2]minor capsid protein; [3]coat protein; [4]green fluorescent protein; [5]tumor inducing; [6]ribulose bisphosphate carboxylase small subunit; [7]homologue of the *Antirrhinum FLORICAULA* gene and the *LEAFY* gene from *A. thaliana* involved in flower development; [8]homologue of the yeast ubiquitin ligase-associated protein; [9]genes mediating disease resistance: *N* gene from *Nicotiana glutinosa* encodes a TIR(toll and interleukin-1 receptor)-NBS(nucleotide–binding site)-LRR(leucine-rich repeat) protein and confers TMV resistance, R_x gene from *Solanum tuberosum* encodes a CC(coiled coil)-NBS-LRR protein that mediates PVX resistance, *Pto* gene from *Lycopersicon esculentum* that encodes a serine-threonine kinase and mediates resistance against *Pseudomonas syringae*); [10]constitutive triple response 1: in *A. thaliana* gene encodes a Raf-like mitogen activating protein kinase kinase kinase-MAPKKK; [11]member of the two-component *Begomovirus* genus which encodes five or six genes on DNA-A (one or two on the virion-sense strand (termed AV) and four in the complementary-sense strand (termed AC). DNA-B encodes two proteins, one on the virion-sense strand BV and one on the complementary strand BC; [12]encodes coat protein, which is required for transmission by insect vector; [13]chlorata 42-encodes a component of the magnesium chelatase complex required for chlorophyll production; [14]luciferase; [15]sulfur allele of magnesium chelatase, an enzyme required for chlorophyll formation; [16]chalcone synthase A gene; [17]large intergenic region V1

or differentiation. Last, VIGS phenotype develops quite fast, usually within 10 to 15 days after virus vector delivery into plant cells.

On the other hand there are two general disadvantages of VIGS. First, it involves genetically modified viruses and it is difficult to predict how an inserted plant gene-derived sequence influences virus photogenecity. Second, in some cases it is not easy to distinguish between effects coming from gene silencing and infection symptoms.

21.4 PRACTICAL ASPECT OF VIGS

21.4.1 Viral Vectors

Although any RNA-based virus is capable of inducing RNA silencing, in practice their application as vectors is limited to those for which a full-length cDNA clone of genomic molecule (or molecules) is available. Furthermore, to create a VIGS vector, viral cDNA has to be modified to enable insertion of the host gene-derived sequence. Despite the introduced changes, the virus vector has to retain its infectivity and ability to replicate and spread in the plant of interest. In addition, the effective methods enabling both the synthesis of infectious viral RNA and its delivery into plant cells should be available. Depending on how these problems are solved, two different types of viral vectors can be distinguished: classical and agrobacterial vectors. To prepare a vector of the first type, viral cDNA is inserted into a regular plasmid (e.g., pUC or pGEM) under the RNA polymerase promoter (e.g., T7 or SP6 polymerase). The linearized plasmid is used as a template for viral RNA synthesis by *in vitro* transcription (see Protocol 21.1). The infectious transcript is introduced into plant cells by mechanical inoculation (Protocol 21.2). For some plants, mechanical inoculation with *in vitro* transcribed RNA is inefficient. To omit this problem, one can propagate the virus vector in another, more susceptible host, isolate viral particles (Protocol 21.3), and then mechanically inoculate the plant of interest with a virus or with viral RNA (Protocol 21.4). However, in this situation vector stability should be analyzed prior to inoculation. To prepare an agrobacterial vector, viral cDNA, accompanied by a promoter (frequently 35S promoter from cauliflower mosaic virus is used) and a transcriptional terminator, is ligated into the Ti plasmid within the T-DNA region (Ti plasmids are routinely used for plant genome transformation). Then the vector is transferred into *Agrobacterium* (for example, by electroporation) (Protocols 21.5 and 21.6), and in this form it is introduced into plant cells by agro-inoculation (Protocol 21.7). Viral genomic RNA is synthesized *in planta*, based on cDNA present in T-DNA.

At present, classical vectors are more frequently used since they are more easily accessible. It seems, however, that in the near future they will be replaced by agrobacterial vectors because the latter make VIGS technology much simpler. In the case of agrobacterial vectors, there is no necessity for large-scale plasmid isolation, plasmid linearization, or *in vitro* transcription. It is especially important if one wishes to screen a large number of genes using a cDNA library. Moreover, recent observations made by Baulcombe and coworkers indicate that, contrary to earlier expectations, agro-inoculation is the most efficient way of introducing cDNA-derived viral RNA into plant cells.[34]

Several examples of viral vectors used in functional genomics of plants are presented in Table 21.1. Two series of classical BMV-based vectors are currently being tested in our laboratory. The first one was obtained by the duplication of the BMV RNA3 sub-genomic promoter just behind the coat protein-coding region.[35] The second series was constructed based on the BMV vector we use to study the RNA recombination process (Figure 21.1).

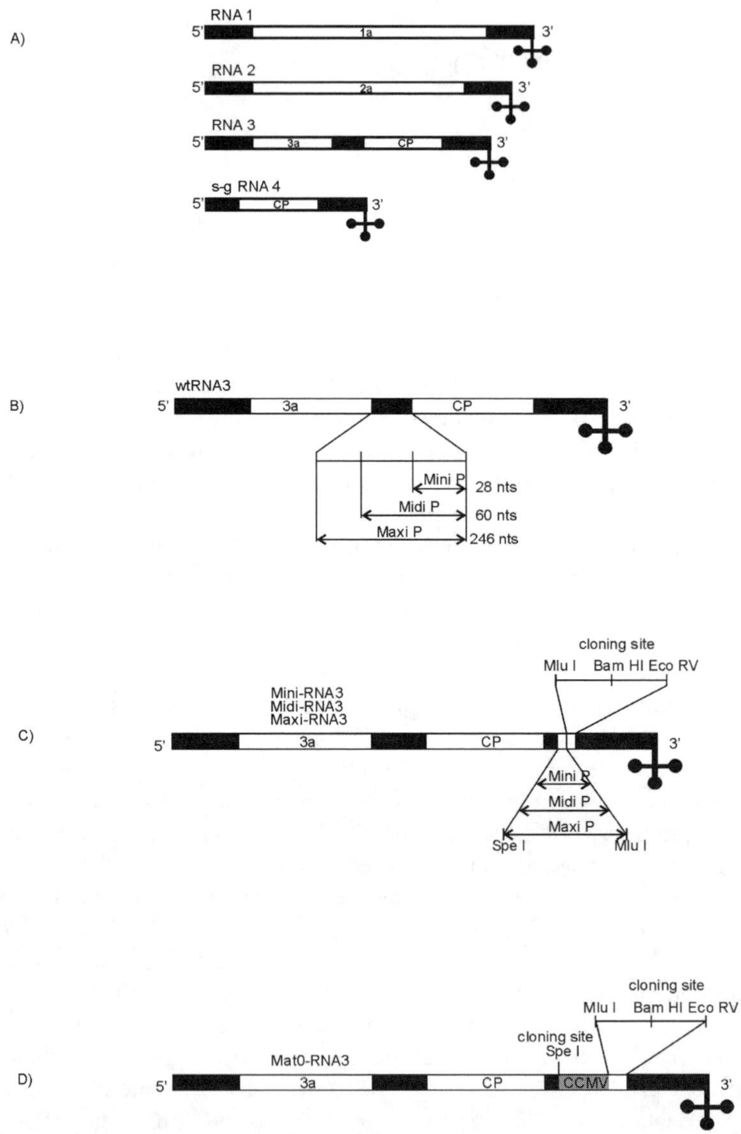

FIGURE 21.1 A schematic description of BMV-based vectors. **A.** The BMV genome. The BMV genome is composed of three RNA molecules called RNA1, RNA2, and RNA3 (open reading frames are boxed and labeled, while 3'-, 5'-, and intergenic noncoding regions are represented by black boxes). RNA1 and RNA2 encode replicase proteins 1a and 2a, respectively, whereas RNA3 encodes the movement protein (3a) and the coat protein (CP). In addition,

21.4.2 Plant Gene-Derived Sequence

There are no exact rules describing precisely what sequence should be inserted into a viral vector to efficiently mediate gene silencing, how long the insert has to be, to which portion of mRNA it should be complementary or homologous, how to select a sequence to target the entire gene family or, conversely, a single gene, and how to avoid the silencing of the closely related genes. An answer to these questions can be found only based on available experimental data and the existing model of RNA silencing. Theoretically, each 21- to 25-nt-long sequence that perfectly matches the gene of interest should be sufficient for its silencing. But the experimental data suggest that usually a 100-nt insert or a longer one is required. If a shorter insert is used, VIGS symptoms are often unclear or transient. On the other hand, too long an insert can destabilize the vector (sequence longer than 500 to 700 nt depending on the vector). There are no evident observations indicating which region of mature mRNA is a better target — coding or noncoding. However, it should be kept in mind that highly structured regions could be especially difficult to access by siRNA. Therefore, any information concerning the secondary and tertiary structure of targeted mRNA would be very helpful. For example, one can try to predict mRNA structure either by using a computer program or by probing it with enzymes or chemical agents specific for single-stranded and double-stranded RNA. Another very complex and still unsolved problem is how sequence similarity between related genes affects their silencing (so- called cross-silencing phenomenon). It seems that cross-silencing may, for the most part, depend on the extent of perfect identity between two mRNAs.

21.4.3 VIGS Assessment

Before one applies VIGS to analyze plant gene function, it is very important to determine how infection with an "empty" virus vector (vector without insert) develops and

FIGURE 21.1 Continued
during viral infection subgenomic RNA4 corresponding to the 3' portion of RNA3 is formed. All genomic molecules have almost identical 3'-noncoding regions, with a tRNA-like structure at the very end. **B.** Wild-type RNA3 molecule. The promoter of subgenomic RNA4 synthesis is located within the intergenic region of wtRNA3 (on the (–) RNA3 strand). The minimal portion of the promoter (28 nucleotides) that is required for RNA synthesis is marked as MiniP. The full-length promoter sequence is called MidiP (60 nucleotides), whereas the entire intergenic RNA fragment (246 nucleotide fragment including the promoter) is marked as MaxiP. **C.** Mini-, Midi-, and Maxi- RNA3 vectors. In Mini-, Midi-, and Maxi- RNA3 vectors, the 3'-noncoding region of RNA3 was modified. The promoter sequence (either MiniP, MidiP, or MaxiP) was inserted just behind the CP coding region. Then the cloning site where VIGS-inducing sequence can be inserted was introduced. **D.** Mat0-RNA3 vector. There is no subgenomic promoter duplication in Mat0-RNA3. Its 3'-noncoding sequence was modified as follows: a cloning site was created just behind the CP coding sequence, then a sequence and from the related bromovirus was inserted (CCMV-derived sequence) together with the second cloning site. The same VIGS-inducing sequence can be inserted into this vector in two different orientations. Alternatively, the CCMV-derived sequence can be removed and the VIGS-inducing sequence can be ligated in one orientation — sense or antisense, e.g., between *Spe*I and *Eco*RV restriction endonuclease sites).

how it influences the plant phenotype. Then optimal experimental conditions should be established, since any environmental changes can strongly affect a VIGS phenotype. For this purpose, constructs inducing apparent silencing symptoms, e.g., vectors targeting the phytoene desaturase (PDS) gene, are used (PDS is involved in the synthesis of carotenoid pigments and its absence results in photobleaching). For the proper assessment of VIGS symptoms, it is also very important to determine if a viral vector encodes any of RNA-silencing suppressors, or not. Only such a well-defined system can be used in functional genomics of plants. Finally, it also has to be admitted that, although very effective, VIGS usually does not cause the gene to be suppressed 100%. The actual changes in the target mRNA and protein levels always need to be established by standard northern or western blot analysis, respectively. In addition, the activation of VIGS should be confirmed by demonstrating target gene-specific siRNA formation (Protocols 21.8, 21.9, and 21.10).

PROTOCOLS

The 10 protocols we use to study the RNA-silencing phenomenon are presented below. Procedures enabling VIGS induction with classical BMV-based vectors are described in Protocols 21.1 to 21.4. They can be easily adapted to other viral vectors. Protocols 21.5 to 21.7 provide information necessary to induce VIGS with Ti plasmid-based vectors. At present, Ti plasmids carrying BMV genomic molecules are not available. That is why we present the typical procedure of plant cell transformation with any Ti plasmid (e.g., PTGS-inducing Ti plasmid bearing the template for dsRNA synthesis). However, one should remember that to induce VIGS, all components of the viral genome have to be delivered into the plant cell. If a virus has a segmented genome, agro-inoculate the plant with a mixture of two or three *A. tumefaciens* strains. Each of them is transformed with a different Ti plasmid. Consequently, all segments of the viral genome are synthesized *in planta*. Finally, Protocols 21.8 to 21.10 present standard methods of siRNA detection and analysis.

PROTOCOL 21.1 *IN VITRO* SYNTHESIS OF INFECTIOUS BMV RNA1, RNA2, AND MODIFIED RNA3

It is important to avoid any RNase contamination. All reagents should be RNase-free. Use sterile disposable plastic ware; nondisposable glass and plastic ware should be treated with 0.2 M NaOH – 0.1% (w/v) SDS and washed with RNase-free water before use.

Materials

- Plasmids carrying cDNAs of BMV, RNA1, RNA2, and modified RNA3 (VIGS-inducing sequence is ligated to the latter plasmid — see Figure 21.1)
- *In vitro* transcription reagents (T7 RNA polymerase 20 U/µl and 5 × transcription buffer, RNasin RNase inhibitor 40 U/µl, RQI DNase 1 U/µl, 100 mM

Dithiothreitol, ATP, CTP, UTP mixture (each in 5-mM conc.)), and 4 mM GTP (all from Promega); 10 mM GpppG (Cap) from New England BioLabs

- *Eco*RI or other appropriate restriction endonuclease (Promega)
- Phenol pH < 7.0
- Chloroform
- 3 M sodium acetate pH 4.8
- 96% (v/v) and 70% (v/v) ethanol
- Centrifuge (Eppendorf 5415R) for standard 1.5-ml Eppendorf tubes
- Incubator

Experimental Procedures

Linearization of the plasmids carrying cDNAs of BMV RNA1, RNA2 and modified RNA3

1. Linearize each plasmid separately. Reaction mixture: 20 μg of plasmid (sufficient for three *in vitro* transcription reactions), 20 to 30 U of appropriate restriction enzyme (*Eco*RI), 10 μl of buffer, water up to 100 μl. Incubate for 4 to 8 h (or overnight, if necessary) at 37°C.
2. Purify DNA by phenol-chloroform extraction and precipitate. Dissolve DNA in 30 μl of water. Alternatively, DNA can be purified on a spin column (e.g., Qiagen's PCR Purification Kit). Check plasmids linearization by agarose gel electrophoresis.

In vitro transcription

3. Synthesize BMV RNA1, RNA2, and modified RNA3 in three separate reactions. Reaction mixture (for 50 μl): 10 μl 5 × transcription buffer, 0.5 μl RNase inhibitor, 5 μl 100 mM DTT, 4 μl 5 mM ATP, CTP, UTP mixture, 1 μl 4 mM GTP, 3 μl 10 mM GpppG (Cap), 5 to 7 μg of linearized DNA, 0.5 μl T7 RNA polymerase 20 U/μl, RNase-free water up to 50 μl.
4. Incubate reaction mixture at 37°C for 60 to 90 min; then add 2.5 μl RQI DNase and further incubate at 37°C for 15 min.
5. Add an equal volume (50 μl) of H_2O, phenol, and chloroform; vortex for 30 sec; centrifuge at 10,000 rpm for 5 min at 4°C; and transfer the upper aqueous phase into a new tube.
6. Add 100 μl of chloroform, vortex for 30 sec, centrifuge at 10,000 rpm for 5 min at 4°C, and transfer the upper aqueous phase into the new tube.
7. Precipitate RNA by adding 10 μl of 3 M sodium acetate, pH 4.8, and 300 μl of 96% (v/v) ethanol. Chill the mixture to –20°C (for 60 min or more) and centrifuge for 10 min at 14,000 rpm at 4°C. Carefully remove the supernatant and dry the pellet in SpeedVac for 3 to 5 min.
8. Dissolve pellet in 45 μl of RNase-free water.

Technical notes. In this form, RNA is ready to be used for mechanical inoculation. It can also be stored at –70°C. Before inoculation, check RNA quality by electrophoresis in a 1% agarose gel (load 2 to 5 μl of transcript solution to each well). As an alternative to this protocol, the mMESSAGE mMACHINE™ Kit from Ambion can be used to synthesize RNA transcripts according to the manufacturer's recommendations.

PROTOCOL 21.2 MECHANICAL INOCULATION

Keep plants in the dark for 6 h before inoculation. The following procedure can also be applied to inoculate plants with a virus or viral RNA.

Materials

- 5 × inoculation buffer, pH 9.2 (50 mM glycine, 30 mM K_2HPO_4, 0.1% bentonit, 0.1% celit (the last two from Sigma))
- Carborund powder (Sigma)
- *In vitro* synthesized BMV RNA1, RNA2, and modified RNA3

Experimental Procedures

1. Prepare the inoculation mix using RNA obtained by *in vitro* transcription (see Protocol 21.1): 40 µl RNA1, 40 µl RNA2, 40 µl RNA3, 40 µl H_2O, 40 µl 5 × inoculation buffer. This amount of RNA is sufficient to inoculate five 2-week *Hordeum vulgare* or two 5-leaf *Chenopodium quinoa* plants.
2. Dust plant leaves with carborund and then, step by step, inoculate each leaf in the following way: pipette 30 to 50 µl of the inoculation mixture onto a leaf and infect the plant by rubbing the leaf with a gloved finger moistened with inoculum. Try not to damage the leaves.
3. After 5 min, spray the inoculated plants with water to remove carborund, keep plants in the dark for 12 h and transfer to a greenhouse.

PROTOCOL 21.3. BMV ISOLATION

Materials

- Extraction buffer (0.5 M sodium acetate, 0.3 M acetic acid, 0.01 M magnesium chloride)
- Storage buffer (0.05 M sodium acetate, 0.01 M acetic acid, 1 mM Na_2EDTA, 1 mM magnesium chloride)
- Chloroform
- 30% (w/v) polyethylene glycol PEG (MW 8000)
- Centrifuge tubes suitable for organic solvents
- Pestle and mortar
- Liquid nitrogen
- Centrifuge (Eppendorf 5415R) for standard 1.5-ml Eppendorf tubes
- Spectrophotometer

Experimental Procedures

1. Collect 1 g of infected leaves (with symptoms) and grind them using pestle and mortar prechilled with liquid nitrogen.
2. Mix the powder with 1 ml of the extraction buffer, transfer into a centrifuge tube, add 0.2 ml of chloroform, and vortex for 30 sec.

4. Centrifuge the suspension for 5 min at 5000 rpm at 4°C (use a precooled rotor).
5. Transfer the upper phase into a new tube. Add 1/3 volume of 30% PEG (c.a. 0.3 ml) and vortex for 1 min. Incubate on ice for 30 min.
6. Precipitate the virus by centrifugation for 10 min at 14,000 rpm at 4°C. Discard the supernatant and dissolve the pellet in 0.2 ml of the storage buffer.
7. Add 0.2 ml of chloroform, vortex for 1 min, and centrifuge for 5 min at 5000 rpm at 4°C.
8. Transfer the upper phase into a new tube. Add 1/3 volume of 30% PEG (c.a. 0.07 ml) and vortex for 1 min. Incubate on ice for 30 min.
9. Pellet the virus by centrifuging for 10 min at 14,000 rpm at 4°C. Remove supernatant and resuspend the virus in 0.2 ml of storage buffer. Leave suspension on ice overnight.
10. Determine virus concentration by measuring UV light absorption at 260 nm. Store the virus at –70°C.

Protocol 21.4 BMV RNA Isolation

Materials

- 10 × RNA extraction buffer, pH 9.0 (0.5 M glycine, 0.5 M sodium chloride, 0.1 M EDTA)
- Phenol pH < 7
- Chloroform
- 10% SDS (w/v)
- 96% (v/v) and 70% (v/v) ethanol
- Centrifuge (Eppendorf 5415R) for standard 1.5-ml Eppendorf tubes
- SpeedVac

Experimental Procedures

1. Pipette 200 µl of purified virus into a 1.5-ml Eppendorf tube; add 25 µl of 10% (w/v) SDS, 25 µl of 10 × RNA extraction buffer, and 250 µl of phenol. Vortex the mixture for 30 sec and separate the phases by centrifuging for 5 min at 14,000 rpm.
2. Transfer the upper (aqueous) phase into a new tube, add 125 µl of phenol and 125 µl of chloroform, vortex for 30 sec, and centrifuge again for 5 min at 14,000 rpm.
3. Transfer the upper phase to a new tube and extract with 250 µl of chloroform, vortex for 30 sec, and centrifuge for 5 min at 14,000 rpm.
4. Transfer the upper phase to a new tube and precipitate RNA by adding 2.5 to 3.0 volumes of 96% (v/v) ethanol. Keep sample on ice for 30 min and then centrifuge for 30 min at 14,000 rpm at 4°C.
5. Discard supernatant and wash the pellet by vortexing for 30 sec with 200 µl of 70% (v/v) ethanol. Centrifuge for 15 min at 14,000 rpm at 4°C. Discard the supernatant, dry the pellet in SpeedVac for 3 to 5 min and dissolve in 30 µl RNase-free water. Use RNA immediately or store at –70°C.

Protocol 21.5 Preparation of Electro-Competent *A. tumefaciens* Cells

Materials

- *A. tumefaciens* strain, e.g., hypervirulent EHA105 strain or any other used for plant cell transformation; can be obtained from any laboratory producing transgenic plants
- Solid AB medium (K_2HPO_4 3 g, NaH_2PO_4 1 g, NH_4Cl 1 g, KCl 0.15 g, agar 15 g; add water to 1 liter and sterilize by autoclaving)
- Liquid YEB medium (beef extract 5 g, yeast extract 1 g, peptone 5 g, sucrose 5 g, $MgSO_4$ 240 mg, pH 7.2, add water to 1 liter and sterilize) with selective antibiotic — rifampicin (final concentration 100 mg/l) from Sigma
- 10% (v/v) glycerol (filter-sterilized)
- 250-ml centrifuge tubes
- Liquid nitrogen
- Spectrophotometer
- Centrifuge Eppendorf 5804 R
- Incubator for bacterial cultures

Experimental Procedures

1. Take *A. tumefaciens* cells from a glycerol stock, spread on solid AB minimal medium containing rifampicin (the *A. tumefaciens* strain we use is resistant to this antibiotic), and incubate at 28°C until single colonies develop.
2. Pick up a single colony from the plate and use it to inoculate 10 ml of liquid YEB medium containing rifampicin. Incubate bacteria overnight at 28°C, with shaking at 280 rpm.
3. Transfer 2.5 ml of the overnight culture to 250 ml of liquid YEB medium with antibiotic and incubate at 28°C with shaking at 280 rpm until its optical density (OD_{600}) reaches 0.5 to 0.8.
4. Place the culture on ice for 30 min and then harvest bacteria by centrifuging for 10 to 15 min at 4500 rpm at 4°C.
5. Discard the supernatant and wash the pellet by resuspending it in the same volume (250 ml) of cold sterile water. Centrifuge the bacteria suspension for 15 min at 5000 rpm at 4°C.
6. Discard supernatant and wash again with the same volume of water. Then resuspend the bacterial pellet in 50 ml of ice-cold 10% (v/v) sterile glycerol. Centrifuge bacteria for 15 min at 5000 rpm at 4°C.
7. Discard supernatant and resuspend pellet in 2 ml of ice-cold 10% (v/v) glycerol. Use bacteria immediately or store at –70°C in 50-μl aliquots.

Protocol 21.6 Transformation of *A. tumefaciens* Cells by Electroporation

Materials

- Ti plasmid (or plasmids) of interest
- Electro-competent *A. tumefaciens* cells

- Liquid SOC medium (tryptone 20 g, yeast extract 5 g, NaCl 0.5 g, 250 mM KCl 10 ml, 2 M $MgCl_2$ 5 ml; add water to 980 ml and sterilize by autoclaving; cool down and add 20 ml of 1 M sterile glucose)
- Solid YEB medium (beef extract 5 g, yeast extract 1 g, peptone 5 g, sucrose 5 g, $MgSO_4$ 240 mg; add water to 1 liter; adjust pH to 7.2; add agar 15 g and sterilize by autoclaving) with selective antibiotics (first antibiotic specific for *Agrobacterium* — rifampicin and the second one specific for Ti plasmid)
- 10-ml tubes
- Electroporator (Bio-Rad Model 1652077 from Bio-Rad) and 2-mm electroporation cuvettes (from Bio-Rad)
- Incubator for bacterial cultures

Experimental Procedures

1. Thaw a 50-μl aliquot of electro-competent *A. tumefaciens* cells on ice for 10 to 15 min and add 50 to 100 ng of the Ti plasmid (not more than 10% (v/v) of the bacterial stock volume). Mix gently and incubate on ice for 5 min.
2. Transfer the mixture of bacteria and the plasmid to a prechilled 2.0-mm electroporation cuvette.
3. Perform electroporation using the following conditions: capacitance 25 μF, voltage 2.5 kV, and resistance 200 ohms.
4. Immediately add 1 ml of room-temperature SOC medium to the cuvette and resuspend bacteria
5. Transfer them to a 10-ml tube and incubate for 3 h at 28°C, with shaking at 50 to 100 rpm.
6. Select transformants by plating 200 to 500 μl of bacterial culture on YEB-agar containing the appropriate antibiotics.
7. Incubate the plate at 28°C in dark. Colonies develop after 48 h.

Protocol 21.7 Agro-Inoculation *N. Benthamiana*

Materials

- *A. tumefaciens* strain transformed with Ti plasmid
- Liquid YEB medium with appropriate antibiotics
- Inoculation buffer (10 mM $MgCl_2$, 10 mM MES, pH 5.7, 150 μM acetosyringon) (from Sigma)
- Incubator for bacterial culture
- Spectrophotometer
- Centrifuge Eppendorf 5804 R
- 50-ml tubes
- 2-ml syringe

Experimental Procedures

1. Pick up a single colony from the plate (see Protocol 21.6) and use it to inoculate 5 ml of liquid YEB medium containing appropriate antibiotics. Incubate the culture overnight at 28°C, with shaking at 250 rpm.

2. Transfer 0.1 ml of the overnight culture to 10 ml prewarmed YEB medium containing antibiotics and acetosyringone (20 μM). Incubate the culture at 28°C with shaking at 250 rpm until its optical density (OD_{600}) reaches 1.0.
3. Harvest the bacteria by centrifuging in a 50-ml tube at room temperature for 10 min at 6000 rpm. Carefully discard the supernatant and gently resuspend the bacterial pellet in the same volume (10 ml) of inoculation buffer (do not use vortex but pipette with a 1-ml tip).
4. Centrifuge the bacterial suspension again for 10 min at 6000 rpm. Discard the supernatant and gently resuspend pellet in the same volume of inoculation buffer (OD_{600} should still be close to 1.0).
5. For agro-inoculation, use four-six leaf *N. benthamiana* plants. Infiltrate 0.5 to 1.0 ml of bacterial culture into a leaf by using a 2- or 3-ml syringe without needle — simply place the leaf between your finger and the syringe (the finger should lie on the upper surface of leaf) and then press the syringe against the lower surface of the leaf. For better results, pierce the inoculated fragment of the leaf with a needle before agro-infiltration. First silencing symptoms are observed 2 to 5 days after infiltration.

Protocol 21.8 Isolation of the Total RNA from Plant Tissue

Materials

- 10 × RNA extraction buffer (0.5 M glycine, 0.5 M sodium chloride, 0.1 M EDTA, pH 9.0)
- Phenol pH < 7.0
- Chloroform
- 10% (w/v) SDS
- 96% (v/v) and 70% (v/v) ethanol
- Liquid nitrogen
- Centrifuge
- Pestle and mortar
- Vortex mixer
- SpeedVac

Experimental Procedures

1. Collect plant material (fragment of leaf). For one extraction, use c.a. 100 mg of plant tissue.
2. Grind leaves using pestle and mortar prechilled with liquid nitrogen.
3. Mix the powder with 200 μl of freshly prepared 1 × extraction buffer (1 ml of 10 × extraction buffer, 1 ml of 10% (w/v) SDS, and 8 ml of water), and transfer the suspension into a tube.
4. Add 200 μl of phenol, vortex for 30 sec, and centrifuge for 5 min at 10,000 rpm at 4°C using a prechilled rotor.

5. Transfer the upper (aqueous) phase into a new tube. Add 100 μl of phenol and 100 μl of chloroform. Vortex and centrifuge for 5 min at 10,000 rpm at 4°C.
6. Transfer the upper phase into a new tube and add 200 μl of chloroform. Vortex and centrifuge for 5 min at 10,000 rpm at 4°C.
7. Transfer the upper phase into a new tube and precipitate RNA by adding 700 μl of 96% (v/v) ethanol. Leave the sample on ice for 30 min. Then centrifuge for 30 min at 14,000 rpm at 4°C.
8. Discard the supernatant, wash the pellet with 500 μl of 70% (v/v) ethanol and centrifuge for 10 min at 14,000 rpm at 4°C.
9. Discard the supernatant, dry the pellet in SpeedVac for 3 to 5 min and dissolve in 30 μl of RNase-free water. Store RNA at –70°C.

N.B. Alternatively, an RNeasy Plant Mini Kit (Qiagen) can be used for total RNA extraction from plant tissue.

Protocol 21.9 siRNA Detection

Materials

- Total RNA extracted from plant tissue (see Protocol 21.8)
- Standard reagents, buffers, and equipment for denaturating agarose and polyacrylamide gel electrophoresis
- Ethidium bromide
- Heating block
- UV transilluminator

Experimental Procedures

Electrophoresis in a denaturating agarose gel

1. Add 1 volume of 5 × RNA loading buffer to 4 volumes of the total RNA samples and mix.
2. Denaturate RNA samples by heating to 65°C for 3 to 5 min and then immediately cool them by putting on ice. Load the samples and run in a 2.5 to 3.0% (w/v) denaturating agarose gel under standard conditions.
3. Visualize RNA by ethidium bromide staining. SiRNAs are usually well visible under UV light if at least 10 to 15 μg of total RNA are loaded.

Electrophoresis in a denaturating polyacrylamide gel

1. Mix 4 volume of total RNA (c.a. 15 mg) with 1 volume of 5 × RNA loading buffer.
2. Denaturate samples by heating to 95°C for 3 min and immediately place on ice.
3. Load the samples on a 15 to 20% (w/v) denaturating polyacrylamide gel and run under standard conditions.
4. After electrophoresis is completed, remove the gel from the glass plates and transfer it to a dish containing 1 × TBE buffer with ethidium bromide for 1 to 2 h at 4°C. Visualize siRNA under UV light.

PROTOCOL 21.10 PREPARATION OF THE PROBE FOR TARGET GENE-SPECIFIC siRNA IDENTIFICATION

To confirm that the detected siRNA was formed due to target mRNA degradation, perform standard Northern blot analysis using *in vitro* synthesized radio-labeled siRNA as a probe. Briefly transfer total RNA separated in agarose or polyacrylamide gel onto a membrane and hybridize with radio-labeled siRNA. To obtain the latter, synthesize radiolabeled dsRNA (by *in vitro* transcription — Protocol 21.1) and digest it with Dicer.

Materials

- *In vitro* transcription reagents (see Protocol 21.1)
- Standard reagents and equipment for Northern blot analysis and polymerase chain reaction (PCR)
- Recombinant Human Dicer Enzyme Kit (Gene Therapy Systems)

Experimental Procedures

1. Synthesize DNA templates for *in vitro* transcription. To this end, amplify the VIGS-inducing insert in two separate PCRs. In the first reaction, the forward primer should contain the T7 polymerase promoter, while in the second reaction, the same promoter should be present in a reverse primer.
2. Prepare separately two complementary RNA molecules by *in vitro* transcription (Protocol 21.1). Set up two reactions — each involving radio-labeled nucleotide triphosphate and one of the PCR products as a template (do not use cap).
3. Obtain radio-labeled dsRNA — mix complementary RNAs in equimolar ratio and anneal them by heating to 95°C and then cooling slowly.
4. Produce radio-labeled siRNA by dsRNA digestion with the commercially available recombinant human Dicer according to the producer's recommendations.
5. Denature *in vitro* prepared siRNA and use them as a probe for northern blot analysis. Reduce hybridization temperature to 45°C.

ACKNOWLEDGMENTS

We are grateful to the Polish Government for support of this work through a grant from the State Committee for Scientific Research (grant number PBZ-KBN-040/P04/2001).

REFERENCES

1. Voinnet, O., RNA silencing as a plant immune system against viruses, *Trends Genet.*, 17, 449, 2001.
2. Plasterk, R.H., RNA silencing: the genome's immune system, *Science,* 296, 1263, 2002.
3. Waterhouse, P.M., Wang, M.B., and Lough, T., Gene silencing as an adaptive defense against viruses, *Nature,* 411, 834, 2001.

4. Tijsterman, M., Ketting, R.F., and Plasterk, R.H., The genetics of RNA silencing, *Annu. Rev. Genet.*, 36, 469, 2002.
5. Szweykowska-Kulińska, Z., Jarmołowski, A., and Figlerowicz, M., RNA interference and its role in the regulation of eucaryotic gene expression, *Acta Bioch. Pol.*, 50, 217, 2003.
6. Waterhouse, P.M, Graham, M.W., and Wang, M.B., Virus resistance and gene silencing in plants can be induced by simultaneous expression of sense and antisense RNA, *Proc. Natl. Acad. Sci. U. S. A.*, 95, 13959, 1998.
7. Tenllado, F. and Diaz-Ruiz, J.R., Double-stranded RNA-mediated interference with plant virus infection, *J. Virol.*, 75, 12288, 2001.
8. Hannon, G., RNA interference, *Nature*, 418, 244, 2002.
9. Ratcliff, F.G., MacFarlane, S.A., and Baulcombe, D.C., Gene silencing without DNA. RNA-mediated cross-protection between viruses, *Plant Cell*, 11, 1207, 1999.
10. Marathe, R. et al., RNA viruses as inducers, suppressors and targets of post-transcriptional gene silencing, *Plant Mol. Biol.*, 43, 295, 2000.
11. Vance, V. and Vaucheret, H., RNA silencing in plants — defense and counterdefense, *Science*, 292, 2277, 2001.
12. Voinnet, O., Pinto, Y.M., and Baulcombe, D.C., Suppression of gene silencing: a general strategy used by diverse DNA and RNA viruses of plants, *Proc. Natl. Acad. Sci. U.S.A.*, 96, 14147, 1999.
13. Voinnet, O., Lederer, C., and Baulcombe, D.C., A viral movement protein prevents spread of the gene silencing signal in *Nicotiana benthamiana*, *Cell,* 103, 157, 2000.
14. Xiang, L. W. and Ding, S.W., Viral suppressors of RNA silencing, *Curr. Opin. Biotech.*, 12, 150, 2001.
15. Schwartz, M. et al., A positive-strand RNA virus replication complex parallels form and function of retrovirus capsids, *Mol. Cell.*, 9, 505, 2002.
16. Ahlquist, P., Bromovirus RNA replication and transcription, *Curr. Opin. Genet. Dev.*, 2, 71, 1992.
17. Ruiz, M.T., Voinnet, O., and Baulcombe, D.C., Initiation and maintenance of virus-induced gene silencing, *Plant Cell*, 10, 937, 1998.
18. Jones, L. et al., RNA-DNA interactions and DNA methylation in post-transcriptional gene silencing, *Plant Cell*, 11, 2291, 1999.
19. Jones, L. et al., RNA-directed transcriptional gene silencing in plants can be inherited independently of the RNA trigger and requires Met 1 for maintenance, *Curr. Biol.,* 11, 1, 2001.
20. Thomas, C. et al., Size constraints for targeting post-transcriptional gene silencing and RNA-directed methylation in *Nicotiana benthamiana* using a potato virus X vector, *Plant J.*, 25, 417, 2001.
21. Wang, M.B. and Waterhouse, P.M., Application of gene silencing in plants, *Curr. Opin. Plant Biol.*, 5, 146, 2002.
22. Holzberg, S. et al., Barley stripe mosaic virus-induced gene silencing in a monocot plant, *Plant J.*, 30, 315, 2002.
23. Lacomme, C., Hrubikova, K., and Hein, I., Enhancement of virus-induced gene silencing through viral-based production of inverted-repeats, *Plant J.*, 34, 543, 2003.
24. Brault, V. et al., Virus-induced gene silencing in transgenic plants expressing the minor capsid protein of beet western yellows virus, *MPMI*, 15, 799, 2002.
25. Burton, R.A. et al., Virus-induced silencing of a plant cellulose synthase gene, *Plant Cell*, 12, 691, 2000.
26. Ratcliff, F., Martin-Hernandez, A.M., and Baulcombe, D.C., Tobacco rattle virus as a vector for analysis of gene function by silencing, *Plant J.,* 25, 237, 2001.

27. Peart, J.R. et al., Ubiquitin ligase-associated protein SGT1 is required for host and nonhost disease resistance in plants, *Proc. Natl. Acad. Sci U.S.A.*, 99, 10865, 2002.
28. Liu, Y., Schiff, M., and Dinesh-Kumar, S.P., Virus-induced gene silencing in tomato, *Plant J.*, 31, 777, 2002.
29. Yu, D. et al., Analysis of the involvement of an inducible *Arabidopsis* RNA-dependent RNA polymerase in antiviral defense, *MPMI*, 16, 206, 2003.
30. Turnage, M.A. et al., Geminivirus-based vectors for gene silencing in *Arabidopsis*, *Plant J.*, 30, 107, 2002.
31. Kjemtrup, S. et al., Gene silencing from plant DNA carried by a geminivirus, *Plant J.*, 14, 91, 1998.
32. Atkinson, R.G. et al., Post-transcriptional silencing of chalcone synthase in petunia using a geminivirus-based episomal vector, *Plant J.*, 15, 593, 1998.
33. Figlerowicz, M., Role of RNA structure in heteroduplex-mediated and site-specific nonhomologous recombination in brome mosaic virus, *Nucleic Acids Res.*, 28, 1714, 2000.
34. Lu, R. et al., Virus-induced gene silencing in plants, *Methods*, 30, 296, 2003.
35. Wojtkowiak, A. et al., RNAi and viral vectors as useful tools in the functional genomics of plants. Construction of BMV-based vectors for RNA delivery into plant cells, *Cell. Mol. Biol. Lett.*, 7, 511, 2002.

22 RNA Silencing in Plants — Biolistic Delivery of RNAi Reagents

Ulrich Klahre and Frederick Meins, Jr.

CONTENTS

22.1 INTRODUCTION

RNA silencing, also known as posttranscriptional gene silencing (PTGS) or RNA interference (RNAi), is an epigenetic form of RNA degradation believed to be an important defense against foreign nucleic acids that is widely used as a tool for inactivating gene expression.[1–3] It is now recognized that RNA silencing occurs in a wide variety of organisms. Examples include transgene-induced PTGS in plants, quelling in *Neurospora Crassa,* and RNA interference (RNAi) mediated by double-stranded RNA (dsRNA) or 21- to 22-nt small interfering RNAs (siRNA) in *Caenorhabditis Elegans, Drosophila Melanogaster*, and mammalian cells. Considerable evidence suggests that the core mechanism has been highly conserved in evolution.

RNA silencing was first described in transgenic plants, where it was shown that *trans*-interactions between multiple copies of entirely foreign transgenes, between foreign transgenes introduced by successive transformation, and between transgenes and homologous host genes could lead to the sequence-specific degradation of mRNAs.[4] Later, it was found that virus-induced gene silencing (VIGS), important for the protection of plants against virus infection,[5] and antisense-mediated gene silencing

0–8493–2141–7/05/$0.00 + $.50

(ASGS)[6] in plants were similar phenomena and share a common RNA-degradation pathway.

This chapter focuses on RNA silencing in plants. We begin with a brief review of its salient features and outline the experimental procedures for triggering silencing by the biolistic delivery of RNA molecules into intact plants.

22.2 THE GENERAL MECHANISM

Figure 22.1 shows a branched model for RNA silencing in plants, emphasizing that diverse sources of silencing signals converge on a common downstream pathway.[1,7,8] The key intermediate is double-stranded RNA (dsRNA) similar in sequence to the transcribed region of target genes, which is generated from transcripts of sense-, antisense-, and inverted repeat transgenes by, as yet, poorly defined mechanisms or as intermediates in viral replication. These dsRNAs undergo endonucleolytic cleavage mediated by the Dicer RNase III to generate one or more classes of 21- to 25-nucleotide-long, double-stranded, small interfering RNA (siRNA) with 2- to 3-nucleotide 3' overhangs, which then promote degradation of cognate RNAs. Studies with *Drosophila* and mammalian cells suggest that antisense siRNAs associated with the RNAi-induced silencing complex (RISC) cleave the target RNA in the region of complementarity.[2,3] Studies with cell-free extracts suggest a similar Dicer-RISC pathway functions in plants.[9] Dicer and Dicer-like enzymes can also process noncoding RNAs with stem-loop structures to generate microRNAs (miRNAs) important in developmental regulation of translation in animals and RNA stability in plants.[10] Interestingly, two size classes of siRNAs, short (21 to 22 nt) and long (24 to 26 nt), are produced in plants[11] and can be generated by different Dicer-like activities present in wheat germ extracts.[9] Evidence suggests these classes could have different functions in silencing.[12,13]

22.3 GENES INVOLVED IN RNA SILENCING

Several *Arabidopsis* genes essential for RNA silencing have been identified by screening for deficiency mutants[14,15] or by their homology to genes important for silencing in animals.[16] *SDE1/SGS2* encodes a protein with homology to RNA-dependent RNA polymerases (RdRP) of tomato and *Neurospora crassa*. The evidence suggests that SDE1/SGS2 is important for the generation of dsRNA from single-stranded, transgene-encoded transcripts and is part of the self-amplification mechanism believed to stabilize the silent state.[3] *Arabidopsis* contains at least three additional genes encoding RdRP homologues, but their involvement in RNA silencing is unclear. SDE3, like SDE1/SGS2, appears to act upstream of VIGS and encodes a putative RNA-helicase homologous to the mouse GB110 protein.[17] SDE4 affects the accumulation of siRNAs and the pattern of DNA methylation associated with RNA silencing.[11] SGS3 encodes a protein of unknown function with little homology to known proteins except for a coil-like domain thought to function in protein–protein interactions.[15]

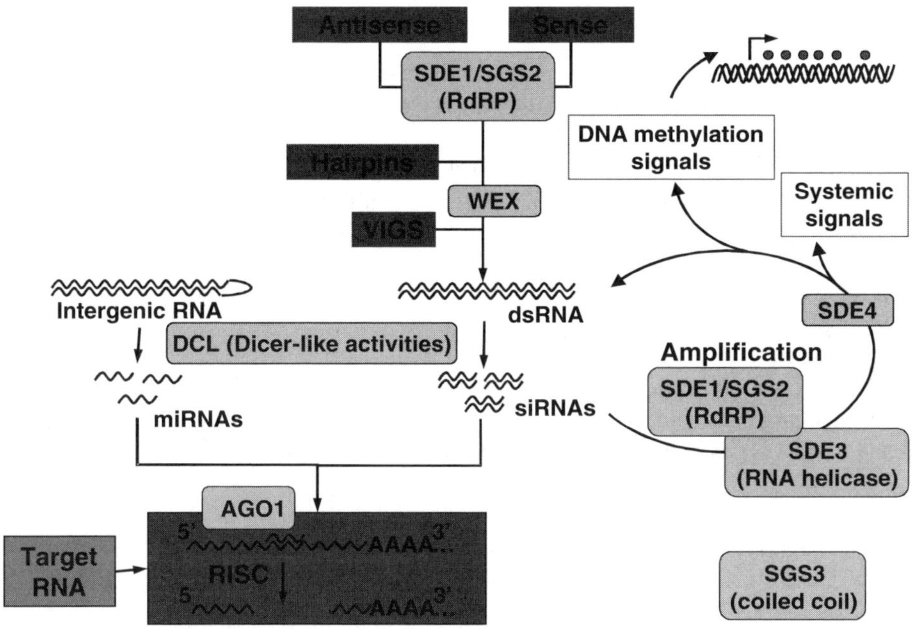

FIGURE 22.1 A branched-pathway model for RNA silencing in plants based on several recent models (1, 3, 7, 8). Sources of upstream silencing signals feeding into the pathway at the dsRNA step (dark grey boxes) and plausible sites of action of several genes important for silencing (light grey boxes) are indicated.

The *Argonaute* gene (*AGO1*), which is implicated in development, is required for RNA silencing[18] in *Arabidopsis* and is homologous to *QDE-1*[19] and *RDE-1*,[20] required for RNA silencing in *N. crassa* and *C. elegans*, respectively. Another member of the AGO protein family, AGO2, is a component of RISC in *Drosophila* cells,[21] suggesting that *AGO1* might act downstream of siRNA formation.

The *Arabidopsis Werner Syndrome-like exonuclease (WEX)* gene[22] encodes a functional RNase D exonuclease domain[23] most similar to that in human Werner Syndrome protein (WRN), which contains functional RecQ helicase and RNase D domains.[24] It is also related to *mut-7*, essential for RNA silencing and RNAi, PTGS, and transposon activity in *C. elegans*.[20,25] Studies with the deficient *wex-1* mutant have shown that WEX is required for RNA silencing but not for transcriptional gene silencing (TGS).[22] The finding that *wex-1* plants develop normally but are blocked in siRNA accumulation suggests that WEX could act upstream of siRNA production in an RNA silencing-specific pathway. Interestingly, although WEX lacks the RecQ helicase domain of WRN, *Arabidopsis* RecQ-like proteins interact with WEX,[26] raising the possibility that WEX and the partner might be the functional equivalent of WRN in mammals.

Although two Dicer activities have been demonstrated in wheat germ extracts,[9] the genes encoding these activities have not been identified. The *Arabidopsis* genome contains four genes encoding Dicer-like proteins (DCL) with structural similarity to the RNase III domain of Dicer.[27] DCL1 and DCL4 have putative nuclear localization signals, and fusions of DCL1 with green fluorescent protein (GFP) are localized in the nucleus, suggesting they may be involved in cleaving nuclear dsRNAs. Mutant alleles of DCL1 have been identified that are embryo-lethal, defective in ovule development, and altered in floral determination. Interestingly, the deficient mutant *dcl1-9* affects production of certain miRNAs but not siRNAs,[28] suggesting that DCL1 is required for the processing of miRNA precursors to miRNAs. Similarly, a mutant deficient in *HEN1*, originally identified as a floral development mutant, shows decreased accumulation of SCARECROW-specific miRNAs but is only partially defective in RNA silencing.[29]

The chromatin mutants, *ddm1* and *met1*, which are deficient in DNA methylation and TGS, show alterations in RNA silencing, as well.[30] Together with the finding that RNA silencing can result in the methylation of the transcribed regions of the affected genes,[31] this suggests that TGS and RNA silencing are in some way linked.

22.4 MAINTENANCE OF THE SILENT STATE: PRODUCTION OF SECONDARY siRNAs AND TRANSITIVITY

While the core mechanism appears to be very similar in plants and animals, the production of secondary siRNAs by a self-amplification mechanism, the elaboration of silencing signals capable of cell-to-cell and systemic movement, and RNA-dependent DNA methylation (RdDM) of transcribed regions occur predominantly in plants.

Once initiated, the silent state shows some degree of stability. It is usually maintained for long periods of time during vegetative growth and persists in cuttings as well as in shoots regenerated from cultured callus tissues. Nevertheless, silencing is a potentially reversible process. Resetting of cells from the silent to high-expressing state regularly occurs after meiosis early in embryonic development. The persistence of the silent state implies that a self-amplification mechanism is involved.[7] One current model is that siRNAs produced by Dicer-like activities serve a dual function: they guide sequence-specific endonucleolytic cleavage of target mRNAs and they prime formation of secondary dsRNAs with target RNAs as a template.[3] This is consistent with the finding that secondary siRNAs representing regions of the target RNA outside the region of homology between the triggering siRNA and the target are produced in both plants[32–35] and *C. elegans*.[36] Silencing can spread along the length of the target RNA, generating secondary siRNAs able to trigger degradation of additional target RNAs in a self-amplifying fashion.[37,38] This phenomenon, called transitivity,[36] can even extend to RNAs encoded by closely related members of the same gene family.[39,40] Silencing of transgenes and transitivity are blocked in deficiency mutants of *SDE1/SGS2*, which encodes a putative RdRP.[14,15,32] QDE-1 is required for RNA silencing in *N. crassa* and has RdRP activity *in vitro*.[41] It supports *de novo* as well as primer-dependent initiation of dsRNA synthesis using single-stranded RNA as a template, suggesting that RdRP could have a key role in the amplification process.

22.5 SYSTEMIC SILENCING SIGNALS

RNA silencing initiated in one region of a plant can spread both locally and systemically.[7,42] The pattern and extent of spread depend on the nature of the transgene, the developmental stage of the plant, and the plant species.[32,43–45] Studies of silencing induced by the local introduction of additional gene copies combined with grafting experiments have shown that silencing signals exist capable of cell-to-cell movement via plasmodesmata and systemic spread via phloem.[46,47] The chemical nature of the silencing signal(s) is unknown, but the analogous spread of viral and endogenous RNAs in plants[48] and the finding that purified RNA molecules can induce local and systemic silencing[33] suggest that the sequence-specific component of the signal is an RNA. At present, it is unclear if there are one or several silencing signals. For example, recent studies suggest that local spread and systemic spread involve different signals.[11,49] Possible RNA candidates include short and long siRNAs, high-molecular-weight dsRNAs, and ill-defined aberrant RNAs that may be generated during the silencing of transgenes.[42] Systemic, but not local, spread of silencing depends on both SDE1/SGS2 and SDE-3, a putative RNA helicase.[49] Thus, it seems likely that the self-amplification mechanism has a role in long-distance movement of silencing signals.

How putative RNA silencing signals are transported via plasmodesmata and the phloem is not known, but it seems likely that this involves the same general mechanisms important for the movement of viral RNAs and certain endogenous mRNAs.[42,50,51] The only other system in which systemic RNA silencing has been documented is *C. elegans*.[52] In contrast to plants, silencing signals appear to move through the extracellular compartment and systemic, but not cell-autonomous, silencing requires the gene *SID-1*.[53] *SID-1* encodes a multispan transmembrane protein that enables passive uptake of dsRNAs by cells. This suggests that the systemic signal is a dsRNA able to enter target cells by a receptor-mediated process.

22.6 RNA-DEPENDENT DNA METHYLATION

RNA silencing in plants is associated with the methylation of cytosines in the transcribed portion of the target gene.[54–56] This phenomenon, called RNA-dependent DNA methylation (RdDM),[57] requires a dsRNA that is cleaved to small 21- to 24-nt RNAs and results in the methylation of cytosines within the region of sequence identity between the triggering RNA and the target DNA.[58] While RdDM of transcribed regions does not appear to reduce transcriptional activity, studies with DNA-methylation mutants[30] and inhibitors of DNA methylation[59] suggest that in some way RdDM helps stabilize the silent state. Interestingly, generation of dsRNA directed against promoters results in methylation of the promoter region and transcriptional silencing of genes regulated by the targeted promoter.[58,60] This raises the intriguing possibility that RdDM may have a more general role in homology-dependent TGS.

At present, it is not clear if dsRNA or siRNAs guide the methylation of target DNA.[61] The finding that the putative RdRP SDE1/SGS2 is required for RdDM[32] implies that production of the guide RNAs is closely linked to self-amplification of silencing signals; however, there is no evidence for the production of systemic RdDM signals.[62]

22.7 EXPERIMENTAL INDUCTION OF RNA SILENCING

Spontaneous RNA silencing in stable transformants is often a stochastic phenomenon; i.e., the frequency of individual plants showing silencing in a genetically identical populations is variable. The incidence of silencing also strongly depends on the nature of the transgene construct, the structure of the transgene insert in the plant genome, and environmental factors.[7] Constructs encoding self-complementary RNAs, i.e., "hairpin" RNAs containing 98- to 853-nucleotide-long arms separated by an intron, have been shown to give efficient silencing in 90 to 100% of independent transformants of diverse plant species.[63]

Silencing can also be initiated locally by introducing additional gene copies. VIGS, which involves the use of RNA-virus vectors encoding copies of target genes, provides a generally applicable method for silencing of resident transgenes and endogenous genes. This method is described in Chapter 21 of this volume. Silencing of individual cells has been reported for a transgene introduced with cognate siRNAs into cultured tobacco protoplasts by electroporation.[64] Silencing has also been achieved by agro-infiltration, i.e., the infiltration of *Agrobacterium tumefaciens* into the intercellular space of leaves,[65] and is discussed in detail in Chapter 23. In brief, strains of *A. tumefaciens* carrying Ti-plasmid binary vectors can deliver T-DNAs containing transgene into plants cells where they are expressed.[66] Agro-infiltration of sense- and antisense-transgenes, as well as promoterless DNA fragments, can efficiently trigger both local and systemic silencing of homologous targets in tobacco and *Nicotiana benthamiana*.[47] A disadvantage of this method is that *A. tumefaciens* induces a pathogen response in the host.[65] The finding that specific RdRP-like genes are induced in the pathogen response[67] raises the possibility of untoward effects of agro-infiltration on the silencing process itself.

Biolistic delivery has been used to trigger silencing in response to additional gene copies, DNA fragments, and RNA molecules.[44,47,68,69] The biolistic delivery of RNA molecules offers several experimental advantages. For example, effects of viral RNA replication and expression of viral RdRPs associated with VIGS; effects of the transcription of delivered DNA, or the delivered DNA itself; and, the activation of plant-defense responses by agro-infiltration are excluded. Biolistic delivery using the particle inflow gun offers the added advantage that macromolecules in liquid phase, such as proteins alone or in combination with nucleic acids, can be delivered into target cells.[70]

Studies with visible markers have shown that expression of biolistically delivered DNA is cell autonomous.[71] Thus, one can use a positive marker for silencing to detect silencing of individual cells and cell-to-cell spread *in planta*. The marker system that we developed[33] is based on the silencing of a transcriptional repressor. Tobacco plants were transformed sequentially with a gene encoding a bacterial tetracycline repressor regulated by a strong promoter and then with the *E. coli β-glucuronidase* (*GUS*) gene regulated by a tetracycline-repressible promoter. In principle, the same strategy using a *green fluorescent protein* (*GFP*) reporter gene could be used for real-time monitoring of silencing in individual cells.

The efficiency of silencing induced biolistically depends on the amount and nature of the nucleic acid delivered as well as on the conditions of bombardment. For example, in the case of tobacco transformed with a tobacco *nitrate reductase* gene, promoter-less DNA fragments as short as 247 bp delivered at a minimum of 0.2 μg DNA per bombardment induced silencing.[68] In contrast, a minimum of 1 μg of siRNA was required to induce systemic silencing of a GFP transgene in *N. benthamiana* plants.[33]

Usually, young plants are bombarded: seedlings 7 to 10 days after germination in the case of *N. benthamiana*[47] and 6-week-old plantlets in the case of tobacco.[68] In a time course study, it was shown that the efficiency of silencing induction in tobacco decreases sharply after the cotyledon stage.[44] Another important factor is the plant species. In our hands, *Arabidopsis thaliana* ecotype Columbia-0 gave a low incidence of local silencing and only rarely of systemic silencing. *Nicotiana* species, in particular, *N. benthamiana*, appear to be particularly susceptible to both induction of silencing and VIGS, which may reflect efficient intercellular and vascular transport of RNAs.[72]

PROTOCOL

Protocol 22.1 Biolistic Induction of RNA Silencing

Several methods for the induction of RNA silencing by biolistic delivery of DNA have been reported.[44,68,73] Here we describe a protocol suitable for the delivery of DNA as well as 21- to 23-nt siRNAs and high-molecular-weight single-stranded and double-stranded RNAs into plants.[33] As judged by the frequency of bombarded plants showing silencing, efficiencies as high as 79% for plasmid DNA and 50% for double-stranded siRNAs have been achieved with this method.

For bombardment of axenic plant materials, place the gun in a sterile flowhood and surface-sterilize with 70% (v/v) ethanol before use. The choice of rupture disk appears not to be too critical, but in our hands 1100 psi gave the best results. An example of the dependence of silencing efficiency on the amount of nucleic acid used to coat particles is shown in Figure 22.2. We routinely use 1 to 2 μg of DNA or dsRNA and 3 to 5 μg of siRNA per bombardment.

Materials

- Helium-driven PDS-1000 particle gun, rupture disks, e.g., 1100 psi, microcarrier gold particles (1-μm diameter), and macrocarriers (Cat. #1652335) (BioRad Laboratories)
- Nucleic acid (DNA, ssRNA, dsRNA, or siRNA)
- Plant seeds (e.g., *Nicotiana benthamiana* or *Nicotiana tabacum* cv. Wisconsin 38) (Institut Expérimental du Tabac, Bergerac, France)
- 3% (w/v) sodium hypochlorite

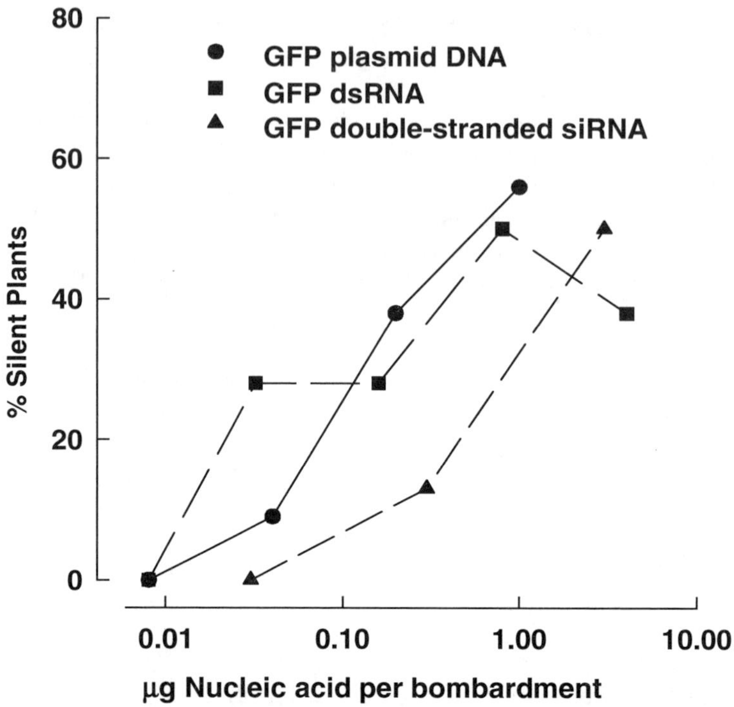

FIGURE 22.2 The effect of nucleic acid dose on the efficiency of biolistically induced RNA silencing. *N. benthamiana* plants transformed with a chimeric *GFP* gene were bombarded at the one–true leaf stage. Gold particles were loaded with the amount of the cognate nucleic acid indicated. Efficiency of silencing is expressed as the percent of > 20 plants that show silencing of at least one leaf ca. 12 days after bombardment. The nucleic acids and plants used are described in Klahre et al.[33]

- Plastic Petri dishes (10-cm diameter, 2 cm deep) (Corning)
- Linsmaier and Skoog medium (DUCHEFA, Haarlem, The Netherlands) solidified with 1% (w/v) purified agar (DIFCO)
- Fluorescent light source (3000 lux) (Tpd 36W 1840, Philips)
- Parafilm M (SPI Supplies, West Chester, PA, USA)
- Sterile water and 50% (v/v) glycerol
- Ethanol
- 2.5 M $CaCl_2$
- 0.1 M spermidine (Sigma)

Experimental Procedures

Plant materials

Raise the plants (in our case, *N. benthamiana* and *N. tabacum* cv. Wisconsin 38) axenically from seed.

1. Surface-sterilize the seeds by incubating them for 3 min in 3% (w/v) sodium hypochlorite, and then rinse 5 times with 10 volumes of sterile water.

2. Sow the seeds at 16 seeds/dish in 10-cm diameter, 2-cm-deep plastic Petri dishes containing 30 ml of autoclaved Linsmaier and Skoog medium[74] solidified with 1% (w/v) purified agar.
3. Close the Petri dishes with Parafilm M and incubate for 10 to 15 days at 26°C with constant fluorescent light (3000 lux).

Coating of gold particles

4. Vortex 60 mg of 1-μm-diameter gold particles in 1 ml of ethanol for 2 min and then collect by centrifugation (14,000 × g, 1 min).
5. Wash the pellet successively by resuspension and centrifugation twice with 70% (v/v) ethanol, twice with ethanol, and twice with sterile water.
6. Suspend the washed gold particles in 1 ml of sterile 50% (v/v) glycerol and store at room temperature.
7. To coat gold particles, mix 25 μl of the washed gold-particle suspension (~1.5 mg) by vortexing at low speed with 5 μl of nucleic acid stock solution, 25 μl of 2.5 M $CaCl_2$, and 10 μl of 0.1 M spermidine.
8. Vortex the resultant suspension at high speed for 2 min, mix well with 200 μl of ethanol, and centrifuge (14,000 × g, 4 sec).
9. Wash the resultant pellet by resuspension and centrifugation in 200 μl 70% (v/v) ethanol, and suspend in 40 μl of ethanol.
10. While continuously resuspending the particles with a pipet, deposit 8-μl aliquots of the suspension onto 5 macrocarriers.

Bombardment

11. Position the plantlets in a Petri dish with the cover removed at the second notch of the particle gun, i.e., ~10 cm from the stopping plate.
12. Deliver the gold particles at 1100 psi with a vacuum of 25 inches of mercury.

Technical notes. 1. Plantlets are bombarded when the first true leaves have emerged. Satisfactory results are also obtained with soil-grown plantlets raised in small pots. 2. It is important to distribute the gold particles uniformly over an area of ~7 mm in diameter at the center of each macrocarrier. Under these conditions, the biolistically delivered particles will be evenly distributed over a ~5-cm-diameter area of target plants. 3. Using the *N. benthamiana* line 16c carrying an mGFP-ER reporter gene with a cauliflower mosaic virus 35S RNA promoter and Nos terminator as the target plant[5] and gold particles coated with 1 to 2 μg of full-length, double-stranded mGFP-ER RNA or double-stranded mGFP-ER siRNA, plants showed local RNA silencing of GFP on bombarded leaves within three days and systemic silencing after 10 to 15 days.[33]

REFERENCES

1. Vance, V. and Vaucheret, H., RNA silencing in plants: defense and counterdefense, *Science*, 292, 2277, 2001.
2. Hannon, G.J., 2002 RNA interference, *Nature*, 418, 244,51.
3. Tijsterman, M., Ketting, R.F., and Plasterk, R.H.A., The genetics of RNA silencing, *Annu. Rev. Genet.*, 36, 489, 2002.
4. Meyer, P. and Saedler, H., Homology-dependent gene silencing in plants, *Annu. Rev. Plant Physiol. Plant Mol. Biol.*, 47, 23, 1996.

5. Ruiz, M. T., Voinnet, O., and Baulcombe, D. C., Initiation and maintenance of virus-induced gene silencing, *Plant Cell*, 10, 937, 1998.
6. Di Serio, F. et al., Sense- and antisense-mediated gene silencing in tobacco is inhibited by the same viral suppressors and is associated with accumulation of small RNAs, *Proc. Natl. Acad. Sci. U.S.A.*, 98, 6506, 2001.
7. Meins, F., Jr., RNA degradation and models for posttranscriptional gene silencing, *Plant Mol. Biol.*, 43, 261, 2000.
8. Béclin, C. et al., A branched pathway for transgene-induced RNA silencing in plants, *Curr. Biol.*, 12, 684, 2002.
9. Tang, G. et al., A biochemical framework for RNA silencing in plants, *Genes Devel.*, 17, 49, 2003.
10. Carrington, J.C. and Ambros, V., Role of microRNAs in plant and animal development, *Science*, 301, 336, 2003.
11. Hamilton, A. et al., Two classes of short interfering RNA in RNA silencing, *EMBO J.*, 21, 4671, 2002.
12. Mallory, A.C. et al., A viral suppressor of RNA silencing differentially regulates the accumulation of short interfering RNAs and micro-RNAs in tobacco, *Proc. Natl. Acad. Sci. U.S.A.*, 99, 15228, 2002.
13. Szittya, G. et al., Low temperature inhibits RNA silencing-mediated defence by the control of siRNA generation, *EMBO J.*, 22, 633, 2003.
14. Dalmay, T. et al., An RNA-dependent RNA polymerase gene in *Arabidopsis* is required for posttranscriptional gene silencing mediated by a transgene but not by a virus, *Cell*, 101, 543, 2000.
15. Mourrain, P. et al., Arabidopsis *SGS2* and *SGS3* genes are required for posttranscriptional gene silencing and natural virus resistance, *Cell*, 101, 533, 2000.
16. Glazov, E.A., A gene encoding an RNase D exonuclease-like protein is required for post-transcriptional silencing in Arabidopsis, Ph,D, thesis, University of Basel, Basel, 2003.
17. Dalmay, T. et al., SDE3 encodes an RNA helicase required for post-transcriptional gene silencing in *Arabidopsis*, *EMBO J.*, 20, 2069, 2001.
18. Fagard, M. et al., AGO1, QDE-2, and RDE-1 are related proteins required for post-transcriptional gene silencing in plants, quelling in fungi, and RNA interference in animals, *Proc. Natl. Acad. Sci. U.S.A.*, 97, 11650, 2000.
19. Catalanotto, C. et al., Gene silencing in worms and fungi, *Nature*, 404, 245, 2000.
20. Tabara, H. et al., The *rde-1* gene, RNA interference, and transposon silencing in *C. elegans*, *Cell*, 99, 231, 1999.
21. Hammond, S.M. et al., Argonaute2, a link between genetic and biochemical analyses of RNAi, *Science*, 293, 1146, 2001.
22. Glazov, E. et al., A gene encoding an RNase D exonuclease-like protein is required for posttranscriptional silencing in *Arabidopsis*, *Plant J.*, 35, 342, 2003.
23. Plchova, H., Hartung, F., and Puchta, H., Biochemical characterization of an exonuclease from *Arabidopsis thaliana* reveals similarities to the DNA exonuclease of the human Werner Syndrome protein, *J. Biol. Chem.*, 2003.
24. Shen, J. C. and Loeb, L. A., The Werner syndrome gene, the molecular basis of RecQ helicase-deficiency diseases, *Trends Genet.*, 16, 213, 2000.
25. Ketting, R.F. et al., *mut-7* of *C. elegans*, required for transposon silencing and RNA interference, is a homolog of Werner Syndrome helicase and RNase D, *Cell*, 99, 133, 1999.
26. Hartung, F., Plchova, H., and Puchta, H., Molecular characterisation of *RecQ* homologues in *Arabidopsis thaliana*, *Nucl. Acids Res.*, 28, 4275, 2000.

27. Schauer, S. E. et al., DICER-LIKE1: blind men and elephants in *Arabidopsis* development, *Trends Plant Sci.*, 7, 487, 2002.
28. Finnegan, E.J., Margis, R., and Waterhouse, P.M., Posttranscriptional gene silencing is not compromised in the *Arabidopsis CARPEL FACTORY (DICER-Like1)* mutant, a homolog of Dicer-1 from *Drosophila*, *Curr. Biol.*, 13, 236, 2003.
29. Boutet, S. et al., *Arabidopsis HEN1*: A genetic link between endogenous miRNA controlling development and siRNA controlling transgene silencing and virus resistance, *Curr. Biol.*,13, 843, 2003.
30. Morel, J.B. et al., DNA methylation and chromatin structure affect transcriptional and post-transcriptional transgene silencing in *Arabidopsis*, *Curr. Biol.*, 10, 1591, 2000.
31. Matzke, M., Matzke, A.J.M., and Kooter, J.M., RNA: guiding gene silencing, *Science*, 293, 1080, 2001.
32. Vaistij, F. E., Jones, L., and Baulcombe, D. C., Spreading of RNA targeting and DNA methylation in RNA silencing requires transcription of the target gene and a putative RNA-dependent RNA polymerase, *Plant Cell*, 14, 857, 2002.
33. Klahre, U. et al., High molecular weight RNAs and small interfering RNAs induce systemic posttranscriptional gene silencing in plants, *Proc. Natl. Acad. Sci. U.S.A.*, 99, 11981, 2002.
34. Sonoda, S. and Nishiguchi, M., Graft transmission of post-transcriptional gene silencing: target specificity for RNA degradation is transmissible between silenced and non-silenced plants, but not between silenced plants, *Plant J.*, 21, 1, 2000.
35. Van Houdt, H., Bleys, A., and Depicker, A., RNA Target Sequences Promote Spreading of RNA Silencing, *Plant Physiol.*, 131, 245, 2003.
36. Sijen, T. et al., On the role of RNA amplification in dsRNA-triggered gene silencing, *Cell*, 107, 465, 2001.
37. Lipardi, C., Wei, Q., and Paterson, B. M., RNAi as random degradative PCR, siRNA primers convert mRNA into dsRNAs that are degradated to generate new siRNAs, *Cell*, 107, 297, 2001.
38. Bergstrom, C. T., McKittrick, E., and Antia, R., Mathematical models of RNA silencing: Unidirectional amplification limits accidental self-directed reactions, *Proc. Natl. Acad. Sci. U.S.A*, 2003.
39. Kunz, C. et al., Developmentally regulated silencing and reactivation of tobacco chitinase transgene expression, *Plant J.*, 10, 437, 1996.
40. Kunz, C. et al., β-1,3-glucanase and chitinase transgenes in hybrids show distinctive and independent patterns of posttranscriptional gene silencing, *Planta*, 212, 243, 2001.
41. Makeyev, E.V. and Bamford, D.H., Cellular RNA-dependent RNA polymerase involved in posttranscriptional gene silencing has two distinct activity modes, *Mol. Cell*, 10, 1417, 2002.
42. Mlotshwa, S. et al., RNA silencing and the mobile silencing signal, *Plant Cell*, 14, S289, 2002.
43. Fagard, M. and Vaucheret, H., Systemic silencing signal(s), *Plant Mol. Biol.*, 43, 285, 2000.
44. Crété, P. et al., Graft transmission of induced and spontaneous posttranscriptional silencing of chitinase genes, *Plant J.*, 28, 493, 2001.
45. Mallory, A.C. et al., The capacity of transgenic tobacco to send a systemic RNA silencing signal depends on the nature of the inducing transgene locus, *Plant J.*, 35, 82, 2003.
46. Voinnet, O. and Baulcombe, D.C., Systemic signalling in gene silencing, *Nature*, 389, 553, 1997.

47. Voinnet, O. et al., Systemic spread of sequence-specific transgene RNA degradation in plants is initiated by localized introduction of ectopic promoterless DNA, *Cell,* 95, 177, 1998.
48. Wu, X., Weigel, D., and Wigge, P.A., Signaling in plants by intercellular RNA and protein movement, *Genes Dev.,* 16, 151, 2002.
49. Himber, C. et al., Transitivity-dependent and -independent cell-to-cell movement of RNA silencing, *EMBO J.,* 22, 4523, 2003.
50. Ueki, S. and Citovsky, V., Inhibition of systemic onset of post-transcriptional gene silencing by non-toxic concentrations of cadmium, *Plant J.,* 28, 283, 2001.
51. Ueki, S. and Citovsky, V., The systemic movement of a tobamovirus is inhibited by a cadmium-ion-induced glycine-rich protein, *Nature Cell Biol.,* 4, 478, 2002.
52. Fire, A. et al., Potent and specific genetic interference by double-stranded RNA in *Caenorhabditis elegans*, *Nature,* 391, 806, 1998.
53. Feinberg, E. H. and Hunter, C. P., Transport of dsRNA into cells by the transmembrane protein SID-1, *Science,* 301, 1545, 2003.
54. Ingelbrecht, I. et al., Posttranscriptional silencing of reporter transgenes in tobacco correlates with DNA methylation, *Proc. Natl. Acad. Sci. USA,* 91, 10502, 1994.
55. Wassenegger, M. et al., RNA-directed *de novo* methylation of genomic sequences in plants, *Cell,* 76, 567, 1994.
56. Jones, A.L., Thomas, C.L., and Maule, A.J., *De novo* methylation and co-suppression induced by a cytoplasmically replicating plant RNA virus, *EMBO J.,* 17, 6385, 1998.
57. Wassenegger, M., RNA-directed DNA methylation, *Plant Mol. Biol.* 43, 203, 2000.
58. Mette, M.F. et al., Transcriptional silencing and promoter methylation triggered by double-stranded RNA, *EMBO J.,* 19, 5194, 2000.
59. Kovarik, A. et al., Drug-induced hypomethylation of a posttranscriptionally silenced transgene locus of tobacco leads to partial release of silencing, *FEBS Lett.,* 467, 47, 2000.
60. Jones, L. et al., RNA-DNA interactions and DNA methylation in post-transcriptional gene silencing, *Plant Cell,* 11, 2291, 1999.
61. Aufsatz, W. et al., RNA-directed DNA methylation in *Arabidopsis*, *Proc. Natl. Acad. Sci. U.S.A.,* 99, 16499, 2002.
62. Mlotshwa, S. et al., RNA silencing and the mobile silencing signal, *Plant Cell,* 14, S289, 2002.
63. Wesley, S.V. et al., Construct design for efficient, effective, and high-throughput gene silencing in plants, *Plant J.,* 27, 581, 2001.
64. Vanitharani, R., Chellappan, P., and Fauquet, C.M., Short interfering RNA-mediated interference of gene expression and viral DNA accumulation in cultured plant cells, *Proc. Natl. Acad. Sci. U.S.A.,* 100, 9632, 2003.
65. Schöb, H., Kunz, C., and Meins, F. Jr., Silencing of transgenes introduced into leaves by agroinfiltration, A simple, rapid method for investigating sequence requirements for gene silencing, *Mol. Gen. Genet.,* 256, 581, 1997.
66. Hooykaas, P. J. J. and Schilperoort, R.A., Agrobacterium and plant genetic engineering, *Plant Mol. Biol.,* 19, 15, 1992.
67. Xie, Z. et al., An important role of an inducible RNA-dependent RNA polymerase in plant antiviral defense, *Proc. Natl. Acad. Sci. U.S.A,* 98, 6516, 2001.
68. Palauqui, J.-C. and Balzerque, S., Activation of systemic acquired silencing by localized introduction of DNA, *Curr. Biol.,* 9, 59, 1999.
69. Schweizer, P. et al., Double-stranded RNA interferes with gene function at the single-cell level in cereals, *Plant J.,* 24, 895, 2000.

70. Finer, J. et al., Development of the particle inflow gun for DNA delivery to plant cells, *Plant Cell Reports,* 11, 323, 1992.
71. Iglesias, V.A. et al., Transient expression of visible marker genes in meristem cells of wheat embryos after ballistic micro-targeting, *Planta,* 192, 84, 1994.
72. Lu, R. et al., Virus-induced gene silencing in plants, *Methods,* 30, 296, 2003.
73. Voinnet, O., Lederer, C., and Baulcombe, D.C., A viral movement protein prevents spread of the gene silencing signal in *Nicotiana benthamiana*, *Cell,* 103, 157, 2000.
74. Linsmaier, B. and Skoog, F., Organic growth factor requirements of tobacco tissue culture, *Physiol. Plant.,* 18, 100, 1965.

23 Agro-Infiltration: A Versatile Tool for RNAi Studies in Plants

Dániel Silhavy

CONTENTS

23.1 INTRODUCTION

Agro-infiltration is an *Agrobacterium*-mediated transient gene expression system that is widely used to express foreign proteins in plants.[1] Importantly, agro-infiltration has also become a key method to study plant RNAi.

The Gram-negative soil bacterium *Agrobacterium tumefaciens* has evolved a unique system to transfer a DNA segment (T-DNA) from its tumor-inducing plasmid (pTi) to plant cells. This natural gene transfer mechanism has been engineered to an efficient plant transformation system. The T-DNA is delimited by two 25-bp direct repeats (called borders), the only *cis*-elements required for T-DNA transfer. Foreign DNA cloned between the borders can also be transferred into plant cells. Vir proteins, the *trans*-factors of T-DNA transfer, achieve the delivery and integration of T-DNA into host cells.[2] Importantly, *cis*- and *trans*-elements of T-DNA transfer can be separated. Therefore, *Agrobacterium*-based plant transformation vector systems consist of two plasmids, a binary plasmid having T-DNA, and a *vir* plasmid expressing Vir proteins. If a plant transgene expression cassette (promoter–transgene-terminator region) is cloned between the T-DNA borders of the binary plasmid, Vir proteins direct mobilization of the transgene cassette into the host cells. In *Agrobacterium* cultures used for agro-infiltration a third helper plasmid could also be present in the bacterium that carries additional copies of certain *vir* genes, helping to improve T-DNA transfer.

0-8493-2141-7/05/$0.00 + $.50

The *Agrobacterium* culture is injected into the leaf through plasmodesmata, and then the bacteria spread in the intercellular spaces. *Agrobacterium* infects the host cells so efficiently that T-DNA is transferred into the nuclei of practically every cell of the infiltrated patch. On entering plant cell nuclei, T-DNAs can either integrate into the host genome or remain present in extra chromosomal form, which can be transcribed. Agro-infiltration results in strong transient expression due to the high copy number of nonintegrated T-DNA.[1,2] High-level transgene expression, however, also triggers RNAi response.[3]

In plants, RNAi operates at two levels, the single-cell level (cell-autonomous silencing) and the whole-plant level (systemic silencing). As a manifestation of transgene-induced cell-autonomous RNAi activity, a few days postinfiltration (dpi) transgene mRNA levels start to decline while transgene-derived siRNAs accumulate in cells of infiltrated patches. Two size classes of transgene-specific siRNAs are generated, small siRNAs (21 to 22 nt) and long siRNAs (24 to 26 nt).[4] If the transgene is homologous to an endogenous gene, mRNAs transcribed from this gene are also targeted by the RNAi machinery, silencing the endogenous gene in the infiltrated cells.

Agro-infiltration-mediated endogenous gene silencing is not cell autonomous but results in systemic silencing signals that confer sequence-specific RNA degradation in the noninfiltrated cells. Short-distance spread of the signal leads to the inactivation of homologous genes around the infiltrated patches, while long-distance signal movement results in silencing of homologous genes in the upper parts of the plant.[5] Importantly, agro-infiltration-mediated systemic silencing occurs only for certain genes, such as integrated transgenes or ectopically expressed plant genes. Agro-infiltration-mediated systemic silencing of a wild-type plant gene has not been reported yet. Perhaps chromatin structure of the integrated gene defines whether systemic silencing can develop or not.[6] The mechanisms of transgene-induced RNAi are poorly understood. It is not known why transient expression triggers RNAi and what features of the endogenous genes determine whether systemic silencing should or should not occur.

23.2 AGRO-INFILTRATION-MEDIATED IDENTIFICATION OF RNAi SUPPRESSOR PROTEINS

Although agro-infiltration has been successfully used to study many aspects of plant RNAi, it is particularly powerful in identifying proteins having RNAi inhibitor activity (RNAi suppressors). A GFP-based agro-infiltration reporter system is widely used in identifying RNAi suppressors. This system is based on the observation that infiltration of the leaves of a GFP transgenic *Nicotiana benthamiana* plant with an *Agrobacterium* culture carrying a binary plasmid that express GFP transgene from the strong 35S-promoter (35S-GFP) leads to strong GFP transient expression and induces cell-autonomous and systemic GFP silencing.[3] Cell-autonomous GFP RNAi manifests as a weakening of green fluorescence, a decrease in the level of GFP mRNA, and an accumulation of both small and long GFP-specific siRNAs in the infiltrated patches. Cell-autonomous RNAi also interferes with the expression of endogen GFP. Consequently, at 8 to 12 dpi the infiltrated patch shows red under UV light, because in the lack of both transient and endogenous GFP activity, the red autofluorescence of chlorophyll is visible. Therefore, by following the appearance and spread of red fluorescence, the development of systemic GFP silencing can be easily monitored. Short-distance systemic silencing results in development of a red fluorescent ring around the infiltrated patch (first appearing at 4 to 6 dpi), while long-distance silencing leads to the appearance and spread of red fluorescence in upper leaves. Red fluorescence first develops around the veins of upper leaves (9 to 12 dpi) and then is spread in the lamina. Three to 4 weeks postinfiltration, upper leaves may become completely red in UV lights. More often, systemic GFP silencing does not develop completely; thus, only cells along veins of upper leaves become red under UV light.

This agro-infiltration-mediated GFP RNAi reporter system has been harnessed to identify RNAi suppressors. If leaves of a GFP transgenic *N. benthamiana* are co-infiltrated with 35S-GFP and with a second *Agrobacterium,* expressing a protein that suppresses RNAi, GFP silencing will be affected.[3] Comparative analyses of GFP RNAi in 35S-GFP infiltrated and in co-infiltrated leaves could reveal whether or not the tested protein acts as an RNAi suppressor. The effect of the co-infiltrated gene on cell-autonomous GFP silencing is analyzed by visual assessment of intensity and duration of transient green fluorescence in the infiltrated patches and by RNA gel blot assays. High-molecular-weight RNA gel blot analyses show the effect of co-infiltrated gene on GFP mRNA accumulation, while low-molecular-weight RNA gel blot experiments show the influence of co-infiltrated gene on GFP-specific siRNA accumulation. The effect of co-infiltrated gene on systemic silencing is tested under UV light by comparing the development of short-distance and long-distance systemic GFP silencing on GFP infiltrated and co-infiltrated plants.

This co-infiltration assay has been used to identify many plant viral RNAi suppressors, to show that insect viral RNAi suppressors could also inhibit silencing in plants[7] and to prove that heterologous dsRNA-binding proteins can interfere with plant RNAi.[8] Moreover, co-infiltration experiments revealed that different suppressors target distinct RNAi steps. For example, co-infiltration of 19-kDa protein of tombusviruses inhibited the development of cell-autonomous and both short- and long-distance systemic GFP silencing, while in similar experiments the P1 suppressor of *Rice Yellow Mottle Virus* barely affected the cell-autonomous and short-distance systemic GFP silencing but inhibited the development of long-distance systemic silencing.[4,5]

23.3 OTHER APPLICATIONS OF AGRO-INFILTRATION IN THE ANALYSIS OF RNAi IN PLANTS

In addition to identification of RNAi suppressors, agro-infiltration has been efficiently used to analyze other aspects of RNAi. Some examples are given below:

i. To compare RNAi inducer activity of transgene-expressing ssRNAs and dsRNAs (inverted repeat construct)[9]
ii. To test the effect of temperature on the efficiency of plant RNAi[10]
iii. To measure the efficiency of different RNAi suppressors to enhance the expression of co-infiltrated genes[11]

iv. To identify targeted steps of RNAi suppressors[4,5,12]
v. To find correlation between the accumulation of small and long siRNAs and RNAi-mediated mRNA degradation, and to observe short- and long-distance silencing[4,5]
vi. To identify potential systemic signal molecules[5]
vii. To analyze miRNA-mediated gene regulation in plants[13] and to test the effect of a silencing suppressor on it[14]

Agro-infiltration is likely to become a key reverse genomic tool to identify plant genes involved in RNAi. In animal systems RNAi-based gene inactivation methods have been used to identify components of the RNAi machinery; a similar strategy could also be employed in plants. The main limitation of agro-infiltration is that although it works very efficiently for *N. benthamiana* and its close relatives, it is inefficient in many other species. However, it may be possible to overcome these limitations by using an alternative method for delivering *Agrobacterium* into plant leaves, such as vacuum infiltration.

PROTOCOLS

Protocols given below deal with testing RNAi repressor activity of a candidate silencing suppressor gene in plants.

PROTOCOL 23.1 GROWING PRECONDITIONED *AGROBACTERIUM* CULTURES

Materials

- *Agrobacterium* cultures
- 35S-GFP *Agrobacterium* C58C1 carrying pGV2260 Ti plasmid, the pCH32 helper plasmid, and a binary plasmid with a GFP expression cassette (35S promoter, GFP transgene, NOS termination region)[3]
- 35S-CS *Agrobacterium* C58C1 carrying pGV2260 Ti plasmid, the pCH32 helper plasmid, and a binary plasmid with a candidate silencing suppressor (CS) expression cassette (35S promoter, CS transgene, NOS termination region)
- LB-preconditioning medium (LB-PM) (LB with 10 mM MES, pH 5.6 with KOH, 5 μg/ml tetracycline, 25 μg/ml kanamycin, 20 μM acetosyringone (all from Duchefa, except acetosyringone))
- Agro-infiltration solution (AS) (10 mM MES, pH 5.6 with KOH, 10 mM $MgCl_2$, 150 mM acetosyringone (Duchefa, except acetosyringone))
- Acetosyringone (Sigma)
- Shaker incubator (Sanyo-Gallenkamp)
- Centrifuge (Sorvall RC-5B)
- Spectrophotometer (HP)

Experimental Procedures

1. Inoculate 10 ml LB-PM with a colony of 35S-GFP and another culture with a colony of 35S-CS. Incubate at 28°C with shaking to OD_{600} 0.6 to 0.8.

2. Collect the cells by centrifuging with 3000 g at room temperature for 10 min. Discard the supernatant and redissolve the cells in 5 ml AS.
3. Incubate the cultures at room temperature for at least 3 h (cultures incubated up to 2 days at room temperature are still active). Adjust the OD_{600} of cultures with AS to 0.5.

Protocol 23.2 Agro-Infiltration of Leaves of GFP Transgenic *N. Benthamiana* Plants

Materials

- 2- to 3-week-old *N. benthamina* GFP transgenic plants (line 16c)
- 2-ml syringe

Experimental Procedures

1. Infiltration is carried out with a 2-ml sterile syringe without needle. Take the following cultures in separate syringes:
   35S-GFP
   35S-CS
   1:1 mixture of the two cultures (35S-GFP + 35S-CS)
2. Infiltrate the abaxial side of leaves of 2- to 3-week-old *N. benthamina* GFP transgenic plants (line 16c) with the *Agrobacterium* cultures. Infiltrate at least 10 plants for each culture.
3. After infiltration, spray the plants with water.

Protocol 23.3 Analyses of Transgene-Induced Cell-Autonomous and Systemic GFP Silencing

Plant RNAi is temperature dependent and, therefore, growth conditions can affect the progression of silencing activity. Keep the infiltrated plants in a well-controlled growth chamber.

Materials

- Handheld, long-wave ultraviolet (UV) lamp (Black Ray model B 100AP, UV Products)
- RNA extraction buffer (0.1 M glycine-NaOH, pH 9.0, 100 mM NaCl, 10 mM EDTA, 2% (w/v) SDS, and 1% (w/v) sodium lauroylsarcosine (Duchefa))
- Phenol and chloroform
- Formaldehyde (REANAL)
- 10 × MAE (0.2 M MOPS, 20 mM sodium acetate, 10 mM EDTA, pH 8.0 (all from Duchefa))
- Agarose (Invitrogen)
- Polyacrylamide gel solution (Ambion)
- 10 × TBE
- Urea (Duchefa)

- Church buffer (1% (w/v) BSA Fraction V (Merck), 1 mM EDTA (Duchefa), 0.5 M phosphate buffer, 7% (w/v) SDS (Duchefa), 100 μg/ml salmon sperm DNA (Amersham Biosciences))
- SiRNA hybridization buffer (50% (v/v) formaldehyde (REANAL), 5 × SSPE, 5 × Denhardt's solution, 0.5% (w/v) SDS (Duchefa) with 100 μg/ml salmon sperm DNA)
- Nytran-N (Schleischer & Schuell)
- UV cross-linker (Stratagene)
- Polyacrylamide gel electrophoresis system (Bio-center)
- Horizontal gel electrophoresis system (Bio-center)
- Storage phosphor screen and PhosphoImager (Amersham Biosciences)

Experimental Procedures

1. Monitor GFP activity and progression of GFP silencing with a handheld UV lamp each day postinfiltration. If the tested protein suppresses cell-autonomous RNAi in plants, the GFP transient expression will be stronger in the co-infiltrated patches than in the leaves infiltrated with 35S-GFP alone. If it affects systemic RNAi, the development of red ring or the GFP silencing in the upper leaves will be delayed or inhibited in co-infiltrated plants.

Total RNA isolation

2. To isolate total RNA from the infiltrated patches of 35S-GFP, 35S-CS, and 35S-GFP + 35S-CS co-injected leaves at 3, 6, and 10 dpi, homogenize 100 mg of leaf tissue in 600 μl of RNA extraction buffer.
3. Extract with equal volume of phenol. Save the aqueous phase.
4. Extract the aqueous phase with equal volume of 1:1 (v/v) phenol:chloroform.
5. Precipitate with 0.1 volume of 3 M sodium acetate and 2.5 volumes of absolute ethanol.
6. Wash the pellet with 70% (v/v) ethanol. Air-dry the RNA pellet and resuspend in 50 μl water.

High-molecular-weight RNA analysis to test GFP mRNA levels

7. Separate 5 μg RNA from each sample on 1.2% (w/w) formaldehyde-MAE agarose gel. To prepare formaldehyde-MAE, dissolve 0.36 g agarose in 27 ml 1 × MAE, boil for a few seconds, cool to 55 to 60°C, and then add 3 ml formaldehyde. Mix and cast the gel.
8. After electrophoresis, soak the gel in 20 × SSC for 15 min. Transfer RNAs onto a Nytran-N membrane and fix by UV cross-linking.
9. Hybridize the membrane with ^{32}P α-CTP labeled DNA probes in Church buffer at 65°C for 12 h.
10. Wash the membrane in 2 × SSC, 0.1% SDS for 2 × 10 min at 65°C.
11. Expose the membrane to a storage phosphor screen and, after overnight exposure, scan in a PhosphoImager. If the tested protein suppresses cell-autonomous RNAi in plant, GFP mRNAs will accumulate to a higher level in the co-infiltrated samples than in the samples infiltrated with 35S-GFP alone.

Low-molecular-weight RNA analysis to test GFP-specific siRNA accumulation

12. Separate 5 to 10 μg total RNA from each sample in 8% (w/v) PAGE with 8.3 M urea prepared in 1 × TBE.
13. Soak the gel in 20 × SSC for 15 min. Transfer RNAs onto a Nytran-N membrane and fix by UV cross-linking.
14. Hybridize the membrane with ^{32}P α-UTP labeled strand-specific RNA probes in siRNA hybridization buffer at 40°C for 12 h.
15. Wash the membrane in 2 × SSC, 0.1% SDS for 2 × 10 min at 40°C.
16. Expose the membrane to a storage phosphor screen overnight and scan in a PhosphoImager. If the tested protein inhibits RNAi in plant, accumulation of GFP-specific siRNAs will be affected in co-infiltrated samples. Depending on the type of suppressor ,only one class, or both small and long GFP-specific siRNA classes, could be influenced in co-infiltrated samples.

ACKNOWLEDGMENTS

I am grateful to David Baulcombe for generously supplying the 35S-GFP vector and *N. benthamiana* GFP 16c line. Our research on temperature dependency of plant RNAi was supported by grants from the Hungarian OTKA (T15 042787).

REFERENCES

1. Kapila, J. et al., An *Agrobacterium*-mediated transient gene expression system for intact leaves, *Plant Science*, 122, 101, 1997.
2. Gelvin, S.B., *Agrobacterium*-mediated plant transformation: the biology behind the "gene-jockeying" tool, *Microbiol. Mol. Biol. Rev.*, 67, 16, 2003.
3. Voinnet, O., Lederer, C., and Baulcombe D.C., A viral movement protein prevents spread of the gene silencing signal in *Nicotiana benthamiana*, *Cell*, 103**,** 157, 2000.
4. Hamilton, A. et al., Two classes of short interfering RNA in RNA silencing, *EMBO J.*, 21, 4671, 2002.
5. Himber, C. et al., Transitivity-dependent and -independent cell-to-cell movement of RNA silencing, *EMBO J.*, 22, 4523, 2003.
6. Mlynarova, L. et al., The presence of a chromatin boundary appears to shield a transgene in tobacco from RNA silencing, *Plant Cell*, 9, 2203, 2003.
7. Li, H., Li, W.X., and Ding S.W., Induction and suppression of RNA silencing by an animal virus, *Science*, 296, 1319, 2002.
8. Lichner, Z., Silhavy, D., and Burgyan, J., Double-stranded RNA-binding proteins could suppress RNA interference-mediated antiviral defences, *J. Gen. Virol.*, 84, 975, 2003.
9. Johansen, L. K. and Carrington, J.C., Silencing on the spot. Induction and suppression of RNA silencing in the *Agrobacterium*-mediated transient expression system, *Plant Physiol.*, 26, 930, 2001.
10. Szittya, G. et al., Low temperature inhibits RNA silencing-mediated defence by the control of siRNA generation, *EMBO J.*, 22, 633, 2003.
11. Voinnet, O. et al., An enhanced transient expression system in plants based on suppression of gene silencing by the p19 protein of tomato bushy stunt virus, *Plant J.*, 33, 949, 2003.

12. Guo, H.S. and Ding, S.W., A viral protein inhibits the long range signaling activity of the gene silencing signal, *EMBO J.*, 21, 398, 2002.
13. Llave, C. et al., Cleavage of scarecrow-like mRNA targets directed by a class of *Arabidopsis* miRNA, *Science*, 297, 2053, 2002.
14. Kasschau, K.D. et al., P1/HC-Pro, a viral suppressor of RNA silencing, interferes with *Arabidopsis* development and miRNA function, *Dev. Cell*, 2, 205, 2003.

Index

D